Springer Theses

Recognizing Outstanding Ph.D. Research

Aims and Scope

The series "Springer Theses" brings together a selection of the very best Ph.D. theses from around the world and across the physical sciences. Nominated and endorsed by two recognized specialists, each published volume has been selected for its scientific excellence and the high impact of its contents for the pertinent field of research. For greater accessibility to non-specialists, the published versions include an extended introduction, as well as a foreword by the student's supervisor explaining the special relevance of the work for the field. As a whole, the series will provide a valuable resource both for newcomers to the research fields described, and for other scientists seeking detailed background information on special questions. Finally, it provides an accredited documentation of the valuable contributions made by today's younger generation of scientists.

Theses may be nominated for publication in this series by heads of department at internationally leading universities or institutes and should fulfill all of the following criteria

- They must be written in good English.
- The topic should fall within the confines of Chemistry, Physics, Earth Sciences, Engineering and related interdisciplinary fields such as Materials, Nanoscience, Chemical Engineering, Complex Systems and Biophysics.
- The work reported in the thesis must represent a significant scientific advance.
- If the thesis includes previously published material, permission to reproduce this must be gained from the respective copyright holder (a maximum 30% of the thesis should be a verbatim reproduction from the author's previous publications).
- They must have been examined and passed during the 12 months prior to nomination.
- Each thesis should include a foreword by the supervisor outlining the significance of its content.
- The theses should have a clearly defined structure including an introduction accessible to new PhD students and scientists not expert in the relevant field.

Indexed by zbMATH.

Gonzalo Álvarez Pérez

Fundamentals of Nano-Optics in Hyperbolic van der Waals Materials

Springer

Gonzalo Álvarez Pérez
Center for Biomolecular Nanotechnologies
Italian Institute of Technology
Arnesano, Lecce, Italy

ISSN 2190-5053 ISSN 2190-5061 (electronic)
Springer Theses
ISBN 978-3-031-75766-2 ISBN 978-3-031-75767-9 (eBook)
https://doi.org/10.1007/978-3-031-75767-9

This Springer imprint is published by the registered company Springer Nature Switzerland AG
The registered company address is: Gewerbestrasse 11, 6330 Cham, Switzerland

A mis padres,
y a Güelita.

Deixarse levar pola luz que desmonta as ideas.

Cedeira
Guadi Galego

Methinks that in looking at things spiritual, we are too much like oysters observing the sun through the water, and thinking that thick water the thinnest of air.

Moby Dick
Herman Mellville

I would like to describe a field, in which little has been done, but in which an enormous amount can be done in principle. This field is not quite the same as the others in that it will not tell us much of fundamental physics (in the sense of "What are the strange particles?") but it is more like solid-state physics in the sense that it might tell us much of great interest about the strange phenomena that occur in complex situations. Furthermore, a point that is most important is that it would have an enormous number of technical applications

---

There is Plenty of Room at the Bottom
Richard P. Feynmann

Sean curiosos: miren donde nadie mira, hurguen donde nadie ve. [...] Tengan paciencia porque todo está ahí: sólo necesitan la complicidad del tiempo. [...] Equivóquense. Sean tozudos. Créanse geniales. Despu´s aprendan.

Zona de obras
Leila Guerrero

Hazte las preguntas, las respuestas llegarán.

Nena
Sexy Zebras

È tutto sedimentato sotto il chiacchiericcio e il rumore. Il silenzio e il sentimento. L'emozione e la paura. Gli sparuti incostanti sprazzi di bellezza. E poi lo squallore disgraziato e l'uomo miserabile. Tutto sepolto dalla coperta dell'imbarazzo dello stare al mondo. Bla. Bla. Bla. Bla. Altrove, c'è l'altrove. Io non mi occupo dell'altrove. Dunque, che questo romanzo abbia inizio. In fondo, è solo un trucco. Sì, è solo un trucco.

La grande bellezza
Paolo Sorrentino

Supervisor's Foreword

It is with great pleasure that we present the doctoral thesis of Dr. Gonzalo Álvarez Pérez entitled *Fundamentals of Nano-Optics in Hyperbolic van der Waals Materials*. This work represents a remarkable contribution to the field of nano-optics, exploring the complex and rapidly evolving field of hyperbolic van der Waals materials and their applications in manipulating light at the nanoscale.

The thesis is exemplary in its scope, depth, and impact. Not only does it significantly advance our fundamental understanding of the interaction between infrared light and anisotropic van der Waals materials at the nanoscale but also presents innovative solutions to challenging related practical problems in nano-optics. It does so by combining experimental exploration, theoretical insights, and computational modeling. The work provides direct demonstrations of novel optical phenomena at the nanoscale—including negative reflection and anomalous refraction—and a series of strategies to manipulate the propagation of infrared light at the nanoscale. These findings are not merely academic; they hold the potential to transform the design of future photonic devices, paving the way for more efficient and compact on-chip optical technologies. Proof-of-concept devices exemplary of this, such as the first planar polaritonic hyperlens or the first the planar nanoresonator for hyperbolic polaritons, are shown in the thesis as well.

Another compelling aspect of this thesis is the introduction of an innovative methodology to extract dielectric functions of anisotropic nanomaterials, like the prototypical mid-infrared anisotropic material, α-MoO_3, by combining far-field spectroscopy and near-field imaging. Such experimental data is augmented by a series of analytical models, together with full-wave electromagnetic simulations and density functional theory calculations. This meticulous approach provides a valuable tool for the scientific community, enabling accurate characterization of complex nanomaterials. In fact, it has already proven to be crucial for further exploration of α-MoO_3, having impacted fields as diverse as infrared optics, sensing, and thermal management.

Gonzalo's thesis also introduces new perspectives in the field of nano-optics, particularly with the concept of polariton canalization and topological transitions in isofrequency curves of polaritons in twisted van der Waals materials. These discoveries have not only contributed to the birth of a new research area, coined as *twist-optics*, but also opened up new avenues for manipulating light at the nanoscale. The potential applications of these findings are vast, ranging from the development of reconfigurable optical elements to the enhancement of light-matter interactions in optoelectronic devices.

Furthermore, Gonzalo's ability to present complex phenomena in a clear and didactic manner makes this thesis both informative and accessible. The work is structured to guide the reader from the most fundamental concepts to more specific and applied aspects, making it an excellent resource for both experts and newcomers to the field. The detailed discussions on the experimental and theoretical techniques used in the research provide a deep understanding of nano-optical tools and instrumentation.

In conclusion, this thesis is not just a culmination of years of dedicated research; it is a testament to the potential of innovative scientific inquiry. Dr. Gonzalo Álvarez Pérez has demonstrated skill, creativity, and perseverance in his work, and his thesis will undoubtedly trigger future developments in nano-optics. The academic and scientific communities will surely benefit from the insights and advancements presented in this thesis, and it is with great confidence that we look forward to the continued impact of Gonzalo's research in the years to come.

Donostia-San Sebastián
Oviedo, Spain
September 2024

Alexey Nikitin
Pablo Alonso González

Preface

Light, a ubiquitous natural resource, plays a central role in our lives. On the most fundamental level, through photosynthesis, light is at the origin of life itself. It is also the reason why we can see, one of our most primary tools for understanding our surroundings and making sense of the world. The study and manipulation of light have led to innumerable technological advancements in diverse scientific fields, including communications, chemistry, security, and disease diagnosis. Optical instruments, such as lenses, microscopes, cameras, radios, routers, optical fibers, or lasers, have become indispensable in our daily lives, as they have transformed the way we communicate, and hence reshaped our reality.

These technologies were developed through centuries of fundamental research on the properties of light. The genesis can be traced back to Euclid's contributions, which explored the principles of reflection and vision in his seminal work "Optica" around 300 BCE, and laid the foundation for geometric optics [1]. Ibn Al-Haytham's seminal opus, "Kitab al-Manazir" (Book of Optics), published in 1015, revolutionized optics and the study of light by introducing a systematic experimental approach and laying the foundation for the understanding of vision, reflection, refraction, and the nature of light itself [2]. Ibn Al-Haytham is often considered one of the early proponents of the scientific method, as he emphasized the importance of experimentation and observation to validate theories. He also stressed the role of light rays entering the eye from external objects rather than the earlier belief, proposed by Euclid, that the eye emitted rays to perceive objects.

Subsequent progress flourished notably between the fifteenth and twentieth centuries. During the Renaissance, and in parallel to his pioneering portrayals of volume, depth, and texture in his artworks, da Vinci's keen observations and insights regarding light's interaction with surfaces and objects contributed significantly to the nascent comprehension of reflective and refractive phenomena [3]. These insights laid the essential groundwork for the evolution of optical theories and techniques across subsequent centuries. Kepler introduced the inaugural coherent explanation of image formation by lenses, thereby laying the bedrock for modern optics and the study of light propagation [4]. Snell elucidated the pivotal relationship between angles of incidence and refraction as light traverses distinct media [5].

This fundamental law, which holds sway in the realm of optics, played a key role in the evolution of lenses and in comprehending the bending of light when transitioning between different media. Snell's law still plays a central role in today's undergraduate courses in Physics, and its generalization to materials investigated nowadays will be explored in Chaps. 5 and 6 of this thesis. Shortly after Snell's studies, Descartes [6] and Newton [7] developed the corpuscular theory, rooted in the notion of discrete particles as the elemental building blocks of matter. This theory stood in contrast to Huygens' groundbreaking wave theory, which proposed that waves propagated through a medium to explain diffraction, and interference phenomena [8]. The different paradigms of the corpuscular and wave nature of light precipitated profound debates and advancements until Young's celebrated double-slit experiment unveiled an interference pattern upon projecting light through closely spaced apertures onto a screen [9]. This interference only found explanation when understanding light as a wave phenomenon. Young's comprehensive exploration of light's interference and diffraction thereby furnished compelling substantiation for its wave nature. Fresnel's groundbreaking work on polarization elucidated the transverse nature of light waves and clarified how light travels in straight lines [10]. He also derived the transmission and reflection coefficients at the interface between two media, known as Fresnel's coefficients, and made significant contributions to the study of biaxial crystals, wherein incident light can split into two beams traveling along different directions. Fresnel formulated the mathematical equations that describe the optical behavior of this type of crystals, known as Fresnel's equations. Biaxial crystals and the propagation of light in them will be extensively discussed across this thesis.

Several decades later, Maxwell's seminal contributions [11], experimentally corroborated by Hertz, culminated in the unification of electricity and magnetism—electromagnetism—leading to the understanding of light as an electromagnetic wave, a form of energy that can travel through vacuum space without the need of a material medium to move: its electric and magnetic fields interact and self-propagate through empty space. All these revelations heralded the dawn of a new era marked by empirical investigation rather than intuitive speculation, with breakthroughs in the unraveling of interference, diffraction, and polarization phenomena, as well as in the understanding of reflection and refraction. The study of these two fundamental optical phenomena plays a central role in this thesis, and two of their most exotic forms will be studied in Chaps. 5 and 6.

The early twentieth century witnessed a cascade of crucial scientific revelations concerning light, untangling its intricate nature. Planck's pioneering work in 1900 on blackbody radiation introduced the concept of quantized energy levels, laying the cornerstone for quantum mechanics, a linchpin in modern physics [12]. Quantized energy levels refer to specific, discrete amounts of energy that particles or systems can have, rather than any value in between. This concept helps explain how atoms and particles behave in a unique way. Building upon this, Einstein's elucidation of the photoelectric effect [13] in 1905 exposed light's dual character as both waves and particles—the latter later named as *photons* in 1926 by Lewis—, underpinning quantum theory. Further expanding on this, de Broglie posited a new hypothesis in

1924: that not only photons but all particles, like electrons, can exhibit wave-like properties, unveiling the wave-particle duality [14]. These revolutionary insights, along with contributions from visionaries like Bohr, Heisenberg, Schrödinger, Pauli, and other contemporaries, marked a departure from classical physics and spurred the birth of quantum mechanics. These studies revealed that, at the atomic and molecular level, physical systems exhibit highly counter-intuitive phenomena, like quantum superposition, entanglement, tunneling, or the aforementioned wave-particle duality, challenging our classical understanding of nature and opening the door for the development of novel technologies like quantum computing, secure communication, and ultra-precise measurements.

Expanding on these advancements, researchers in the latter half of the century delved not only into studying but also into controlling and manipulating materials at the atomic and molecular scale—a dimension billions of times smaller than a meter, commonly referred to as the *nanoscale*. This exploration gave rise to nanotechnology, a field intricately intertwined with the foundational discoveries of the twentieth century, particularly quantum mechanics. Nanotechnology is dedicated to the precise control and shaping of matter and devices at the nanometer scale. The conceptual groundwork for nanotechnology was laid by Feynman's 1959 lecture "There's Plenty of Room at the Bottom," envisioning the potential to manipulate and engineer matter atom by atom [15]. During the 1980s, a groundbreaking decade for nanotechnology, these visionary theoretical concepts materialized into a tangible reality with transformative applications across diverse fields. Binnig and Rohrer pioneered the scanning tunneling microscope [16], while Binnig, Quate, and Gerber spearheaded the creation of the atomic force microscope [17]. These revolutionary instruments allow the imaging of samples at the nanometer scale with unprecedented precision, achieved through the use of extremely small and sharp probes and advanced electronics. Consequently, scientists gained the ability to visualize and manipulate individual atoms for the first time, significantly advancing the development of nanotechnology.

However, a significant challenge arises when trying to bridge nanotechnology and the study and manipulation of light at the nanometer scale. Light cannot be easily controlled at scales much smaller than its wavelength due to the so-called diffraction limit, which had been postulated by Abbe [18] in 1873. This limit arises from light's wave nature and restricts its ability to resolve details smaller than its own wavelength. Attempting to shrink light to nanoscale dimensions involves grappling with the inherent wave nature of photons. These photons naturally disperse and resist compression beyond a certain limit, analogous to attempting to fit a bulky object through a narrow funnel. This is in fact the reason why the human eye cannot perceive objects at the nanometer scale, such as atoms or molecules, as its wavelength detection range is in the hundreds of nanometers. As a result, most of the existing optical devices are largely constrained by diffraction. Bridging the precise control of nanoscale structures with the unique properties of light could trigger advancements in optical computing, high-resolution imaging, and sensing. This convergence is vital not only for our understanding of light and nature but also for the development of more compact, integrated, efficient, environmentally friendly, and high-performance technologies.

Nevertheless, while we cannot physically shrink light itself, there are solutions to overcome the diffraction limit. An approach that has drawn significant attention in the last decades was introduced in a pioneering work by Hopfield [19] in the 1950s. It consists on exploiting the synergistic interaction of light with matter. When light interacts with materials which support collective dipolar excitations, like collective movements of free electrons in metals [20] or collective vibrations of atoms in polar crystals [21], light and the dipolar excitations start to exchange energy. It is said that light *couples* to the dipolar excitations. Depending on how much energy they exchange, the coupling can be weaker or stronger. When the coupling is strong enough, this is, when they are able to exchange enough energy, they act like a single entity and a light-matter hybrid, called *polariton*, forms.

Due to their hybrid light-matter nature, polaritons have a larger momentum than that of light, and thus, offer superb field confinement and enhancement compared to traditional electromagnetic radiation. The confinement of optical fields below the diffraction limit in metals initially sparked the rise of plasmonics [22], unleashing a realm of fascinating physics and giving birth to the fields of nano-optics [23,24] and important applications of metamaterials [25], materials engineered to exhibit properties not usually found in natural materials. In the last few decades, polaritons have also been widely studied in polar crystals. Moreover, since the optical properties of noble metals and polar crystals can be tailored, polaritons can be controlled and customized in these types of materials, opening up opportunities for controlling light at the nanometer scale and developing technological applications. In fact, these advancements have already led to a myriad of discoveries such as ultrasensitive sensing and spectroscopy, improved chemical detection, super-resolution imaging, or enhanced quantum efficiency for detectors across the ultraviolet, visible, and infrared spectral ranges, to name a few.

More recently, advancements in nano-optics have been driven by the advent of van der Waals (vdW) materials, compounds made up of thin layers which are held together by weak forces. Due to the weakness of such interlayer forces, these materials can easily be separated into sheets as thin as a single atom, like taking a page out from a pile. Novoselov and Geim successfully isolated and characterized in 2004 the first atomically thin material: graphene [26], a single layer of carbon atoms arranged in a two-dimensional honeycomb lattice. The discovery of graphene initiated a cascade of research studies that showcased the exceptional properties that can arise from two-dimensional structures. The possibility of manually peeling off atomically thin layers from a bulk material using adhesive tape, together with the remarkable mechanical, optoelectronic, and thermal properties of vdW materials, opened up further possibilities for developing extremely compact nanodevices with tailored properties with relative easiness, holding promise for advancements in electronics, energy storage, sensing, and other cutting-edge technologies. It turns out that some of these extremely thin layers can also have remarkable properties when it comes to the confinement and control of light at the nanoscale, as they can support polaritons with nanoscale dimensions, allowing for record-high confinement at the nanoscale [27,28]. With the parallel development of state-of-the-art imaging techniques such as near-field optical microscopy [29], a field of research on polaritons in vdW materials was sparked over the last decade, allowing the demonstration of proof-of-concept nano-optical devices, like nanolenses and nanoprisms.

In vdW materials, the inherent differences between the forces that hold the layers together and those within each layer result in anisotropic crystal lattices. This anisotropy sparked a great interest in nanophotonics [30]. In fact, anisotropy has been a key property employed in the design of optical components for hundreds of years. In a crystal, the anisotropy is originated by the atoms being arranged differently along different directions, mainly due to the weak forces that hold the layers together, as opposed to the much stronger covalent bonds that keep the atoms joint within each of the layers. This anisotropy affects the properties of each layer, like its conductivity, strength, or its optical properties. It can also have a significant impact on the collective vibrations of atoms within a material, known as *phonons*. This can lead to varying speeds and patterns of vibration in different directions, resulting in anisotropic phonons. When coupled to light, these anisotropic phonons can interact with electromagnetic waves in distinct manners along different directions, giving rise to what are known as *anisotropic phonon polaritons*. In particular, when the material is strongly anisotropic, phonon polaritons can propagate only along some directions, while their propagation is forbidden along others, opening the door for a directional control of light at the nanometer scale. In addition, the wavelength varies depending on the direction of propagation, and it can actually be incredibly small, allowing for an extremely high confinement. This type of extremely directional propagation, known as *hyperbolic* propagation, can lead to intriguing optical phenomena, such as negative reflection and refraction or an enhanced density of optical states. These phenomena are at the core of extraordinary applications, such as superlenses or invisibility cloaks. Excitingly, in some of these vdW materials, such as the semiconductor alpha-phase molybdenum trioxide (α-MoO_3), the crystal anisotropy can occur not just due to the weak interlayer forces, but along every direction, even within each individual layer [31]. As a result, hyperbolic propagation of phonon polaritons can also occur along the surface of the material. This provides an unprecedented platform not only to study and visualize counter-intuitive optical phenomena, now directly accessible in experiments, but also to precisely manipulate the energy flow at the nanoscale. In recent years there has been a growing interest in the quest for materials with ever higher optical anisotropy for their ability to compress light to smaller dimensions and display exotic optical phenomena. This is currently a key direction in the field of optics and nano-optics, as evidenced by the intensive study of strongly anisotropic phonon polaritons supported by vdW slabs, heterostructures, and bulk crystals [32,33].

Interestingly, most of these strongly anisotropic phonon polaritonic excitations take place within the mid-infrared (mid-IR) and terahertz (THz) spectral bands, as collective vibrations in solids have energies in such ranges. The nature of mid-IR or THz light is similar to that of visible light, with the only difference that they possess a lower energy than the one that can be detected by our eyes. The mid-IR range is significant for spectroscopy and chemical sensing because many important molecules have characteristic vibrational modes at these frequencies. By analyzing the absorption and emission of light in the mid-IR range, chemical compounds and their properties can be detected and identified. In addition, the mid-IR is also important for thermal applications, since the blackbody emission reaches its maximum in this range, allowing for the detection and measurement of heat signatures and thermal patterns. On the other hand, the THz range is important for imaging, communication, and security applications. THz light can penetrate materials like clothing, plastics, and papers. This capability makes it suitable for medical imaging, where it can provide detailed information about biological tissues, as well as for wireless communication and security scanning, where it can detect concealed objects. Therefore, learning to control, steer, and manipulate strongly anisotropic polaritons at the nanometer scale in the mid-IR and THz ranges is important for deepening our fundamental understanding of light-matter interactions, and it is also a promising avenue for the development of novel nano-optical technologies with unprecedented functionalities.

The work that comprises this thesis started in 2018, soon after the discovery of α-MoO_3 as the first natural material supporting strongly in-plane anisotopic phonon polaritons [31]. Part I lays the groundwork for the subsequent chapters. Namely, Chap. 1 provides an overview of propagating polaritons in metals, polar crystals, graphene, and the vdW polar crystals hBN and α-MoO_3. We put special emphasis on the role played by optical anisotropy, particularly on naturally hyperbolic materials. Chapter 2 gives an introduction to the main analytical, numerical and experimental methods used throughout the thesis. After these two introductory chapters, Part II focuses on how to predict the polaritonic response of strongly anisotropic media and, in particular, α-MoO_3. Specifically, the dispersion relation for polaritons in slabs of α-MoO_3 and other strongly anisotropic materials is derived in Chap. 3. In Chap. 4, we report on the experimental extraction of the dielectric permittivity of α-MoO_3 through a combination of far- and near-field methods. Building on these predictive capabilities and taking advantage of the unique properties of strongly anisotropic phonon polaritons, exotic instances of two very fundamental optical phenomena—reflection and refraction—are discussed in Part III. We will also discuss the potential technological advancements that such studies enable for nanophotonics. Namely, in Chap. 5, we discuss the first visualization of negative reflection of polaritons at the nanoscale. Based on our analysis, we design the first in-plane cavity for strongly anisotropic polaritons. In turn, in Chap. 6, we discuss the visualization of anomalous refraction of polaritons at the nanoscale and design the first planar lens for polaritons, improving the focus size by a factor of 50. In the final part of the thesis, Part IV, we develop strategies to control, direct, and guide hyperbolic phonon polaritons, crucial for their implementation in nanotechnologies. Chapter 7 shows how to steer hyperbolic phonon polaritons along directions that are, in principle, forbidden for propagation. Chapter 8 is devoted to the birth and latest research of twist-optics: an emerging field dedicated to engineer and control of polaritons by stacking strongly anisotropic slabs with a relative angle between them. Chapter 9 introduces a device that combines graphene and a strongly anisotropic slab to dynamically manipulate hyperbolic phonon polaritons via an applied voltage. Finally, Chap. 10 summarizes the research carried out in Prof. Basov's laboratory during a Fulbright stay at Columbia University in 2022, aiming to develop mechanically in-operando rotatable bilayers.

Altogether, the findings presented in this thesis deepen our understanding of the propagation of strongly anisotropic phonon polaritons in vdW materials, and thus of the fundamental aspects of nano-optics in hyperbolic media. Encompassing a comprehensive spectrum, these results span from the analytical framework and accurate characterization of one of the most archetypal mid-infrared hyperbolic materials, α-MoO_3, to the conception and development of technologies specially promising for practical applications and implementation, including the rise of a new research field: twist-optics. Our work also delves into the investigation and elucidation of three fundamental yet intriguing phenomena—negative reflection, anomalous refraction and canalization—, their intrinsic properties, and their potential applications. In fact, this thesis provides the first visualization of some of these effects, and thus its impact goes beyond the realm of mid-infrared nanophotonics. In sum, this work provides a very broad exploration into the fundamentals of nano-optics in hyperbolic two-dimensional materials.

Mieres, Spain
2023

Gonzalo Álvarez Pérez

Acknowledgments/agradecimientos

At first, you'll have to learn to use your new body, to move your arms and legs, to pull yourself upright. You will learn to walk and run, to use your hands to make sounds and form words. Slowly, you will learn to take care of yourself. [...] the colors, sensations and sounds will wash over you constantly. You will see so many living things, plants and animals, beyond imagining. [...] everything is in motion, everything is always changing. There will be so much for you to learn and so much for you to feel, pleasure and fear, joy and disappointment, sadness and wonder. In your confusion and delight, you will forget where you came from.

Star child
Claire A. Nivola

Im Grunde sind es immer die Verbindungen mit Menschen, die dem Leben seinen Wert geben.

Wilhelm von Humboldt

It takes a village es un proverbio africano que dice que se necesita toda una comunidad para criar a un niño. Creo que aplica a todos los logros de la vida: casi nada es mérito únicamente propio. Y esta tesis no es una excepción. Empecé esta aventura hace ya más de cinco años, lleno de dudas e incertidumbre. Durante este tiempo he sido feliz a nivel personal y he disfrutado haciendo ciencia y trabajando con gente a la que admiro y aprecio. Y, aunque el trabajo lleve mi nombre, en realidad pertenece a todas las personas que me han estado enseñando, ayudando, rodeando y queriendo de mil maneras diferentes durante los últimos años. Mi más sincero y profundo agradecimiento a todos.

Esta tesis se la debo a mis supervisores, Pablo y Alexey. Ha pasado ya tiempo desde el 27 de julio de 2017, cuando acudí por primera vez al despacho de Pablo, por aquel entonces en la Facultad de Ciencias en Oviedo. Allí hablamos sobre lo que podría ser (y al final fue) el tema de mi TFM y mi tesis de doctorado. Pablo me habló de polaritones, de grafeno, de transformadas de Fourier que daban como resultado una hipérbola y de algo que tenía «muchos *k*'s». La verdad es que entendí muy poco de todo aquello. De hecho, tardé unos cuantos meses en entender lo que era *realmente* un polaritón, si es que a día de hoy *realmente* lo entiendo. Pero decidí unirme por la confianza, la pasión y el buen rollo que Pablo me transmitió.

Echando la vista atrás, definitivamente fue una buena decisián. Durante todo este tiempo he aprendido un montón y hemos estado haciendo física muy interesante. Gracias, Pablo. No solo me convenciste para elegir la ciencia, sino que además me diste la oportunidad de hacerlo en tu equipo. Ahora puedo decir que entiendo un poco mejor los conceptos de los que me hablaste aquel primer día, y eso no habría sido posible sin tu experiencia, conocimientos, esfuerzo y paciencia. Gracias por enseñarme a analizar datos, a diseñar figuras, a preparar diapositivas, a dar charlas, a escribir artículos y respuestas a revisores. Por remangarte y ponerte manos a la obra conmigo para hacer un ajuste, repasar una y otra vez un abstract o una imagen, paciente y constructivamente. Por tomarte el tiempo para leer esta tesis en detalle. En definitiva, gracias por enseñarme nano-óptica y mostrarme cómo se hace la ciencia, día a día. Aunque no solo me has enseñado eso, también me has aconsejado cuando tenáa dudas sobre la vida, sobre el futuro. Como decía Pedro Etxenike en sus *Consejos a un joven científico*, la ciencia requiere sosiego. Y tú me has dado sosiego, pero también me has animado y corregido cuando ha hecho falta. Siempre he sentido un fuerte respaldo de tu parte, sabiendo que buscabas lo mejor para mí: me has dado todas las facilidades para ir a conferencias, hacer estancias, solicitar becas, establecer colaboraciones y buscar postdocs. Gracias también por la buena onda, por las bromas, las experiencias en viajes y congresos y las anécdotas fuera del trabajo. Gracias por compartir tu pasión por la ciencia y exponerme a un mundo que ha definido tanto mi vida. Por cierto, me enorgullece mucho que hayas apostado por Asturias y el trabajo que estás haciendo aquí. Ha sido un auténtico placer ser tu alumno de doctorado. Presumiré, no solo de director de tesis, sino también de haber tenido a un mentor como tú.

Alexey, tengo tantas cosas que agradecerte. Para empezar, guiarme en mis primeros pasos en investigación. Aún recuerdo cuando llegué a nanoGUNE en octubre de 2017. Aquel día empezaba mi primer trabajo serio, donde me pagaban por estudiar. Tú me decías que derivase la relación de dispersión en láminas biaxiales, que no hacía falta que me fijase tanto en cómo lo hacían en otros papers, que simplemente me pusiese frente a una hoja en blanco y la derivase. Y yo no sabía ni por dónde empezar. Me costó un poco, pero al final (gracias a ti y a la ayuda clave de Kirill) lo conseguimos. Has sido muy paciente conmigo. Me has enseñado mucho, sobre todo a ser analítico. De ti he aprendido a buscar las soluciones a problemas difíciles partiendo desde los casos más sencillos, una enseñanza que va mucho más allá de lo científico. También a quedarme solo con los términos importantes y a despreciar los que importan menos, no solo en las ecuaciones. Gracias por tu apoyo continuo e incondicional, por tu profesionalidad y buen hacer. Por invitarme a Donosti para poder trabajar juntos en los proyectos teóricos. Gracias por acogerme siempre que me ha hecho falta, me has hecho sentir uno más en tu grupo con las excursiones, a la txakolindegia, a Lukas. Seguro que estos últimos años no han sido fáciles, y te agradezco de corazón haber estado siempre ahí.

Al igual que he tenido mucha suerte con los supervisores de mi tesis, también la he tenido con los compañeros de trabajo en Oviedo. Javi Martín, casi todo lo que sé de COMSOL me lo has enseñado tú. Gracias por todo tu apoyo y consejos. Gracias por todas las discusiones, las científicas y las no científicas. Queda pendiente una conversación sobre la superficialidad para dentro de diez o quince años. Aitana, gracias por el buen humor y buen rollo que contagias. Duan, your help has been unvaluable during these years. To be honest, I wouldn't have even faced many interesting problems without your ideas, energy and help. You're a true scientist, and you spread the passion you feel for science to others. Some results of this thesis wouldn't be here if it weren't for you. By the way, I really hope we can make that trip to Datong together in the next years!

Javi Taboada, Ana, Abel, Christian, Adam, José, Enrique, Julia, Lucía, Jorge, Patri. Os merecéis un puesto especial aquí. No puedo sentirme más afortunado y agradecido de haber formado parte de este grupo, esta familia (de peña algo rara, un poco personajes, hemos de reconocerlo). Javi Taboada, compañero de doctorado desde el primer día. Por fuera, duro como una marañuela, pero por dentro mucho más blando. Siempre dispuesto a ayudar, sincero como nadie y más bueno que el pan. Muchas gracias. Christian, primer compañero de *teoricadas* en el grupo, listo para una birra en todo momento. Llegaste a este grupo y fuiste el pegamento que nos unió a todos un poco más. Ha sido un placer haber discutido contigo de ciencia, y, sobre todo, de tantos temas fuera de la ciencia, empezando por los Dire Straits. Muchos buenos momentos fuera de la oficina. Algunos viajes. También dentro de la oficina. Muchas risas. Muchas estupideces. Mucho tiempo *perdido*, pero con mucho gusto. Ah, y mil gracias por las correcciones en el prefacio. Ojalá un día no sea necesario discutir la necesidad de defender y proteger la cultura propia. Ana, gracias por tantas muestras y medidas que forman parte de esta tesis, y por el apoyo, los consejos, las confidencias, y todo lo que hemos vivido juntos estos años, que no ha sido poco. Abel, Adam, me alegro mucho de haberos tenido al lado durante esta etapa. José, qué bueno parecias antes de empezar, cuando todavía eras alumno de TFG, pero, ¡ay, amigo!, luego se descubrió el pastel. Eres un crack, Una pena no haber coincidido más tiempo. Para hacerte un par de caños en las pachangas, y para seguir aprendiendo de tus masterclasses de empatía y psicología. Te echaré de menos. Enrique y Julia, viene bien un soplo de aire fresco y de organización en el laboratorio, llegáis fuerte las nuevas generaciones. Lucía, qué poco hemos coincidido, pero ha bastado para ver la persona tan guay que eres, comprometida con lo que de verdad importa. Estoy seguro de que te va a ir genial en esta nueva etapa. Patricia, qué bueno llevarme de estos años a una persona con la que comparto tanto: tantos intereses, tantas emociones. Te comes el mundo si quieres. Jorge, compañero de analógicas, crack del beerpong, el DFTista más fan del motxo y seguramente la persona con menos fairplay sobre la faz de la tierra. Ha sido muy guay haber compartido este tiempo con todos vosotros. Dentro del trabajo, donde habéis creado un ambiente tan cómodo que ir a currar cada mañana se haya sentido como si no fuera trabajo, pero sobre todo fuera. Estos años ha habido historias y anécdotas para rellenar varias temporadas de una sitcom o de una telenovela. De las cutres. De las que enredan la trama, con dramas retorcidos y absurdos entre los personajes, giros de guion totalmente injustificados, reapariciones y cameos de personajes antiguos. Me da pena irme del Aumente, del Peluche, dejar las pachangas, dejar el bar de Fran, porque estoy seguro de que las siguientes temporadas no defraudarán.

Gracias también a todos los alumnos de grado que habéis pasado por el laboratorio durante esta etapa y que habeís contribuido a generar un ambiente tan familiar: Lucas, Ana de Castro, Miriam, Ceci, Andrés, Samuel.

Gracias a los miembros de la Comisión Académica. A María, Carlos y Víctor, por escucharnos atentamente durante largas sesiones y por hacer siempre preguntas tan acertadas. Por recordarnos los plazos y los trámites burocráticos. María, gracias también por tutorizarme durante esta tesis. Gracias también a Jaime y a José Ignacio, y a todas las personas del Departamento de Física que habéis estado siempre dispuestas a echarnos una mano. Gracias a Bea, Marta, Eduardo y María, de los SCTs, siempre con una sonrisa, da gusto trabajar con gente así.

En estos años, he tenido la enorme suerte de trabajar con un montón de científicos en varios grupos y lugares diferentes. Empezando por el grupo de Alexey en Donosti. Kirill, you are the most talented person for calculations and simulations I've ever met. It's been a privilege to work with you so closely on so many projects and to learn so much from you. Chapter 3 and some other parts of this thesis wouldn't be here if it weren't for your talent and key contributions. Thank you. Javi Alfaro, quién nos iba a decir cuando nos conocimos en nanoGUNE que viviríamos unos meses en Nueva York a la vez. Gracias por tu ayuda y por resolverme tantas dudas, desde que empecé mi TFM hasta la escritura de la introducción de esta tesis. Arturo y Nathaniel, ha sido un placer trabajar con vosotros. Sois unos cracks. Arturo, el capítulo 9 no estaría aquí sin tu trabajo. Nathaniel, la primera parte del capítulo 8 tampoco sin el tuyo. Andrei, Kate, Olga, it's been a pleasure too. Rainer, thanks so much for your suggestions and comments, always on point, on our works. I am fascinated by your ability to find the weak and strong points of a paper at first glance. I've learned a lot from you, and your insights have substantially improved our papers, and therefore some parts of this thesis. Thank you. Many thanks to all the people in your group too: Lisa, Marta, Andrei, Irene, and the rest, it's been great to work with you. Ion Errea, mil gracias por tus cálculos DFT para el Capítulo 4, por la buena onda siempre. Al resto de la gente de Donosti con la que he coincidido durante el máster y después, en las estancias de doctorado y conferencias: Gabriele, Carlos, Jon, Nodar, Antton, Irene, Txemikel, Cris, y los que me dejo. Todos vosotros, y los pintxos/pizza potes, habéis convertido a Donosti en mi segunda casa durante este doctorado. Mila esker.

Thanks a lot to the Dresden team: Thales, Susanne, Max, Tobias, Lukas, Mike, and the rest. It's been a pleasure to work with you and to expand some of the findings on hyperbolic polaritons in the mid-IR to the THz regime. And a privilege getting to do experiments at the FELBE in Dresden. Thanks also to Qingdong and Qiaoliang from Monash. Gracias a Iván PG por la fabricación de las muchas antenas de oro utilizdas en esta tesis.

I've had the luck to undertake many short—and not-so-short—stays abroad. The first one, back when I had barely started my PhD, in Nashville. Josh, you've always been very attentive and generous with me. I appreciate the trust and hospitality, especially for hosting a total newbie like me in your lab at Vanderbilt. It's also been great catching up with you in different parts of the world (Nashville, DC, Dresden) during these years and learning new physics every time we chat. Thanks a lot for taking the time to have a look at this thesis and for your insights and comments, too. Tom, thank you for the help with the FTIR and the W-VASE software. You've taught me a lot and you've been crucial for our work on MoO_3. Chapter 4 wouldn't be here without you. Joseph, thanks for your help too. And thanks for showing me Nashville around—never not missing those hot biscuits from The Loveless Cafe. Ryan, Katja, Aditah, Saurabh, it's been great to work with you these last few months in Dresden (except for the lack of sleep in the shifts). Coincidentally this (long) thesis has been kind of enclosed in time in between collaborations with the Vandie group and I've gotten to see the evolution of its members. Ryan, thanks for your help with the FTIR pics, too. (I'll miss the German chocolate waffle snacks!)

Si mi primera experiencia en EEUU fue buena, la segunda fue inolvidable. En algún sitio leí que «nadie puede rechazar una invitación a vivir en Nueva York». No imaginaba yo cuando comencé esta tesis que terminaría viviendo allí durante nueve meses. Dmitri, thank you for hosting me and being so supportive. You have sponsored one of the best experiences of my life. I treasure all the time I stole from you to talk about Physics. Thanks also to Cory Dean and James Hone for allowing me to use their labs. To Frank, Sam, Maëlle, thank you for your help in the lab trying to rotate those very stubborn bilayers. Frank, always grateful for showing me some of the best spots in NY, too. To all the members of the groups: Shuai, Rocco, Bjarke, Yinan, Brian, Aaron, and the rest. Gracias a Lucía Herrer y José Mª de Teresa por tantas muestras. Sigo confiando en que podamos rotarlas. Gracias infinitas a las personas detrás de la Comisión Fulbright—Sara, Patricia, Victoria, Alberto—por el apoyo, la confianza y la cercanía. Por hacerlo todo tan fácil y ser tan atentos, siempre con una sonrisa. Sin duda, mucho más que una beca. Gracias a los que me ayudasteis leyendo una y mil veces la solicitud, practicando la entrevista: Pablo, Alexey, Jesús Blanco, Javi Martín, Alba, papá, Jesu, Marta Lomba, Miguel, Christina. A todo el equipo que hicisteis que me sintiera como en casa durante esos nueve meses en NY: Sandra, Nico, Małgosia, Laura, Pauliña, Thales, David, Juan, Chris, Wiebke, Lars, Helena, Sydney, Patrick, Albert, Marco, Sofía, Alex, Guille, Marta, Miguel. Algunos de vosotros sois lo mejor que me llevo de NY. Repetiría mil veces esos meses llenos de risas y planes con todos. Meses también de transformación, de cambiar perspectivas y descubrirme más a fondo.

Al poco de regresar de Nueva York, tuve la oportunidad de pasar algo más de tres meses en Berlín. Alex, right from the start you were incredibly kind and generous; it has been an absolute pleasure working with you. I can't thank you enough for letting me join your group and for always looking out for me, whether inviting me to collaborations, beamtimes, having a look at this thesis, or discussing Physics while you were having a smoke on the benches outside the Fritz Haber Institute. I feel like I've learned way more about free electron lasers, crystal structures and phonons in three months in your group than ever before. I would like to thank everyone in the group: Giulia, Dory, Richarda, Niclas. Thank you for welcoming me so warmly and for being so hospitable and attentive; you create an awesome atmosphere that feels like a wonderful family, where you take care of each other. Giulia, it's been awesome to revisit the basics with you without any fear of judgment. I already miss all our late-evening conversations, whether in the FHI or in the miserable room, and whether about the reality of LO phonons, the scientific system, or politics. You're great. The world would be a better place if more people were as genuinely good, kind and sympathetic as you. Gabi, por supuesto estas palabras tambien aplican para ti. Niclas, thank you so much for having a look at this thesis, for your feedback and suggestions. I've learned a lot from you. Muchas gracias por abrirme las puertas de vuestra casa, por la hospitalidad (y el mole). Also, my gratitude to the whole team at the Physical Chemistry Department for creating such an exceptional scientific and familial atmosphere that one longs to work in. Michael, Ben, Álvaro, Max, Alex, Ugaitz, Alkisti, Joanna, Luis, Daniel, Sebastian, Marcel, Melanie, Martin. Manuel and Daria, from the administration team, you are fantastic. Vielen Dank an alle und bis bald.

Gracias a los profesores que he tenido a lo largo de mi vida y que han sido una auténtica inspiración. Jesús Blanco, creo que mi verdadera curiosidad por la ciencia y por la física de la materia condensada nació gracias a ti. Recuerdo todas las charlas que mantuvimos durante tu supervisión de mi TFG, muchas veces más allá de la ciencia, los consejos impagables, tu ayuda con la Fulbright. Profesor y científico de vocación, de esos que se recuerdan siempre. Gracias a Gerardo Domenech, un apoyo, especialmente durante mis primeros pasos en la física. Los cafés, algunas veces más regulares, otras menos, pero siempre continuos, ayudan a uno a ser consciente de los cambios, a evaluar el paso del tiempo. Gracias a Edu Portugués por enseñarme estadística y a hacer ciencia en otro campo. Ha sido un placer trabajar juntos. También me gustaría agradecer profundamente al Dr. Castro, por hacerme ver hace veintinueve años. Tu valentía y buen hacer fueron absolutamente determinantes en mi vida. Sin ti, no habría llegado hasta aquí, ni a muchos otros puntos antes, ni por supuesto sería quien soy a día de hoy. Gracias también a Victoria por escucharme durante estos últimos meses, por los consejos, el apoyo, y el esfuerzo. Haces que me conozca y entienda un poquito mejor.

La investigación de esta tesis ha sido mi trabajo durante estos años. He defendido fervientemente entre familia, amigos y ante quien fuera que el doctorado es un trabajo. Pero un trabajador no puede trabajar sin un salario. Gracias al apoyo económico de la Consejería de Educación, Universidades e Investigación del Gobierno de Asturias. La ayuda (que no debería llamarse ayuda) Severo Ochoa ha sido mi sustento desde febrero de 2020. Agradezco también a Pablo y a la ERC Starting Grant 2DNANOPTICA el contrato que me permitió iniciarme en la ciencia. Gracias a este contrato pude tener un sueldo hasta conseguir la beca. Qué pena que estas oportunidades no sean la norma y muchos desistan de seguir en ciencia y otros tengan que currar gratis los primeros meses, si se lo pueden permitir. A veces olvidamos la larga lista de privilegios sociales y económicos sobre los que se suele asentar la realización una tesis doctoral, y a los que no todo el mundo tiene acceso, desde aspectos generales como educación y salud (públicas) de calidad hasta otros más específicos como un entorno social apropiado o la tranquilidad necesaria para hacer investigación.

Por ello, en esta parte final quiero agradecer a los míos, a los de siempre. Me siento muy afortunado de tener los amigos y familia que tengo. Esta tesis tiene mucho de vosotros. Gracias a los que veo a menudo y a los que, por desgracia, no tanto, aunque hay amistades que no se rompen nunca. Jesu (el que siempre estuvo, está y estará, un hermano), Miguel (qué bueno haber compartido viajes y anécdotas juntos estos años, y haberte visto crecer tanto), Alper, Nuo Alicates. Peleteiro (ejemplo de superación y de ética), Marta y Álvaro (lo que unió Reflexu que no lo separe nadie), Estela, Sandra (El Nano siempre presente), Leti (qué orgullo haber visto los primeros ensayos de Comala, sois unos artistas), Borja y Francesca, Alessandro, Fer, Natalia, Donato. Gracias a Irene Faza, María Abril y Christian por la ayuda con la traducción del prefacio al asturiano. Irene Estébanez, gracias infinitas, por todo lo que compartimos y porque también te debo un poco de esta tesis: sin tu ayuda y recomendación no la habría comenzado. Alba, los primeros años de esta etapa siempre estarán unidos a ti. Gracias por el tiempo que vivimos juntos, por lo que me has enseñado. Contigo guardo momentos que son ya para siempre. Y qué bien que les hayas demostrado a todos, incluso a los que no confiaron, que eres una científica de los pies a la cabeza. Henar, ojalá hubiese más personas tan genuinamente buenas y comprometidas con lo que de verdad importa: cuidar a la gente que uno quiere, lo local y la comunidad, los ideales, los libros, la música. Gracias, por el tiempo juntos, y por descubrirme libros y artistas y lugares a los que luego he vuelto una y otra vez.

A mi familia. A mis tías y tíos: Belén, Luisa, Merce, Juan, Mari, Min, Jose. Gracias por tanto y tanto cariño, por ser unos segundos padres para mí, por tratarme como si fuese vuestro propio hijo. Belén y Juan, gracias por descubrirme tantos rincones en Asturias, por los findes en la cabaña. Gracias a vosotros he aprendido a querer y a apreciar la naturaleza y lo nuestro. Luisa, siempre me has consentido mucho. Ya se notaba que disfrutaba con los experimentos cuando *operaba* naranjas en tu casa. A Zulima, Charo, y el resto. Aquí no puedo poner la lista entera porque ya me quedo sin páginas. A Bárbara, Sandra, Mª José, Maite, Colm, Thierry. A Marta, lo más cercano que tengo a una hermana, sabes lo mucho que te admiro. Eres un ejemplo de fuerza, de resistencia, de supervivencia. Qué bueno haber compartido tanto estos últimos años. Te quiero millones. Gracias a todos. Sé que el tema de esta tesis no es el más accesible, y no es fácil sentir su valor. Por eso me conmueve profundamente cuando siento que apreciáis mi trabajo.

Ya termino. Las últimas líneas son para alguien que se fue durante esta tesis, para alguien que llegó y para los que siempre han estado:

Güelita, ojalá siguieses aquí para celebrar esto. Nos dejaste a todos queriéndote y llenos de buenos recuerdos. Sin ti, esta familia no sería ni una pequeña parte de lo que es. Siento que me faltaba muchísimo aún por aprender contigo y de ti. Eras el mejor ejemplo de ganas de vivir, de cuidado de los tuyos, de bondad, sencillez, humildad y sacrificio. Un modelo a seguir para mí. Tu personalidad es todo a lo que aspiro.

Ahora, la más importante en estos últimos años, literalmente la mejor persona que conozco. No sé ni por dónde empezar. Ojalá tener la habilidad con las palabras de Bolaño o Zambra para describir lo feliz que me haces. *Got the music in you baby, tell me why.* Estos años han sido una aventura y tú has hecho que lo bueno fuese muy bueno y que lo malo fuese un poco menos malo. Contigo me siento en casa donde sea: Asturias, Nueva York, Berlín, Lecce y tantos otros sitios tienen tu huella y eso significa que son hogar para mí. Me das paz y tranquilidad, me entiendes mejor que yo mismo, me haces más autoconsciente, me cuidas hasta niveles inimaginables. Te admiro y estoy orgulloso de ti, de tu empatía, de tu bondad genuina, de tu creatividad, de tu inteligencia, de lo cool y goofy y genial que eres. Conocerte ha sido lo mejor de estos años y, con que lo que está por venir signifique para mí la mitad de lo que hemos vivido hasta ahora, ya querré vivirlo. *Keep hitting the iron bell.* Y, bueno, mil gracias también a Manolita y Jose, me hacéis sentir como en casa.

Finalmente, mamá, papá. Sois, y siempre habéis sido, mi verdadero sustento. Sin vosotros, nunca habría terminado, ni siquiera llegado a empezar, esta tesis. Papá, gracias por fomentar siempre una actitud crítica en mí, desde que me enseñaste a leer; por insistir una y otra vez en que vaya por el buen camino de la vida, aunque no siempre estemos de acuerdo en cuál es ese buen camino; por cuestionarlo todo, creo que eso me ha enseñado a argumentar mejor; por empujarme a ser autoexigente; por la confianza que sé que tienes en mí. Mamá, gracias por apoyar todas las decisiones que he ido tomando; por enorgullecerte más que yo de mis pequeños logros; por lo que te has esforzado para hacerme aprender desde que me enseñaste a leer la hora y a atarme los cordones. A los dos, gracias de corazón por cuidarme siempre, por vuestro amor tan puro y genuino, por vuestro apoyo incondicional y desinteresado, por hacerme sentir tan querido y estar siempre a mi lado, y por haber dado absolutamente todo lo que estaba en vuestra mano para hacerme llegar hasta aquí. Estoy muy orgulloso de los padres que tengo y jamás encontraré la forma de agradeceros todo lo que habéis hecho por mi. Esta tesis es un pequeño reconocimiento a vuestro esfuerzo y cariño. Es vuestra.

> «Dicen que eso es la felicidad: nunca sentir que sería mejor estar en otra parte, nunca sentir que sería mejor ser alguien más. Otra persona. Alguien más joven, más viejo. Alguien mejor.»

En *Poeta chileno*, de Alejandro Zambra, se dice que esta es una idea perfecta e imposible. Pero describe muy bien mi sentir durante estos años, donde ha habido un montón de momentos en los que he sido enteramente consciente de ser feliz en el mismo momento en que lo estaba siendo. Todos vosotros habéis hecho de este doctorado una etapa muy plena, y yo solo puedo seguir diciendo gracias.

Mieres, Spain Gonzalo Álvarez Pérez
2023

List of Publications

Parts of this thesis have been published in the following journal articles. Equally contributing authors are shown:

Articles Contained in This Thesis

Chapter 3

- G. Àlvarez-Pérez, K. V. Voronin et al. "Analytical approximations for the dispersion of electromagnetic modes in slabs of biaxial crystals". Physical Review B 100, 235408 (2019) [34].

Chapter 4

- G. Álvarez-Pérez et al. "Infrared Permittivity of the Biaxial van der Waals Semiconductor α-MoO_3 from Near-and Far-Field Correlative Studies". Advanced Materials 32, 1908176 (2020) [35].

Chapter 5

- G. Álvarez-Pérez, J. Duan et al. "Negative reflection of nanoscale-confined polaritons in a low-loss natural medium". Science Advances 8, eabp8486 (2022) [36].

Chapter 6

- J. Duan, G. Álvarez-Pérez et al. "Planar refraction and lensing of highly confined polaritons in anisotropic media". Nature Communications 12, 4325 (2021) [37].

Chapter 7

- J. Duan, G. Álvarez-Pérez et al. "Enabling propagation of anisotropic polaritons along forbidden directions via a topological transition". Science Advances 7, eabf2690 (2021) [38].

Chapter 8

- J. Duan et al. "Twisted Nano-optics: Manipulating Light at the Nanoscale with Twisted Phonon Polaritonic Slabs". Nano Letters 20, 5323–5329 (2020) [39].
- J. Duan, G. Álvarez-Pérez, C. Lanza, K. V. Voronin et al. "Multiple and spectrally robust photonic magic angles in reconfigurable α-MoO_3 trilayers. Nature Materials 22, 867–872 (2023) [40].

Chapter 9

- G. Álvarez-Pérez, A. González-Morán et al. "Active Tuning of Highly Anisotropic Phonon Polaritons in Van der Waals Crystal Slabs by Gated Graphene". ACS Photonics 9, 383–390 (2022) [41].

Articles Not Contained in This Thesis

- J. Taboada-Gutiérrez, et al. "Broad spectral tuning of ultra-low-loss polaritons in a van der Waals crystal by intercalation". Nature Materials 19 (9), 964–968 (2020) [42].
- Y. Wu, Q. Ou, et al. "Chemical switching of low-loss phonon polaritons in -MoO by hydrogen intercalation", Nature Communications 11 (1), 2646 (2020) [43].
- T. V. A. G. de Oliveira et al. "Nanoscale" Confined Terahertz Polaritons in a van der Waals Crystal", Advanced Materials 33 (2), 2005777 (2021) [44].
- P. Aguilar-Merino, G. Álvarez-Pérez, J. Taboada-Gutiérrez, et al. "Extracting the Infrared Permittivity of SiO2 Substrates Locally by Near-Field Imaging of Phonon Polaritons in a van der Waals Crystal", Nanomaterials 11 (1), 120 (2021).
- E. García-Portugués et al. "A goodness-of-fit test for the functional linear model with functional response" Scandinavian Journal of Statistics 48 (2), 502–528 (2021).
- J. Martín-Sánchez, J. Duan, et al. "Focusing of in-plane hyperbolic polaritons in van der Waals crystals with tailored infrared nanoantennas", Science Advances 7 (41), eabj0127 (2021) [45].
- J. Duan, F. J. Alfaro-Mozaz, et al. "Active and Passive Tuning of Ultranarrow Resonances in Polaritonic Nanoantennas", Advanced Materials 34 (10), 2104954 (2022) [46].
- T. Nörenberg, G. Álvarez-Pérez, et al. "Germanium Monosulfide as a Natural Platform for Highly Anisotropic THz Polaritons", ACS Nano 16 (12), 20174–20185 (2022) [47].
- M. Obst, et al. "Terahertz Twistoptics–engineering canalized phonon polaritons". ACS Nano 17 (19), 19313–19322 (2023) [48].
- E. Galiffi, G. Carini, X. Ni, et al. "Extreme light confinement and control in low-symmetry phonon-polaritonic crystals". Nature Reviews Materials, 9, 9–28 (2024) [33].
- K. V. Voronin, et al. "Fundamentals of polaritons in strongly anisotropic thin crystal layers". ACS Photonics, 11 (2) 550–560 (2024) [49].

- A. I. F. Tresguerres-Mata, C. Lanza, et al. "Observation of Naturally Canalized Polaritons in LiV_2O_5 Thin Layers". Nature Communications, 15, 2696 (2024) [50].
- J. Taboada-Gutiérrez, et al. "Unveiling the Mechanism of Phonon-Polariton Damping in α-MoO_3. ACS Photonics, 11(9), 3570–3577 (2024) [51].

Book Chapters Not Contained in This Thesis

- E. García-Portugués et al, "Goodness-of-fit Tests for Functional Linear Models Based on Integrated Projections", Functional and High-Dimensional Statistics and Related Fields, 107–114 (2020).

References

1. Euclid, O. Byrne, *The First Six Books of the Elements of Euclid* English (William Pickering, 1847)
2. Alhazen, *Kitab al-Manazir [Book of Optics]* (circa 1011–1021 CE)
3. L. Da Vinci, *The Notebooks of Leonardo da Vinci* English, ed. by J.P. Richter (1883)
4. J. Kepler, *Dioptrice* Latin (1611)
5. W.S. Snellius, *Eratorum Chilias* Latin (1621)
6. R. Descartes, *La Géométrie* French (1637)
7. I. Newton, *Opticks: or, A Treatise of the Reflections, Refractions, Inflections and Colours of Light* English (1704)
8. C. Huygens, *Traité de la Lumière* French (1690)
9. T. Young, Experiments and calculations relative to physical optics. English. Philos. Trans. R. Soc. Lond. **94**, 1–16 (1804). https://www.jstor.org/stable/107135
10. A.-J. Fresnel, *Oeuvres complètes d'Augustin Fresnel* (Imprimerie Impériale, Paris, 1866)
11. J.C. Maxwell, *A Treatise on Electricity and Magnetism English* (Clarendon Press, 1873)
12. M. Planck, Über das Gesetz der Energieverteilung im Normalspektrum. German. Ann. Phys. **4**, 553–563 (1901). https://link.springer.com/chapter/10.1007/978-3-663-13885-3_16
13. A. Einstein, Über einen die Erzeugung und Verwandlung des Lichtes betreffenden heuristischen Gesichtspunkt (Photon Theory of Light). Ann. Phys. **17**, 132–148 (1905)
14. L. De Broglie, Recherches sur la théorie des quanta. French. Ann. Phys. **3**, 22 (1925). https://theses.hal.science/tel-00006807/document

15. R. Feynman, There's plenty of room at the bottom. *Caltech Eng. Sci.* **23**, 22–36 (1960). https://www.zyvex.com/nanotech/feynman.html
16. G. Binnig, H. Rohrer, Scanning tunneling microscopy. Surf. Sci. **126**, 236–244 (1983). https://www.sciencedirect.com/science/article/pii/0039602883907161
17. G. Binnig, C.F. Quate, C. Gerber, Atomic force microscope. Phys. Rev. Lett. **56**, 930–933 (1986). https://link.aps.org/doi/10.1103/PhysRevLett.56.930
18. E. Abbe, Beiträge zur Theorie des Mikroskops und der mikroskopischen Wahrnehmung. Arch. Mikrosk. Anat. **9**, 413–468 (1873). https://doi.org/10.1007/BF02956173
19. J.J. Hopfield, Theory of the contribution of excitons to the complex dielectric constant of crystals. Phys. Rev. **112**, 1555–1567 (1958). https://link.aps.org/doi/10.1103/PhysRev.112.1555
20. R.H. Ritchie, Plasma losses by fast electrons in thin films. Phys. Rev. **106**, 874–881 (1957). https://link.aps.org/doi/10.1103/PhysRev.106.874
21. M. Born, K. Huang, *Dynamical Theory of Crystal Lattices* (Clarendon Press, 1954)
22. S.A. Maier, *Plasmonics: Fundamentals and Applications* (Springer New York, New York, 2007). https://link.springer.com/book/10.1007/0-387-37825-1
23. L. Novotny, B. Hecht, *Principles of Nano-Optics* (Cambridge University Press, 2006)
24. M.L. Brongersma, P.G. Kik, *Surface Plasmon Nanophotonics*. Springer Series in Optical Sciences (2007)
25. N. Engheta, R.W. Ziolkowski, *Metamaterials: Physics and Engineering Explorations* (Wiley and Sons, 2006)
26. K.S. Novoselov et al., Electric field effect in atomically thin carbon films. Science **306**, 666–669 (2004). https://www.science.org/doi/abs/10.1126/science.1102896
27. D.N. Basov, M.M. Fogler, F.J.G. de Abajo, Polaritons in van der Waals materials. Science **354**, 195 (2016). https://science.sciencemag.org/content/354/6309/aag1992
28. T. Low et al., Polaritons in layered two-dimensional materials. Nat. Mater. **16**, 182–194 (2017). https://www.nature.com/articles/nmat4792
29. X. Chen et al., Modern scattering-type scanning near-field optical microscopy for advanced material research. Adv. Mater. **31**, 1804774 (2019). https://onlinelibrary.wiley.com/doi/abs/10.1002/adma.201804774
30. S. Dai et al., Tunable phonon polaritons in atomically thin van der Waals crystals of boron nitride. Science **343**, 1125–1129 (2014). http://science.sciencemag.org/content/343/6175/1125
31. W. Ma et al., In-plane anisotropic and ultra-low-loss polaritons in a natural van der Waals crystal. Nature **562**, 557 (2018). https://www.nature.com/articles/s41586-018-0618-9
32. M. He et al., Anisotropy and modal hybridization in infrared nanophotonics using low-symmetry materials. ACS Photon. **9**, 1078–1095 (2022). https://doi.org/10.1021/acsphotonics.1c01486

33. E. Galiffi et al., Extreme light confinement and control in low-symmetry phonon-polaritonic crystals. Nat. Rev. Mater. (2023). https://doi.org/10.1038/s41578-023-00620-7
34. G. Álvarez-Pérez, K.V. Voronin, V.S. Volkov, P. Alonso-GonzáLlez, A.Y. Nikitin, Analytical approximations for the dispersion of electromagnetic modes in slabs of biaxial crystals. Phys. Rev. B **100**, 235408 (2019). https://link.aps.org/doi/10.1103/PhysRevB.100.235408
35. G. Álvarez-Pérez et al., Infrared permittivity of the biaxial van der Waals semiconductor α-MoO_3 from near- and far-field correlative studies. Adv. Mater. **32**, 1908176 (2020). https://onlinelibrary.wiley.com/doi/abs/10.1002/adma.201908176
36. G. Álvarez-Pérez et al., Negative reflection of nanoscale-confined polaritons in a low-loss natural medium. Sci. Adv. **8**, eabp8486 (2022). https://www.science.org/doi/abs/10.1126/sciadv.abp8486
37. J. Duan et al., Planar refraction and lensing of highly confined polaritons in anisotropic media. Nat. Commun. **12**, 4325 (2021). https://doi.org/10.1038/s41467-021-24599-3
38. J. Duan et al., Enabling propagation of anisotropic polaritons along forbidden directions via a topological transition. Sci. Adv. **7**, eabf2690 (2021). https://www.science.org/doi/abs/10.1126/sciadv.abf2690
39. J. Duan et al., Twisted nano-optics: manipulating light at the nanoscale with twisted phonon polaritonic slabs. Nano Lett. **20**, 5323–5329 (2020). https://doi.org/10.1021/acs.nanolett.0c01673
40. J. Duan et al., Multiple and spectrally robust photonic magic angles in reconfigurable α-MoO_3 trilayers. Nat. Mater. **22**, 867–872 (2023). https://doi.org/10.1038/s41563-023-01582-5
41. G. Álvarez-Pérez et al., Active tuning of highly anisotropic phonon polaritons in van der Waals crystal slabs by gated graphene. ACS Photon. **9**, 383–390 (2022). https://doi.org/10.1021/acsphotonics.1c01549
42. J. Taboada-Gutiérrez et al., Broad spectral tuning of ultra-low-loss polaritons in a van der Waals crystal by intercalation. Nat. Mater. **19**, 964–968 (2020). https://doi.org/10.1038/s41563-020-0665-0
43. Y. Wu et al., Chemical switching of low-loss phonon polaritons in α-MoO_3 by hydrogen intercalation. Nat. Commun. **11**, 2646 (2020). https://doi.org/10.1038/s41467-020-16459-3
44. T.V.A.G. De Oliveira et al., Nanoscale-confined terahertz polaritons in a van der Waals crystal. Adv. Mater. **33**, 2005777 (2021). https://onlinelibrary.wiley.com/doi/abs/10.1002/adma.202005777
45. J. Martín-Sánchez et al., Focusing of in-plane hyperbolic polaritons in van der Waals crystals with tailored infrared nanoantennas. Sci. Adv. **7**, eabj0127 (2021). https://www.science.org/doi/abs/10.1126/sciadv.abj0127
46. J. Duan et al., Active and passive tuning of ultranarrow resonances in polaritonic nanoantennas. Adv. Mater. **34**, 2104954 (2022). https://onlinelibrary.wiley.com/doi/abs/10.1002/adma.202104954

47. T. Nörenberg et al., Germanium monosulfide as a natural platform for highly anisotropic THz polaritons. ACS Nano **16**, 20174–20185 (2022). https://doi.org/10.1021/acsnano.2c05376
48. M. Obst et al., Terahertz twistoptics–engineering canalized phonon polaritons. ACS Nano **17**, 19313–19322 (2023). https://doi.org/10.1021/acsnano.3c06477
49. K.V. Voronin, G. Álvarez-Pérez, C. Lanza, P. Alonso-González, A.Y. Nikitin, Fundamentals of polaritons in strongly anisotropic thin crystal layers. ACS Photonics **11**(2), 550–560 (2024). https://doi.org/10.1021/acsphotonics.3c01428

Contents

Part I Introduction

1 Introduction to Nanoscale-Confined Propagating Polaritons 3

2 Experimental and Modelling Techniques 47

Part II Enabling Predictive Capabilities for Hyperbolic Polaritons in van der Waals Materials

3 Dispersion of Polaritons in Biaxial Slabs................................ 93

4 Infrared Permittivity of α-MoO_3 from Near- and Far-Field Correlative Studies 117

Part III Anomalous Optical Phenomena at the Nanoscale in Strongly Anisotropic Media

5 Negative Reflection of Nanoscale-Confined Hyperbolic Polaritons ... 143

6 Anomalous Refraction and Lensing of Nanoscale-Confined Hyperbolic Polaritons 157

Part IV Controlling, Directing and Guiding Hyperbolic Polaritons at the Nanoscale

7 Enabling Propagation of Hyperbolic Polaritons Along Forbidden Directions 175

8 Twist-Optics: Controlling the Propagation of Phonon Polaritons with Twisted van der Waals Stacks 189

9 Active Tuning of Hyperbolic Polaritons in van der Waals Materials by Integrating a Gated Graphene Layer..................... 213

10 Twistable Polaritonics with In-Operando Rotatable van der Waals Bilayers 225

11 Conclusions and Outlook ... 235

A Infrared Dielectric Functions of AgCl and BaF_2 ... 243

B Ab Initio Infrared Dielectric Function of α-MoO_3 ... 247

C Analytical Shape of a Total Hyperbolic Retroreflector and a Planar Hyperlens for Polaritons ... 251

D Rotation of the Propagation Direction of In-Plane Hyperbolic Polaritons in a Slab Embedded Between Two Semi-infinite Media ... 259

E Dyadic Green's Function for a Two-Dimensional Conductivity Biaxial Layer ... 265

F Derivation of the Dispersion Relation of Polaritons in Twisted Anisotropic Conductivity Bilayers ... 269

G Phonon Polaritons in Twisted Biaxial Trilayers: Further Analytical Considerations ... 277

H Active Tuning of Hyperbolic Polaritons in van der Waals Materials by Gated Graphene: Further Analytical and Numerical Considerations ... 283

Part I
Introduction

In this part, we present the fundamental concepts that will establish the foundational framework for the subsequent chapters, which will present original research conducted within the scope of this thesis.

Chapter 1 describes how polaritons can confine light at the nanometer scale and beat the diffraction limit of light at visible and infrared frequencies. It then furnishes a comprehensive outline of the characteristics of propagating polaritons in metallic media, polar crystals, and some of the most studied van der Waals (vdW) materials within the domain of nano-optics. We put special emphasis on the mid-infrared (mid-IR) frequency range, as well on the crucial role of anisotropy in shaping the characteristics of propagating polaritons. The chapter culminates with an introduction to hyperbolic phonon polaritons and their properties, particularly in the material that will be featured prominently in this thesis: the vdW semiconductor alpha-phase molybdenum trioxide.

Chapter 2 provides a detailed explanation of the main experimental and theoretical techniques utilized in this thesis, with a focus on the aspects related to the investigation of propagating polaritons in van der Waals (vdW) materials. These encompass mechanical exfoliation and the dry-transfer technique, Fourier-transformed infrared spectroscopy, scattering-type scanning near-field microscopy, as well as various analytical and numerical modeling approaches. The chapter also delivers a detailed overview of the state-of-the-art in the field, serving as a motivation for the subsequent chapters.

Chapter 1
Introduction to Nanoscale-Confined Propagating Polaritons

Abstract In recent decades, there has been an increasing interest in the search for structures that enable confining and manipulating light at the nanometer scale. However, the diffraction limit of light poses a fundamental challenge in achieving resolution beyond the wavelength of the illuminating light, restricting the fabrication and imaging of sub-wavelength structures and thus impeding progress in the field of nano-optics. The initial section of this chapter illustrates this limitation by demonstrating that waveguides—structures designed to direct light along specific directions—made of traditional dielectric materials are not suitable for confining light in nanometric volumes at visible or longer wavelengths.

In the subsequent sections, we discuss considerable advances that have been made to overcome this limitation. All are based on leveraging polaritons—hybrid light matter excitations—to achieve nanoscale confinement and control of light. We will show that polaritons can propagate along interfaces between materials with permittivities of different signs as a result of the strong interaction between light and collective dipolar matter oscillations. Then, we offer a comprehensive overview of propagating polaritons. Initially, these were explored in bulk metals and polar crystals, and subsequently in graphene and hexagonal boron nitride with the advent of van der Waals (vdW) and two-dimensional (2D) materials. We also put a special focus on the mid-infrared (mid-IR) frequency range, a technologically important spectral region, and we highlight the role that anisotropy plays in the properties of propagating polaritons. Finally, we introduce alpha-phase molybdenum trioxide, the main material studied in this thesis. This vdW semiconductor supports the propagation of strongly anisotropic polaritons along its surface, promising unprecedented control of light and the flow of energy at the nanoscale.

G. Álvarez Pérez, *Fundamentals of Nano-Optics in Hyperbolic van der Waals Materials*, Springer Theses, https://doi.org/10.1007/978-3-031-75767-9_1

1.1 Guided Light in Dielectric Slabs and the Diffraction Limit of Light

Guided Light in Dielectric Slabs

The precise control and manipulation of light at scales smaller than the wavelength of light are essential for the development of miniaturized optical circuits, innovative biosensors, and efficient light-matter interactions. However, conventional dielectric materials have limitations in localizing light at sub-wavelength dimensions [1]. In this section, such limitation is exemplified through the study of light propagation in a slab waveguide.

A dielectric slab can guide light in the form of discrete electromagnetic modes, known as waveguide modes, which arise as a result of the difference in refractive indices between the waveguide medium and the surrounding cladding material. This causes the electromagnetic field to bounce back and forth between the boundaries of the waveguide due to the phenomenon of total internal reflection (Fig. 1.1a). This effect creates standing waves inside the slab, and consequently only specific wavelengths or frequencies can form stable guided modes within the waveguide, while other wavelengths or frequencies will not be guided and will be attenuated. These allowed modes are classified as transverse electric (TE) modes and transverse magnetic (TM) modes, where the electric (magnetic) field is perpendicular to the direction of propagation. To illustrate how light propagates in such waveguides, we calculate the dispersion of TM modes in a dielectric waveguide made of silica (SiO_2) (Fig. 1.1a), a typical dielectric material, commonly used in optical communications. Yet, an analogous analysis can be carried out for TE modes as well. The dispersion of TM modes in a dielectric waveguide, which relates their momentum or wavevector k to their energy or frequency ω, is described by the following equation [2–4]:

$$\tanh(q_{\mathrm{SiO_2},z} k_0 d) = -\frac{2 q_{\mathrm{SiO_2},z} \cdot \varepsilon_{\mathrm{SiO_2}} \cdot q_{\mathrm{air},z}}{q_{\mathrm{air},z}^2 \cdot \varepsilon_{\mathrm{SiO_2}}^2 + q_{\mathrm{SiO_2},z}^2}. \tag{1.1}$$

Here, $q_{\mathrm{SiO_2},z} = \sqrt{q^2 - \varepsilon_{\mathrm{SiO}}}$ and $q_{\mathrm{air},z} = \sqrt{q^2 - 1}$, where $q = k/k_0$ is the normalized wavevector and $k_0 = \omega/c$. In the visible and near-infrared (NIR) frequencies, the permittivity of SiO_2 depends weakly on the frequency. Neglecting losses for simplicity, it can be approximated as $\varepsilon_{\mathrm{SiO_2}} = 2.28$. Figure 1.1b illustrates how light propagates in a SiO_2 waveguide for a fixed thickness of $d = 500\,\mathrm{nm}$ surrounded by air, as a function of the incident frequency ω. The dispersion of the first three TM modes is plotted in Fig. 1.1b. For each mode, the normalized wavevector approaches one for small frequencies and increases thereafter, tending to $\sqrt{\varepsilon_{\mathrm{SiO_2}}} k_0$, which corresponds to the wavevector of light propagating through bulk SiO_2 [5]. This implies that the minimum wavelength of the mode is determined

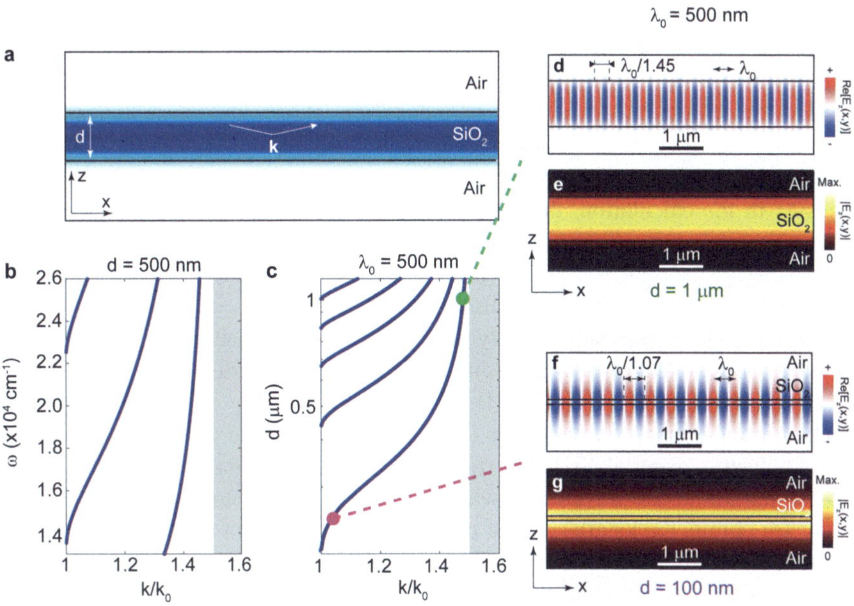

Fig. 1.1 Guided light in a SiO_2 slab at visible frequencies. (**a**) Schematics of the SiO_2 waveguide surrounded by air. (**b**,**c**) Normalized wavevector of the first-order modes as a function of the incident frequency (**b**) and the waveguide thickness (**c**), for a fixed thickness $d = 500$ nm and a fixed incident wavelength $\lambda_0 = 500$ nm, respectively. The asymptote at large k/k_0, is marked by the gray area. (**d**,**f**) Snapshot of the vertical component of the electric field of the fundamental TM mode (TM0), calculated for waveguides with thickness of 100 nm (**d**) and 1 μm (**f**), marked by the green and magenta dots in panel (**c**). (**e**,**g**) Corresponding field amplitude of the TM0 mode for waveguides with thickness of 100 nm (**e**) and 1 μm (**g**), marked by the green and magenta dots in panel (**c**)

by the wavelength of light in the material, and thus the confinement of light in traditional dielectric waveguides is limited by diffraction.

In Fig. 1.1c, we analytically calculate the normalized wavevector as a function of the SiO_2 slab thickness for a fixed incident wavelength of $\lambda_0 = 500$ nm. For very small thicknesses, the normalized wavevector is close to one and increases for thicker layers. At high frequencies, the modes' wavevectors tend to $\sqrt{\varepsilon_{SiO_2}}k_0$, corresponding to the wavevector of light propagating through bulk SiO_2 [5]. Once again, the minimum wavelength of the mode is determined by the wavelength of light in the material, and hence limited by diffraction.

On the other hand, to illustrate the mode wavelength, as well as its transverse confinement, as a function of the waveguide thickness, we simulate the field distribution of the lowest-order TM mode (TM0) for two thicknesses,[1] $d = 1$ μm

[1] We note here that, since the decay length into vacuum is mostly governed by the momentum, the transverse confinement of modes with the same momentum but different thicknesses can also

and 100 nm (marked by the green and magenta dots, respectively, in Fig. 1.1c). To do so, we use a finite-element-method simulation, carried out using COMSOL Multiphysics (more details about the modelling will be discussed in the next chapter, see Sect. 2.5). Figure 1.1d,f shows the simulated snapshots of the vertical component of the electric field, $\mathrm{Re}(E_z)$, for the two aforementioned thicknesses, while the field amplitudes, $|E_z|$, are shown in Fig. 1.1e,g. We note that, due to the absence of losses in the calculation, the waveguide modes propagate without dissipation. For a waveguide with thickness of 1 μm, the electric field of the mode is mainly confined inside the SiO_2 slab (Fig. 1.1d,e), and the fields rapidly decay with distance to the SiO_2/air interfaces. However, for the thin waveguide ($d = 100$ nm) the electric field of the mode extends significantly into the surrounding medium [6] (Fig. 1.1f,g). Moreover, the calculation also shows that the mode wavelength is reduced in thicker waveguides (compare Fig. 1.1d and f).

The presented analysis illustrates that, in SiO_2 (and more generally in any low index dielectric material [7]), the transverse field confinement cannot be much smaller than the mode wavelength which, in turn, is similar to that of light in free space. This limits the length scale of potential light circuits in novel optoelectronic devices. Therefore, in order to exploit the premises of nano-optics, it is imperative to find an alternative mechanism of interaction between light and matter that beats the diffraction limit. In order to do so, we start by revisiting the diffraction limit itself in the next subsection.

The Diffraction Limit of Light

The results of the analysis conducted above, namely the impossibility of confining light into deeply sub-wavelength dimensions in dielectric waveguides, stem from the diffraction limit of light, a concept formulated by Abbe in 1873 [8]. Abbe postulated that the minimum distance d to optically distinguish two objects, for instance in a microscope, is determined by the wavelength of the employed electromagnetic radiation, as expressed by the equation:

$$d = \frac{\lambda}{2n \cdot \sin\theta} \tag{1.2}$$

Here, λ/n stands for the wavelength of light in a material with refractive index $n = \sqrt{\varepsilon}$, and θ represents half the angle of the focusing cone. The phenomenon of diffraction causes the image of a single point dipole source to appear as an interference pattern, termed the *point spread function*, which relies on the numerical

be studied. Nevertheless, for non-zero losses, which is accurate for realistic structures and will be applicable to the structures explored in this thesis, the lowest-order modes dominate. Therefore, in this analysis we restrict ourselves to illustrating the transverse confinement of the lowest-order mode.

aperture (NA $= n \cdot \sin\theta$) of the focusing objective [9]. Visible light has a wavelength ranging from 400 to 700 nm, allowing conventional microscopes to resolve entities such as cells and bacteria with dimensions on the order of microns. Nevertheless, this resolution falls short for smaller objects like viruses, proteins, or individual molecules. With the emergence of nanotechnology, both natural phenomena and artificial structures have necessitated alternative imaging techniques due to their very small scale. One straightforward method to enhance resolution is to decrease the employed wavelength, as predicted by Eq. (1.2), for instance using X-ray high resolution imaging. This approach applies not only to electromagnetic radiation but also to massive particles, thanks to the particle-wave duality of quantum mechanics. This principle has led to the routine use of electron microscopes [10], achieving resolutions at the nanometer scale. As the De Broglie wavelength is inversely proportional to the particle mass, even greater resolutions are attainable by utilizing massive ions [11]. However, these solutions present substantial technological challenges compared to conventional optical microscopes, since they often operate in high vacuum conditions and the samples under study usually are required to be conductive or must undergo laborious preparation routines. Additionally, the utilization of specific wavelengths of light to probe distinct excited states of a sample is frequently constrained. For instance, molecular vibrational modes and solid-state phonons lie in the infrared (IR) range, while electronic transitions typically occur at visible and ultraviolet frequencies. Consequently, surmounting the diffraction limit has remained a central focus in optics, yielding fundamental advancements through both far- and near-field techniques.

In order to discuss the possibilities to beat the diffraction limit and achieve sub-diffractional field confinement, we start by considering the fact that this limit can be interpreted as an embodiment of the Heisenberg uncertainty principle [12], one of the essential pillars of quantum mechanics:

$$\Delta r \Delta p \geq \frac{\hbar}{2}. \tag{1.3}$$

In the case of a photon, the momentum is given by $\Delta p = \Delta k\hbar/2\pi$. In this context, we consider a simplified scenario where a beam propagates in the z direction within vacuum and is focused in the x direction by an objective, with $k_0^2 = k_x^2 + k_z^2$. The dispersion of momentum in the x direction is linked to the numerical aperture of the objective, as $k_x^{\min} = 0$ and $k_x^{\max} = k_0\mathrm{NA}$. In the ideal case of an infinitely extended objective, NA equals 1. Thus, $k_x^{\max} = k_0 = 2\pi/\lambda$. Substituting into Eq. (1.3) yields:

$$\Delta x \geq \frac{\lambda}{2}. \tag{1.4}$$

This result aligns with the diffraction limit presented in Eq. (1.2), as well as with the calculations and simulations carried out in the preceding subsection. Surpassing the diffraction limit necessitates a broader distribution in photon momentum. But

$k_0^2 = k_x^2 + k_z^2$, and as long as k_x and k_z remain real numbers, which holds true for propagating free-space light, the diffraction limit persists. Conversely, if k_z takes on imaginary values, the spatial extent of the field in the x direction can be confined beneath the diffraction limit, as $k_x > k_0$ is possible. In general, an imaginary $\mathbf{k}$ vector[2] corresponds to an exponentially damped field, as expressed by $\mathbf{E} \propto \exp(i\mathbf{k} \cdot \mathbf{r})$. Crucially, the existence of these evanescent waves is generally associated with the presence of a material boundary. The region in proximity to the surface, where these exponentially decaying fields manifest, is referred to as the *near field*. These field components do not propagate and remain unobservable in the far field, farther from the interface.

The easiest way to produce an evanescent wave is to consider two dielectric media exhibiting total internal reflection (TIR) when light comes from the medium with the higher refractive index $n_2 > n_1$ above the critical angle $\theta_c = \arcsin(n_1/n_2)$. Nevertheless, the low confinement inherent to the waveguide modes shown in Fig. 1.1 clearly illustrates their limitations in effectively surpassing the diffraction limit and enabling nanoscale confinement of light. A suitable alternative can be achieved through the excitation of surface waves that propagate along the interface of two materials with permittivities of opposite signs and feature exponentially decaying fields in the perpendicular direction. The study of these surface waves will be a central focus of this thesis, and the remaining part of this chapter is devoted to the detailed discussion on their origin and their properties in different types of materials.

1.2 Surface Polaritons at Material Interfaces

The existence of surface states at the interface between two media can be derived analytically from Maxwell's equations applying continuity relations for the electric and magnetic fields. Without loss of generality, we consider the interface to be located at $z = 0$, and the media 1 and 2 to be at $z > 0$ and $z < 0$, respectively. The surface wave condition is enforced by demanding the field components to be exponentially decaying in the direction z, perpendicular to the interface. To have exponentially decaying fields along the z directions, the wavevectors must be imaginary, with positive imaginary part, $\mathrm{Im}(k_{iz}) > 0$. $i = 1, 2$. Thus, the fields are given by $E_{ij}(z), H_{ij}(z) \propto \exp(-\mathrm{Im}(k_{iz})|z|)$, where the index j runs over the field components, and i over the two media 1 and 2, $i = 1, 2$. From this condition[3] and the continuity of the fields, one gets that no transverse electric (TE) surface wave can exist at the boundary [13]. Only transverse magnetic (TM) surface waves can exist, provided that the real parts of the dielectric functions of the two media have

[2] Along this thesis, we will use bold letters to indicate vectors.

[3] If $\mathrm{Im}(k_{iz}) < 0$, the oscillations could exponentially grow, which is unrealistic for a system in which energy is not amplified.

opposite signs [13]:

$$\frac{k_{1z}}{k_{2z}} = -\frac{\varepsilon_1}{\varepsilon_2}.$$

Therefore, surface waves within a single boundary can only occur between a dielectric with $\text{Re}(\varepsilon_1) > 0$ and a material with $\text{Re}(\varepsilon_2) < 0$. The two **k** vectors in the z direction satisfy the following equations [13]:

$$k_{iz}^2 = k_0^2 \varepsilon_i - k_p^2, \qquad i = 1, 2,$$

where k_p is the wavenumber of the surface wave in the propagation direction and is given by:

$$k_p(\omega) = \frac{\omega}{c}\sqrt{\frac{\varepsilon_1(\omega)\varepsilon_2(\omega)}{\varepsilon_1(\omega) + \varepsilon_2(\omega)}}. \tag{1.5}$$

We have made explicit the frequency dependence of the permittivities and the momentum in Eq. (1.5), which constitutes the dispersion relation of evanescent surface states in the simplest case of an interface between two semi-infinite isotropic media. From a classical electrodynamics perspective, this equation determines many of the fundamental properties of these evanescent surface states, which are highly dependent on the characteristics of media 1 and 2 through the permittivities ε_1 and ε_2. In particular to those of medium 2, the one with negative permittivity, which is generally driven by strong optical resonances stemming from collective dipolar oscillations in the material. As will be discussed in detail in the coming subsections, these can be oscillations of the electron density (plasmons) in metals, lattice vibrations (phonons) in polar dielectrics, or bounded electron-hole pairs in semiconductors (excitons), to mention a few.

The evanescent surface states described by Eq. (1.5) are more often called surface *polaritons*, referring to their nature as hybrid light-matter excitations that arise from the *strong coupling* between light and such collective matter dipolar oscillations. The meaning of *strong coupling* can be understood from a quantum-electrodynamics perspective. Collective matter oscillations in the material can be described as quantum harmonic oscillators, which then interact with the modes of the quantum electromagnetic field, the photons. Both systems can exchange energy rapidly and strongly, leading to significant transfers of energy between them. When this exchange is faster than the energy decay in both systems separately, we refer to strong coupling. When strongly coupled, photons and collective excitations can no longer be separated or understood individually, but in turn they are intertwined, making it customary to refer to polaritons as hybrid light-matter excitations. This is reflected in the perturbation of both the linear dispersion relation of photons and the constant dispersion of the collective modes, resulting in avoided crossing between the two dispersion lines of polaritons [14], as will be shown for different systems along this chapter.

In a more rigourous way, polaritons are bosonic quasiparticles, the eigenstates of the system formed by the collective modes of the material and the radiation field [14]. However, while this very nature of polaritons comes from a treatment based on quantum field theory, in this thesis we will restrict ourselves to a classical treatment, as it will suffice to predict and interpret the phenomena under study. Notably, the description up to Eq. (1.5) constitutes the simplest form of the linear classical electromagnetic theory for surface polaritons [14–16]. We will follow this level of description throughout this thesis, where charge oscillations, crystal symmetries and phonon anisotropies effectively enter into the permittivity function. We note, however, that higher-order optical response theory might be more appropriate in some cases [17], in particular when intense light fields are employed, as higher-order susceptibility tensors typically posses higher-order symmetry than the linear response [18]. Furthermore, we will assume polariton wavelengths much larger than the unit cell of the crystal, constituting a local field approximation. Non-local effects [19, 20] manifest in an explicit wavevector-dependence of the permittivity, and can also affect the dispersion relation of polaritons, for example in thin films of polar materials [21] or in atomic-scale heterostructures [22].

Surface Plasmon Polaritons in Metals at Visible Frequencies

Metals such as gold, silver, and copper can support surface polaritons at the interface with a dielectric at visible and IR frequencies, where their permittivity is usually negative [23].

To understand why, let us start by considering that, in metals, the outermost electrons are not strongly bound to individual nuclei, but are instead shared by the entire metal lattice and can move freely across it. In a state of equilibrium, the negative charge of the electrons is balanced by the positive charge of the ionized metal atoms. When this equilibrium is disrupted by altering the electron density in a region, the electrons are pushed away from that region due to their mutual repulsion, causing it to become positively charged. However, equilibrium is not instantly restored because the electrons possess inertia, causing them to continue moving away from the region even after the system becomes neutral. Consequently, after a certain period, the region becomes negatively charged. This attracts electrons, causing inward motion and repeating the cycle, creating collective oscillations of the charge carriers (Fig. 1.2a) until energy dissipates.

Drawing parallels between electronic dynamics in a metal and the motion of charged ions in a plasma, the aforementioned charge oscillations are termed *plasma oscillations*. They possess a characteristic frequency ω_p, which is often referred to as the plasma frequency and depends on the carriers' mass m, density n, and charge e:

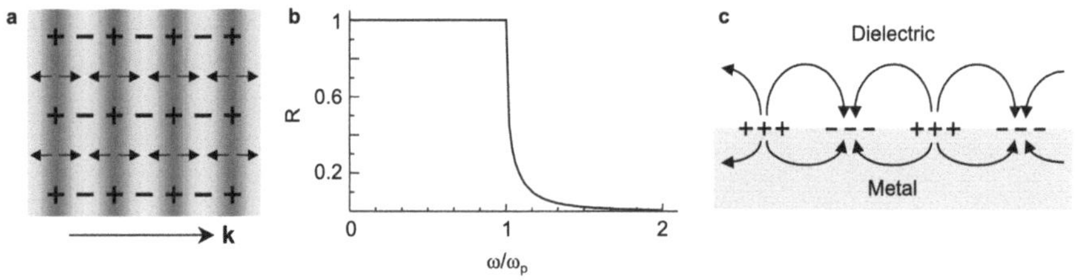

Fig. 1.2 **Plasma oscillations in metals.** (**a**) Charge fluctuations in a free carrier plasma oscillation. The lighter regions denote areas with excess electron densities. The small arrows indicate the direction of the electric fields, which are parallel to the direction of propagation of the wave, as indicated by its wave vector **k**. (**b**) Reflectivity of an undamped free carrier gas as a function of frequency. (**c**) Electric fields associated with electron charge density fluctuations at the surface of a metal. Adapted from [24]

$$\omega_p = \sqrt{\frac{ne^2}{m\varepsilon_0}},$$

where ε_0 is the vacuum permittivity. The quantized modes of plasma oscillations are known as *plasmons*, with an average energy $E_p = \hbar\omega_p$.

Since the restoring force arises from the electrostatic interactions along the direction of the electric field, these oscillations are longitudinal in nature [24]. For a transverse plasma mode to exist, there would need to be a mechanism by which the electrons could be displaced perpendicular to the applied electric field but, in a plasma, there is no restoring force acting in the perpendicular direction.

Metals typically have a high electron density, and the (effective) mass of electrons in metals is low. As a result, the plasma frequency ω_p is relatively high, and the corresponding wavelength is in the UV or near-UV range. In addition, at the surface of a semi-infinite metal, only half of the charges is in the metal, while the other half is outside and do not induce an electric field. The current flow across the surface is then cut in half, and the charges undergo an oscillation at the so-called surface plasma frequency ω_{sp}:

$$\omega_{sp} = \frac{\omega_p}{\sqrt{2}}.$$

These effects can be described by the permittivity function of a metal $\varepsilon_m(\omega)$ at visible frequencies. $\varepsilon_m(\omega)$ is mainly influenced by two factors: (1) the aforementioned presence of conduction electrons that can move freely through the metal and (2) interband excitations that occur between energy bands in the electronic structure [9]. These factors contribute to the permittivity function in different ways, with the free electrons influence being usually described by a Drude-like permittivity function—a simple classical theory that describes the behavior of electrons in a metal as if they were a free electron gas—, and the interband excitations contributing according to the Drude-Sommerfeld model. For simplicity, here we consider a

Drude model to describe the permittivity of a metal in the visible and IR range. By writing the equations of motion for a free electron gas, the permittivity can be expressed as [24]:

$$\varepsilon_m(\omega) = 1 - \frac{\omega_p^2}{\omega^2 + i\gamma\omega}, \tag{1.6}$$

where ω_p is the plasma frequency of the metal, i is the imaginary unit and γ is the damping rate. which is inversely proportional to the average time between collisions of electrons. A larger damping rate implies that the electrons collide more frequently, which leads to a decrease in electrical conductivity. The permittivity function in Eq. (1.6) does not accurately describe the optical response of all metals, as it does not capture some effects that are important in real materials, such as interband excitations [25]. To describe the behavior of materials at high frequencies, where quantum mechanical effects become significant and the classical assumptions of the Drude model might not fully apply, the first term in the summation in Eq. (1.6) if often replaced by a constant ε_∞, helping to extend the model's applicability:

$$\varepsilon_m(\omega) = \varepsilon_\infty - \frac{\omega_p^2}{\omega^2 + i\gamma\omega}. \tag{1.7}$$

To further account for resonant absorption in the interband region, Lorentz or Gaussian oscillators can be added to Eq. (1.6) in an empirical manner, but this is not explored here since, despite this limitation, the Drude model can still provide an intuitive description of the optical properties of many metals.

Neglecting losses, we can see from Eq. (1.6) that, for frequencies below ω_p, $\varepsilon_m(\omega)$ becomes negative. To understand the role of collective oscillations in this negative permittivity region, let us consider the following. When light of a frequency near ω_p impinges on a metal surface, its alternating current (AC) electric field exerts a force on the free electrons of the metal, distorting equilibrium and causing them to collectively oscillate in response to the incident field's frequency. The reflectivity of the metal as a function of the incident frequency can be calculated as:

$$R(\omega) = \left| \frac{\sqrt{\varepsilon_m(\omega)} - 1}{\sqrt{\varepsilon_m(\omega)} + 1} \right|^2 .$$

By substituting the frequency dependence of $\sqrt{\varepsilon_m(\omega)}$ into this formula, we can see that $R(\omega)$ approaches unity for $\omega \leq \omega_p$, and then decreases for $\omega > \omega_p$ (Fig. 1.2b). This is because, below ω_p, the permittivity is negative, meaning that free or unbound electrons can respond quickly to the incident electromagnetic field and begin to coherently oscillate in a direction opposing it. Therefore, the material does not *permit* the electromagnetic radiation, and becomes highly reflective. Above ω_p, metals start to exhibit some absorption and especially transmission of light due to the inability of electrons to respond rapidly to the changing field. The exchange of

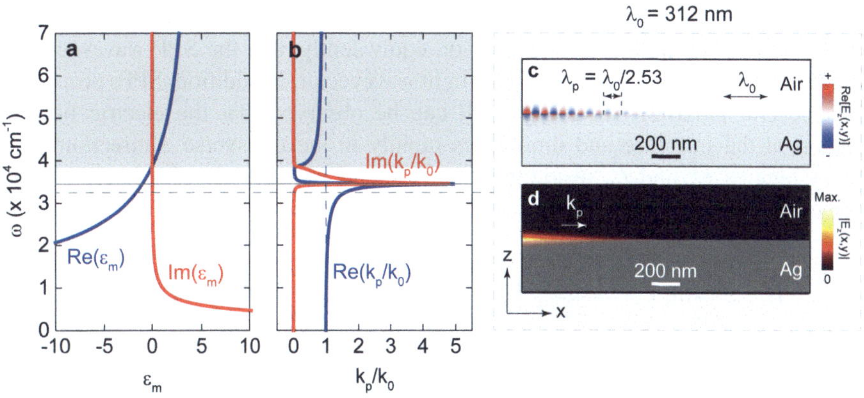

Fig. 1.3 **Surface plasmon polaritons at an Ag/air interface.** (**a**) Real and imaginary parts of the permittivity of Ag. (**b**) Real and imaginary parts of the SPP wavevector at the Ag/air interface. (**c**) Field amplitude of the SPP, propagating along the x direction as a function of x and z. (**d**) Snapshot of the vertical component of the electric field of the SPP. The solid gray line indicates the frequency at which $\mathrm{Re}(\varepsilon_m) = -1$. The white background in (**a**,**b**) indicates the region where $\mathrm{Re}(\varepsilon_m) < 0$. The blue dashed vertical line in (**b**) shows the dispersion of light in free space. The horizontal dashed line in (**a**,**b**) marks the frequency at which the fields distributions in (**c**,**d**) were calculated

energy between the collective oscillations and the incoming field for $\omega \leq \omega_p$ can be very high, reaching the strong coupling regime and giving rise to a surface plasmon polariton (SPP), a light-plasmon hybrid that propagates along the metal surface.

To illustrate the properties of SPPs, we consider a typical material in plasmonics: silver (Ag). The permittivity of Ag at visible frequencies is given by the following parameters [26]: $\varepsilon_\infty = 4$, $\omega_p = 7.74 \cdot 10^4\,\mathrm{cm}^{-1}$ and $\gamma = 183.9\,\mathrm{cm}^{-1}$. The real and imaginary parts of the permittivity of the Drude metal are shown in Fig. 1.3a, illustrating that, below ω_p (compensated by the effect of ε_∞), the permittivity becomes negative. The dispersion of SPPs at the interface between Ag and air can be calculated by setting $\varepsilon_1(\omega) = 1$ and $\varepsilon_2(\omega) = \varepsilon_m(\omega)$ in Eq. (1.5), with $\varepsilon_m(\omega)$ given by Eq. (1.7). Figure 1.3b shows that the SPP wavevector is close to the light line at low frequencies, this is, in the mid-IR and terahertz (THz) spectral ranges. This is indicated by $\mathrm{Re}(k_p/k_0)$ approaching the blue dashed vertical line in Fig. 1.3b, which depicts the dispersion of light in free space. As the frequency increases, the normalized wavevector also increases and can exceed several times the free-space wavevector of light, particularly close to the frequency where $\mathrm{Re}[\varepsilon_m(\omega)] = -1$, which is also located in the near-UV range. This frequency is marked by a solid gray line in Fig. 1.3a,b, and determines the transition from normal to anomalous SPP dispersion. In the anomalous dispersion regime, the group velocity, given by $v_g = \partial k_p/\partial\omega$, is negative. As the frequency exceeds that at which $\mathrm{Re}[\varepsilon_m(\omega)] = 0$, the SPP dispersion gradually approaches the light line.

Figure 1.3c,d displays the vertical component of the electromagnetic field of the SPP that the Ag/air interface supports at $\lambda_0 = 312\,\mathrm{nm}$ ($\omega = 3.4 \cdot 10^4\,\mathrm{cm}^{-1}$).

The images clearly show that the SPP wavelength is substantially smaller than that or free space light, $\lambda_p = 2.53\lambda_0$ or, equivalently, that the SPP wavevector is significantly larger than the free-space light wavevector. In addition, SPPs propagate over several polariton wavelengths. It can be observed that the electric field is highest at the interface and diminishes rapidly in the transverse z direction. The decay lengths L_z^0 and L_z inside the dielectric and Ag, respectively, determine the rate of decay, and are given by:

$$L_z^0 = \frac{1}{k_0}\mathrm{Im}\left(\sqrt{\frac{\varepsilon_1 + \varepsilon_m}{\varepsilon_1^2}}\right), \qquad L_z = \frac{1}{k_0}\mathrm{Im}\left(\sqrt{\frac{\varepsilon_1 + \varepsilon_m}{\varepsilon_m^2}}\right). \tag{1.8}$$

At the frequency $\omega = 3.4 \cdot 10^4\,\mathrm{cm}^{-1}$, we obtain $L_z^0 = 20\,\mathrm{nm}$ and $L_z = 17\,\mathrm{nm}$. This calculation shows that the decay length is smaller in Ag than in the dielectric and that both of them are significantly smaller than the free-space wavelength ($\lambda_0 = 294\,\mathrm{nm}$). This property holds generally for SPPs and is in contrast to waveguide modes in dielectrics.

Furthermore, the electromagnetic field of the SPP decays exponentially as it propagates (as shown in Fig. 1.3c). This is due to the ohmic losses in Ag, resulting in the heating of the material. The propagation length L_p, defined as the distance for the SPP electric field to decay by a factor of $1/e$, can be calculated in general from the dispersion as $L_p = 1/\mathrm{Im}(k_p)$. In terms of the polaritonic wavelength, which can also be calculated from the dispersion as $\lambda_p = 2\pi/\mathrm{Re}(k_p)$, L_p can be expressed by means of the following figure of merit (FOM):

$$\mathrm{FOM} = \frac{L_p}{\lambda_p} = 2\pi\frac{\mathrm{Im}(k_p)}{\mathrm{Re}(k_p)}. \tag{1.9}$$

For $\omega = 3.4 \cdot 10^4\,\mathrm{cm}^{-1}$, we find a value of $L_p = 0.3\,\mu\mathrm{m}$. In relative terms to the SPP wavelength, this can be expressed by means of the FOM. Evaluation of Eq. (1.9) yields a FOM value of 2.5. Therefore, the SPP propagates 2.5 wavelengths before decaying a factor $1/e$. For slightly lower frequencies, L_p significantly increases. At $\omega = 3 \cdot 10^4\,\mathrm{cm}^{-1}$, $L_p = 9\,\mu\mathrm{m}$ and FOM $= 34$. At $\omega = 2.5 \cdot 10^4\,\mathrm{cm}^{-1}$, we get $L_p = 42\,\mu\mathrm{m}$ and FOM $= 116$. One should note, however, that the increased propagation length comes at the cost of reduced confinement. Conversely, within the anomalous dispersion regime, calculations yield values down to FOM ≤ 0.35. A propagation length in the range of microns is common for SPPs in metals at visible frequencies [9, 27].

Hence, SPPs in metals, such as gold, silver and copper, at visible frequencies can have sub-diffractional wavelengths ($\lambda_p < \lambda_0$), high lateral confinement ($L_p < \lambda_0$ and $L_z < \lambda_0$), and propagation lengths in the order of dozens of microns. Although using more complex geometries can increase the propagation length of SPPs (up to the millimeter range) [28–31], their propagation length is still much shorter compared to that of visible light in a dielectric waveguide [3, 7, 32]. This limitation makes it challenging to employ SPPs for long-distance communication systems.

The discovery of SPPs on metal surfaces [33], along with subsequent findings such as the surface enhanced Raman scattering (SERS) effect [34, 35] and extraordinary optical transmission through sub-wavelength holes in metals [36], led to the emergence of plasmonics [13]. This field has since focused on the study of both *localized* modes, confined in three spatial dimensions, such as nanoparticles or nanoresonators, and *propagating* modes, confined in one dimension and travelling along the other two, as in metal/dielectric interfaces. The potential for manipulating light at the nanoscale and the rich physics of surface plasmons spurred the development of nanophotonics [37] and many applications of metamaterials [38]. Research in plasmonics has since then led to various advancements, including improved chemical detection [39–42] and increased quantum efficiency for detectors in the UV, visible, and NIR spectral ranges [43–48]. However, high optical losses in metals at optical frequencies have prevented the full realization of the potential of plasmonics for many applications [49–51].

Surface Plasmon Polaritons in Metals at Mid-infrared Frequencies

Metal-based plasmonics has proven to be effective in the UV to NIR spectral range, but its usefulness is limited beyond the NIR due to the very large negative permittivity of metals at longer wavelengths. In the mid-IR range, the permittivity of metals such as gold or silver has large negative values, resulting in SPPs at semi-infinite metal/dielectric interfaces having a wavevector similar to the free-space wavevector (see Fig. 1.3 at small values of ω), and subsequently large transverse decay lengths in the lateral direction. For instance, evaluation of Eq. (1.8) at $\omega_0 = 2000\,\text{cm}^{-1}$ yields $L_z^0 = 30\,\mu\text{m}$. Therefore, SPPs are not suitable for developing subwavelength-scale mid-IR to THz photonics.

This is a crucial limitation, since the mid-IR spectral band (400–4000 cm^{-1}) and the THz spectral range offer numerous potential applications [52]. For instance, the region known as the *atmospheric window*, which spans from 8 to 12 μm, is particularly advantageous for imaging through atmospheric obscurants and opaque objects, and utilizing free space optics. Additionally, the *fingerprint region*, generally considered to be between 2.5 and 6.7 μm, is associated with the vibrational motion of molecules. This region provides a unique pattern or *fingerprint* that can be used to identify the types of chemical bonds and functional groups present in a sample. Within the fingerprint region, IR-active vibrational fingerprints can be used for detecting trace levels of chemical compounds. Moreover, the blackbody emission reaches its maximum in this range (at room temperature), making it interesting for thermal applications, such as thermal emission control, monitoring, and imaging.

Therefore, alternative optical materials that support plasmon-like modes and/or exhibit optical properties conducive to sub-diffraction confinement in the IR to THz spectral ranges have been extensively studied. In particular, advances in efficient and coherent sources could unlock new opportunities in imaging, sensing, and detection technologies. While dielectric materials can significantly reduce optical losses, resonators with sub-diffractional optical confinement cannot be made out of positive permittivity materials. Therefore, the search for alternative materials for nanophotonic applications focused into identifying novel lower-loss free-carrier-based plasmonic materials. One possibility is to use artificial metamaterials that sustain propagating surface polaritons, such as metal-insulator-metal waveguides [53–56], transmission lines [57, 58], and dielectric stripes on metallic surfaces [59]. Heavily doped semiconductors with carrier densities around $n = 10^{19} - 10^{21}\,\text{cm}^{-3}$, such as germanium (Ge), silicon (Si), III-V alloys and cadmium oxide (CdO) [60–62], can also support SPPs with large wavevectors and lateral confinement at mid-IR frequencies [63–65]. However, the damping rate of free electrons in doped semiconductors, which is key in determining characteristics such as lifetime and propagation length [66, 67], is typically in the order of $10^{13} - 10^{15}\,\text{s}^{-1}$ [61, 68, 69], which is relatively fast and therefore limits the efficiency of polaritons in these materials. Hence, materials combining smaller damping rates and small negative-valued permittivities could exhibit surface polaritons with better characteristics than SPPs in metals and doped semiconductors at mid-IR frequencies.

Surface Phonon Polaritons in Polar Crystals at Mid-IR Frequencies

Polar dielectrics, such as NaCl, CaF_2, SiC, GaAs, AlN or Al_2O_3 can also support surface polaritons at their interface with an insulator. In this case, the existence of surface polaritons is due to an uneven distribution of electron density that creates a dipole moment between atoms within the unit cell of the polar dielectric.

To explain how this uneven distribution can lead to the existence of surface polaritons, we start by considering that atoms in solids are bound to their equilibrium positions by forces that hold the solid together. However, atoms are not completely rigid, but in turn they are being constantly displaced from their equilibrium positions. When atoms move, they experience restoring Coulomb forces. Usually, the displacements are assumed to be small, and thus the motion can be assumed to be harmonic, meaning that a Taylor expansion can be applied on the force, which is then proportional to the displacements. Due to the interaction between atoms, the displacement of one or more atoms from their equilibrium position gives rise to a set of vibrations propagating through the lattice at characteristic frequencies, which are determined by the structure of the crystal itself. The resonant frequencies of the harmonic vibrations are given by $\omega = \sqrt{K/\mu}$, where K is the spring constant and μ is the reduced mass. Since μ here is related to the mass of the atoms, much higher than those of free electrons discussed in the previous subsection, it becomes clear

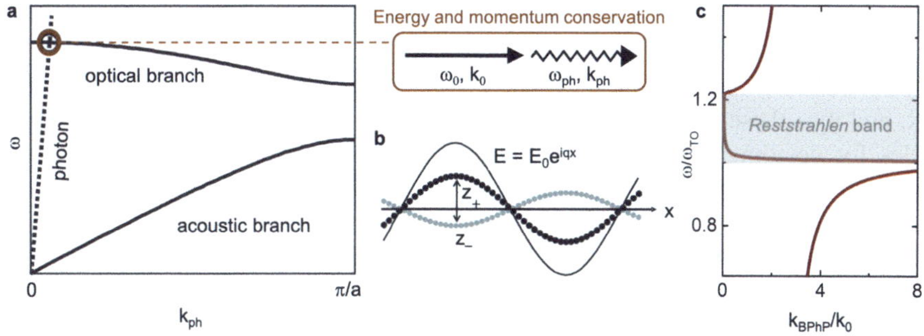

Fig. 1.4 Lattice absorption process by infrared active phonons and bulk phonon polaritons. (**a**) Dispersion curves for the acoustic and optical phonon branches in a typical crystal with a lattice constant a. The dispersion of the photon modes in the crystal is shown by the dotted line. In the brown square, the straight arrow represents the photon that is absorbed, while the wiggly arrow represents the phonon that is created. Adapted from [24]. (**b**) Interaction of a TO phonon mode propagating in the x direction with an electromagnetic wave of the same wavevector. The black circles represent positive ions, while the gray circles represent the negative ions. The solid line represents the spatial dependence of the electric field of the electromagnetic wave. Adapted from [24]. (**c**) Calculated normalized bulk phonon polariton (BPhP) dispersion in a prototypical phonon-polaritonic material. The area shadowed in gray shows the *Reststrahlen* band

that these vibrations occur at lower frequencies than the plasma oscillations, more specifically in the IR to THz regime, with $\omega_0/2\pi \approx 10^{12} - 10^{13}$ Hz [24].

Any arbitrary lattice vibration can be considered to be a superposition of vibration normal modes, which can be divided into acoustic and optical modes. The corresponding quantized modes of the lattice vibrations are called *phonons*. Acoustic phonons are in-phase vibrations of the atoms in the lattice, while optical phonons are out-of-phase vibrations [15, 70]. Both the acoustic and optical phonon modes of a crystal are then sub-divided into transverse and longitudinal. Figure 1.4a shows generic dispersion curves for the acoustic and optical phonons in a simple crystal, adapted from Ref. [24], and where the fact that the longitudinal and transverse modes tend to have different frequencies is neglected for simplicity. The angular frequency ω_{ph} of the acoustic and optical phonons is plotted against the wavevector k_{ph} in the positive half of the first Brillouin zone. At small wavevectors, the slope of the acoustic branch is equal to v_s, the velocity of sound in the medium, while the optical modes are essentially dispersionless near $k_{ph} = 0$. The figure also shows the dispersion of the light waves in the crystal, which have a constant slope of $v = c/n_d$, where n_d is the refractive index of the polar dielectric. The refractive index has been highly exaggerated here in order to make the dispersion of the photon noticeable on the same scale as the phonon dispersion.

For a mode to be able to interact directly with light, conservation laws require that the photon and the phonon have the same energy and momentum, which happens when the dispersion curves intersect (brown square in Fig. 1.4). Since $c/n_d \gg v_s$, the only intersection point for the acoustic branch occurs at the origin, which corresponds to the response of the crystal to a static electric field. The situation is different for the optical branch: there is an intersection at finite ω_{ph}, which is identified with the brown circle in Fig. 1.4a. These modes can generally couple to light, rendering them optically active. As discussed above, ω_{ph} is usually in the IR, and thus these modes are generally referred to as IR active. Detailed selection rules for deciding which phonon modes are IR active can be derived by using group theory, but this is beyond the scope of this thesis. In addition, as the optical branch is essentially flat for small k_{ph}, the frequency of this resonance is equal to the frequency of the optical mode at $k_{ph} = 0$, the Γ point of the Brillouin zone. Therefore, all along this thesis, it will be typically sufficient to consider only Γ-point phonons, owing to the small momentum of free-space light.

The IR-active character of optical modes can be intuitively understood from the following picture: when an electromagnetic field is applied, photons couple to phonons through the driving force exerted on the atoms by the AC electric field. Electromagnetic waves are transverse, and thus they apply driving forces to the transverse vibrations of the crystal, which means that they couple to the transverse optical (TO) phonon modes.

Since optical phonons involve out-of-phase vibrations, it becomes clear that they typically exist if the lattice consists of two or more elements with an asymmetric charge distribution, with positive and negatively charged ions moving along opposite directions and creating such out-of-phase movements. Therefore, in order to support optically active phonons, the crystal must have an asymmetric charge distribution, usually referred to as a polar character. In a simple picture, this charge separation can arise from differing electronegativities of the atoms within the unit cell. Higher electronegativity then leads to a partial negative charge, while lower electronegativity results in a partial positive charge. This causes atoms with an asymmetric charge distribution to oscillate out of phase, creating transverse electric dipoles. For instance, ionic crystals, such as alkali halides (e.g., Na^+Cl^-), exhibit large electronegativities, leading to a complete electron transfer between the atoms, and thus to large permanent dipoles. However, when these atoms vibrate, the induced separation of charge is minimal because the electron transfer has already occurred. Therefore, the dipoles associated with vibrational modes in ionic compounds are generally weaker compared to those in covalent materials, giving rise to a low polarizability. On the other hand, materials with covalent bonds involve shared electrons, resulting in partial positive and negative charges. These partial charges, and thus the dipoles they originate, depend on properties such as the strength of the bonds, the masses of the atoms or the symmetry of the lattice. For instance, AlN and hBN exhibit significantly strong dipoles despite smaller differences in electronegativity than ionic crystals, while molybdenum disulfide (MoS_2) displays notably weaker dipoles. It therefore becomes necessary to consider crystal properties such as the bond strength and the masses of the atoms in the discussion of charge separation. In general, lighter masses are associated with more easily displaced atoms, which respond readily to external forces or perturbations, leading to a larger change in polarization. All these factors are usually captured by a partial or effective charge, the Born effective charge [15], a measure of how the charge distribution of an ion responds to atomic displacements, which is traditionally calculated using Density Functional Theory (DFT).

To examine the interaction between a TO phonon mode with $k_{ph} \approx 0$ and an infrared light wave, we consider a linear chain, as shown in Fig. 1.4b. The solid line depicts the AC electric field's spatial variation in the infrared light wave. The chain comprises alternating positive (black circle) and negative (gray circle) ions, with motion along the x axis. In this transverse mode, the atom displacements occur along the z axis. The displacements of the positive and negative ions in a TO mode are in opposite directions and are given the symbols z_+ and z_- respectively. The phonon wavelength significantly exceeds the size of a crystal unit cell, typically $< 10^{-9}$ m (Fig. 1.4b exaggerates atom size for clarity). Therefore, for long-wavelength TO modes ($k_{ph} \approx 0$), the motion of the atoms in different unit cells is nearly identical and we can focus on vibrations within a single unit cell. The out-of-phase displacements of oppositely charged atoms generate an AC electric field that is in phase with the external light. This is crucial, since a strong interaction can then occur between TO phonons and light when their wavevectors and frequencies coincide.

As a result of the strong interaction between light and the IR active phonons, phonon-photon coupling can occur. When the magnitude of the phonon-photon coupling is large enough, hybrid phonon-photon modes can form and propagate through the bulk of the material, thereby referred to as *bulk phonon polaritons* (BPhPs). We plot in Fig. 1.4c the BPhP dispersion $\omega = k_0\sqrt{\varepsilon_d(\omega)}$ for an archetypal phonon polaritonic material, normalized to ω_{TO} and k_0. The BPhP dispersion shows an anti-crossing, a clear signature of strong coupling. In the anti-crossing regime, no BPhP state is allowed to propagate inside the polar crystal. In turn, in non-polar crystals, the lack of a permanent electric dipole moment reduces the phonon-photon coupling, making the formation of BPhP less likely or even non-existent.

The interaction between the TO phonon and the light wave, and thus the optical response of the crystal, is captured by the dielectric permittivity. The permittivity function of a polar crystal at IR frequencies can be modelled by writing down the equations of motion for the displaced ions taking into account the restoring force, yielding a Lorentz model [24, 71, 72]:

$$\varepsilon_d(\omega) = \varepsilon_\infty + (\varepsilon_0 - \varepsilon_\infty)\frac{\omega_{\mathrm{TO}}^2}{\omega_{\mathrm{TO}}^2 - \omega^2 - i\gamma\omega}, \tag{1.10}$$

where ε_∞ represents the dielectric constant at frequencies well above the phonon resonance, but below the next natural frequency of the crystal due, for example, to the bound electronic transitions in the visible/ultraviolet spectral region. ε_∞ is a characteristic feature of polar dielectrics [70] and can be observed experimentally through techniques such as Raman or IR spectroscopy. In turn, ε_0 is the static permittivity.[4] Dielectric spectroscopy yielding ε_0 reports on the total polarity of the system. Finally, γ is the damping rate of TO phonons. One of the most attractive features of polar crystals is the typically small phononic damping rate ($\gamma \approx 10^{12}\,\mathrm{s}^{-1}$), up to two orders of magnitude smaller than that of free electrons in metals and doped semiconductors, as they are governed by the scattering of optical phonons which occurs on longer time-scales (typically in the picosecond range) compared to their plasmonic counterparts (typically in the femtosecond range) [52]. Therefore, these materials exhibit optical properties similar to metals below the plasma frequency, but with reduced damping rates and vastly different electrical properties [73].

Additionally, Eq. (1.10) tells us that, at a certain frequency which we label ω', the dielectric constant can fall to zero (supposing that we have a lightly damped system so that we can set $\gamma = 0$). The condition for this to happen can be solved to obtain $\omega' = \sqrt{\varepsilon_0/\varepsilon_\infty}\omega_{\mathrm{TO}}$. We can understand the meaning of $\varepsilon = 0$ by considering Gauss's law. If the medium has no free charges, the total charge density is zero, and we have $\nabla \cdot \mathbf{D} = \nabla \cdot (\varepsilon \mathbf{E}) = 0$, where we have made use of $\mathbf{D} = \varepsilon \mathbf{E}$ to relate the electric displacement $\mathbf{D}$ to the electric field $\mathbf{E}$ in a dielectric medium. If $\varepsilon(\omega) \neq 0$, it follows that $\nabla \cdot \mathbf{E} = 0$, which implies that the waves are transverse. This is the usual case that applies when considering the propagation of electromagnetic waves in a dielectric medium. However, if $\varepsilon(\omega) = 0$, we can satisfy $\nabla \cdot (\varepsilon \mathbf{E}) = 0$ with waves for which $\nabla \cdot \mathbf{E} \neq 0$, i.e. longitudinal waves, as bulk plasmons, which were discussed in the previous subsection [24]. These longitudinal electromagnetic waves are called longitudinal optical (LO) phonons. The waves at $\omega = \omega'$ therefore correspond to LO phonons, and we identify ω' with the frequency of the LO mode at $k_{ph} = 0$, namely ω_{LO}. Therefore, we can rewrite the relation between the TO and LO phonon frequencies in the following form: $\omega_{\mathrm{LO}}^2/\omega_{\mathrm{TO}}^2 = \varepsilon_0/\varepsilon_\infty$, known as the Lyddane-Sachs-Teller (LST) relation. The LO phonon emerges at higher energies than the TO phonon because the associated macroscopic longitudinal polarization makes the atoms experience a restoring force that not only depends on their displacements but also on the strength of the electric field, as opposed to transverse vibrations. IR light does not couple or interact with LO phonons in most cases, as this is a longitudinal field.

Within the frequency region between ω_{TO} and ω_{LO}, the reflectivity[5] approaches 100%, but unlike in metals, this is realized without free carriers. The frequency range between the TO and LO phonon frequencies is called the *Reststrahlen* band (RB). *Reststrahlen* is the German word for *residual rays*. Light cannot propagate into the medium in the RB [75]. As a result, polar crystals can exhibit negative permittivities in the IR to THz spectral range.

---

[4] The symbol ε_0 is also often used to represent the vacuum permittivity, even in the context of this thesis. However, the static permittivity will not be appearing in the remaining of this text, eliminating any potential confusion.

[5] The high reflectivity characteristic of this region can evolve into a single peak at the TO phonon frequency as the material enters the optically thin limit [74].

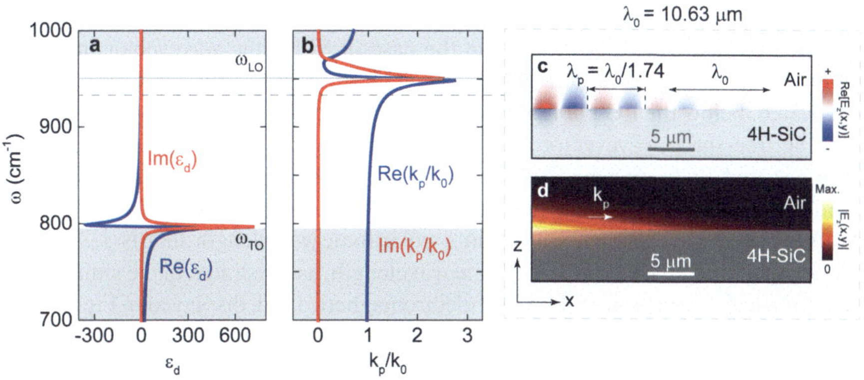

Fig. 1.5 Surface phonon polaritons at a 4H-SiC/air interface (**a**) Permittivity of 4H-SiC in the mid-IR. (**b**) Dispersion of the surface phonon polaritons on a 4H-SiC/air interface. (**c**) Field amplitude of the SPhP as a function of x and z. (**d**) Snapshot of the vertical component of the electric field. The dashed line in (**a–b**) marks the frequency at which the fields in (**c,d**) were calculated. The solid gray line in (**a,b**) marks the frequency at which $\mathrm{Re}(\varepsilon_d) = -1$. The white region in (**a,b**) represents the mid-IR RB of 4H-SiC

Therefore, when incident fields of frequencies inside the RB—typically within the region from the mid-IR down to the THz—interact with a polar material, they stimulate surface excitations mediated by the lattice vibrations, the so-called surface phonon polaritons (SPhPs) [52]. These electromagnetic modes screen out the incident field and propagate at the interface between the polar crystal and the dielectric.

In order to explore in detail these properties, we choose a representative phonon-polaritonic material. 4H-SiC is a polar crystal that has a RB in the mid-IR [76, 77]. Its permittivity function is well described by the Lorentz permittivity model:

$$\varepsilon_d(\omega) = \varepsilon_\infty \left(1 + \frac{\omega_{\mathrm{LO}}^2 - \omega_{\mathrm{TO}}^2}{\omega_{\mathrm{TO}}^2 - \omega^2 - i\gamma\omega}\right). \qquad (1.11)$$

where we have taken into account the LST relation into Eq. (1.10).[6] The permittivity function of 4H-SiC is given by Eq. (1.11), with the parameters [77, 78]: $\varepsilon_\infty = 6.7$, $\omega_{\mathrm{TO}} = 969\,\mathrm{cm}^{-1}$, $\omega_{\mathrm{LO}} = 793\,\mathrm{cm}^{-1}$, $\gamma = 4.76\,\mathrm{cm}^{-1}$. The permittivity function of 4H-SiC is shown in Fig. 1.5a.

The dispersion relation of SPhPs at the air/4H-SiC interface, normalized to the free-space wavevector is given by Eq. (1.5), with $\varepsilon_1(\omega) = 1$ and $\varepsilon_2(\omega) = \varepsilon_d(\omega)$, as given by Eq. (1.11). Figure 1.5b shows that the SPhPs exist in the range between the

[6] A more general expression for the dielectric permittivity of polar phonons is obtained by allowing different damping rates for the TO and LO phonons, γ_{TO} and γ_{LO}, respectively [75], Along this thesis, we will assume that $\gamma_{\mathrm{TO}} = \gamma_{\mathrm{TO}}$.

TO phonon frequency, at which the SPhP wavevector is similar to the wavevector of light in free-space, and the LO. As in the case of SPPs, the wavevector increases up to the frequency at which $\mathrm{Re}(\varepsilon_d) = -1$, depicted by the solid gray line in Fig. 1.5a,b, below the LO phonon frequency. SPhPs can possess wavevectors higher than the free-space wavevector (up to $2.5k_0$ for 4H-SiC), differently from SPPs in metals at mid-IR frequencies, which have wavevectors close to k_0. At $\lambda_0 = 10.63$ μm ($\omega_0 = 940\,\mathrm{cm}^{-1}$, where the wavevector of the SPhP is well separated from the free-space wavevector), we obtain a polaritonic wavelength that is 1.74 times smaller than free-space light at the same wavelength, as illustrated in the simulations of the vertical component of the SPhP electromagnetic field, displayed in Fig. 1.5c,d.

Interestingly, due to the small range of frequencies that the RB of polar crystals typically covers (see the area shadowed in white in Fig. 1.5a for 4H-SiC), the permittivity function of a polar crystal is more dispersive than that of a typical plasmonic material [52, 69]. Hence, SPhPs are also more dispersive than SPPs and, as a consequence, the group velocity of SPhPs, $v_g = \partial\omega/\partial k_p$, is typically smaller than that of SPPs.

In addition, the SPhP electromagnetic fields decay exponentially in the lateral direction, similar to the SPP fields, as seen in Fig. 1.5c,d. The SPhP decay lengths into the dielectric and polar material are given by Eq. (1.8). Plugging the permittivity of the polar dielectric, defined in Eq. (1.11), into $\varepsilon_m(\omega)$ in Eq. (1.8), we obtain L_z and L_z^0. At $\omega_0 = 940\,\mathrm{cm}^{-1}$ ($\lambda_0 = 10.63\,\mu\mathrm{m}$), the SPhP decay length into 4H-SiC is $L_z = 819\,\mathrm{nm}$, and the decay length into air is $L_z^0 = 1.3\,\mu\mathrm{m}$, both being much smaller than λ_0. The propagation length of the mode at the same frequency is $L_p = 9\,\mu\mathrm{m}$, which is 0.76 times the free-space photon wavelength or 1.5 times the SPhP wavelength. At higher frequencies, larger confinement rates can be achieved at the expense of higher optical losses. For instance, at $\omega_0 = 945\,\mathrm{cm}^{-1}$ ($\lambda_p = 10.58\,\mu\mathrm{m}$), $\lambda_p = 4.7\,\mu\mathrm{m}$ $L_z = 819\,\mathrm{nm}$ and $L_z^0 = 1.3\,\mu\mathrm{m}$. However, the propagation length is just $L_p = 2.5\,\mu\mathrm{m}$, almost half of the polaritonic wavelength $\lambda_p = 4.7$ μm. In addition, it is worth noting that, although the damping rate of polar phonons in 4H-SiC is two orders of magnitude smaller than that of free electrons in doped semiconductors [61], the normalized propagation lengths of SPhPs in 4H-SiC are not much longer than those of SPPs in doped semiconductors [52]. This is related to the fact that the propagation length, L_p, approximately depends on the damping rate, γ, and the group velocity, v_g, as $L_p \propto \gamma/v_g$ [52], i.e., the smaller damping is largely compensated by the smaller group velocity in 4H-SiC.

In sum, polar crystals, such as 4H- SiC, provide subwavelength confinement and long propagation lengths in the mid-IR. Nevertheless, fabricating nanostructures from 4H-SiC, as well as other bulk polar crystals, proves to be a challenging task. In 4H-SiC, this is primarily due to its robust covalent bonds, which make it highly resistant to thinning methods. This limits the applicability of these materials in nano-optics. In fact, the progress in nano-optics significantly relies on the advancement of materials that are more user-friendly and facilitate the fabrication of nanostructures out of them.

1.3 Polaritons in van der Waals Materials

In the past decade, there has been a surge of research activity focused on van der Waals (vdW) materials, more easily manipulable for nanofabrication and nanostructuring purposes than traditional bulk crystals. This easiness of use stems from the fact that vdW materials consist of individual atomic planes bonded by weak vdW attraction. Due to the weakness of such interlayer forces, these materials can easily be separated into sheets as thin as a single atom, which as thus referred to as two-dimensional (2D). Moreover, owing to their low dimensionality, these 2D sheets exhibit unique electronic, optical, and mechanical properties, turning the advent of vdW materials into a transformative breakthrough in the fields of nano-optics and polaritonics.

The first 2D material discovered was graphene. Scientists had theorized the potential existence and production of graphene for decades. In 1962, Boehm et al. [79] looked for the thinnest possible fragments of reduced graphite oxide and identified some of them as monolayers relying on a relative transmission electron microscope contrast, which strongly depends on focusing conditions. Boehm et al. also [80] introduced the term graphene in 1986, derived from *graphite* and the suffix *-ene*, which reflects the fact that this graphite allotrope of carbon contains numerous double bonds. Overall, graphene received little attention until 2004, when the material was rediscovered, isolated and investigated [81]. Graphene proved to be surprisingly easy to isolate and to display intriguing electronic properties, such as a linear electronic band dispersion [82, 83] and large electron mobilities, which held promise for applications in nanoelectronics. However, it was soon realized that graphene's optical properties were just as fascinating. Furthermore, the discovery of graphene led to the identification of numerous vdW materials with distinct physical properties, including metallic, semiconducting, insulating, and superconducting compounds. Since then, a vast field of research emerged focused on exploring the properties of vdW materials that are atomically thin [84]. The ability to stack and combine vdW materials in a precisely controlled manner has opened up new possibilities for designing and fabricating complex heterostructures with tailored properties.

In particular, vdW materials display nearly all electronic and optical phenomena found in solids, including plasmonic oscillations of free electrons characteristic of metals, light emission/lasing and excitons encountered in semiconductors, and intense phonon resonances typical for insulators. As a result, a broad range of polaritonic modes have been either experimentally observed or theoretically predicted in vdW materials [85, 86]. A brief summary, shown in Fig. 1.6, includes:

- Plasmon Polaritons: These arise from the interaction between conduction electrons and light in materials like graphene [87, 88] and black phosphorous [89, 90]. In addition, natural hyperbolic plasmon polaritons have been observed in thin films of WTe_2 [91]. A detailed discussion on hyperbolicity will follow in the next subsections.

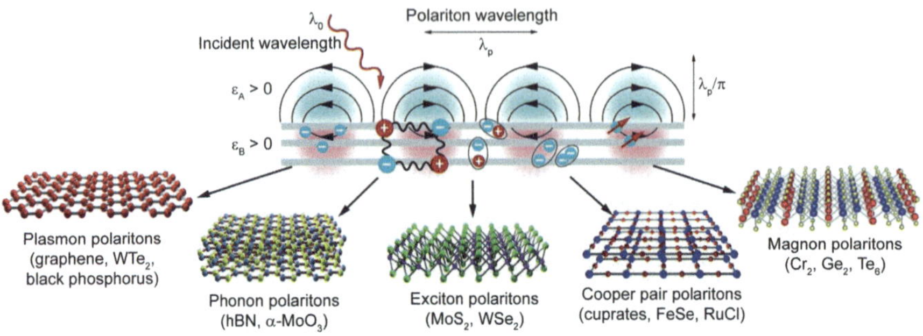

Fig. 1.6 Polaritons in vdW materials. The family of vdW materials supports polaritons arising from different physical phenomena: conduction electrons (leading to plasmon polaritons), infrared-active phonons (phonon polaritons), excitons (exciton polaritons), superfluidity in superconductors with high critical temperature T_c (Cooper-pair polaritons), and magnetic resonances (magnon polaritons). The matter oscillation component results in negative permittivity ($\varepsilon_2 < 0$) of the polaritonic material, giving rise to optical-field confinement at the interface with a positive-permittivity ($\varepsilon_1 > 0$) environment. vdW polaritons exhibit strong confinement, as defined by the ratio of incident light wavelength λ_0 to polariton wavelength λ_p. Adapted from [85]

- Phonon Polaritons: Formed by the coupling of light with bound atomic charges at mid-infrared (mid-IR) frequencies in hexagonal boron nitride (hBN) [92, 93], alpha-phase molybdenum trioxide (α-MoO_3) [94, 95], and alpha-phase vanadium pentoxide (α-V_2O_5) [96], among others. Phonon polaritons will be a pivotal topic of this thesis and, as such, a more comprehensive discussion on phonon polaritons in these two materials will also follow in this chapter.
- Exciton Polaritons: Arising from the coupling of light with bounded electron-hole pairs, exciton polaritons manifest in materials such as MoS_2 and WSe_2 at visible and near-infrared (near-IR) frequencies [97].
- Magnon polaritons: They are expected to originate in magnetic materials, such as antiferromagnetic and ferromagnetic materials [98]. Magnetic materials can also support *hybrid polaritons*, such as the coupling between magnetic excitons and photons, as recently shown in the antiferromagnet CrSBr [99]. Magnetic fields can also have profound effects on the collective motion of free electrons and excitons, including hybrid magnetoexcitons, as recently shown in graphene [100].
- Further, Cooper pair polaritons [101, 102] are expected to arise in high-temperature superconductors.

A more extensive discussion on the physics and technology of all types of polaritons can be found in Ref. [103].

Importantly, the strongly anisotropic crystal structure of vdW materials (in-plane covalent bonding and out-of-plane weak vdW bonding) can cause a highly anisotropic optical response. This optical anisotropy will lead, as commented below, to so-called elliptic and hyperbolic polaritonic dispersion, which has been theoretically predicted [89, 104, 105] and experimentally observed [57, 58, 106] in many vdW materials. The role of anisotropy, crucial for directing and steering light at the nanoscale, will be central to this thesis, and will be introduced in this chapter through the paradigmatic cases of hBN and α-MoO_3. Prior to that, our attention will be directed towards graphene due to its distinction as the initial atomically thin material to be successfully isolated. Additionally, it served as the pioneer in showcasing the experimental observation of polaritons within atomically thin materials. Consequently, it played a pivotal role in instigating investigations on polaritons within 2D and vdW materials, which constitutes one of the primary topics of this thesis.

Plasmon Polaritons in Graphene

Graphene is an allotrope of carbon consisting of a single layer of atoms arranged in a hexagonal lattice, see inset in Fig. 1.7. In a graphene sheet, each atom is bonded to its three closest neighbors by a strong σ-bond. Electrons are delocalized in the valence band, meaning they do not belong to a specific atom but are spread out over the entire graphene sheet. This is a fundamental property of the electronic structure in graphene. In addition, the valence band is touched by a conduction band at a single point in reciprocal space, the Dirac point, with a linear, rather than quadratic, dependence of energy on momentum of the charge carriers, making graphene a semimetal with unusual electronic properties that are best described by theories for massless relativistic particles [82, 83]. Importantly, field-effect transistors (FET) can be made based on graphene, showing bipolar conduction. This is central to

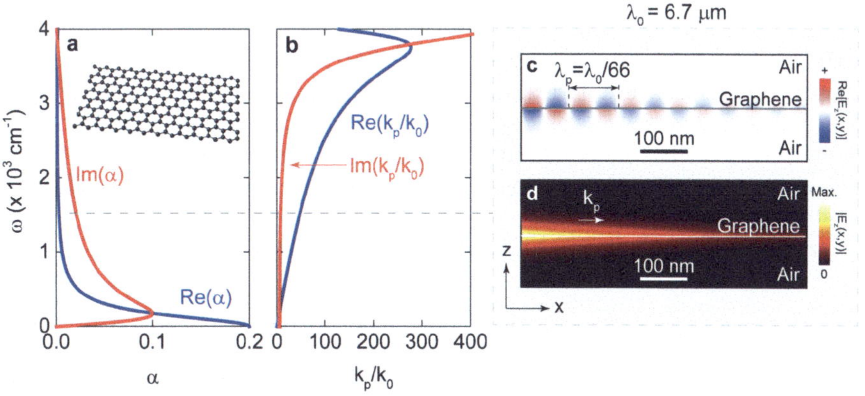

Fig. 1.7 Ultraconfined graphene plasmon polaritons. (**a**) Normalized conductivity of graphene at mid-IR frequencies for a Fermi energy of 0.3 eV, a relaxation time of 0.1 ps, and a temperature of 300 K. (**b**) Real and imaginary parts of the normalized wavevector of the GPPs. (**c**) Snapshot of the vertical component of the electric field of a GPP. (**d**) Field amplitude distribution of a GPP. The dashed line in (**a**,**b**) marks the frequency at which the fields in (**c**,**d**) were calculated

the application of this material in nanophotonics, as will be discussed below. Additionally, graphene conducts heat and electricity very efficiently along its plane. It strongly absorbs light of all visible wavelengths [107, 108], which accounts for the black color of graphite; yet a single graphene sheet is nearly transparent because of its extreme thinness (although its absorption is remarkably high for a single atom thick sheet). It is also about 100 times as strong as would be the strongest steel of the same thickness [109, 110].

The unique structural, electronic and optical properties of graphene make it an ideal platform for optical, electronic and mechanical studies and applications. In the following, we discuss the optical response of graphene, described by its conductivity. Many optical phenomena in graphene can be described by the classical electromagnetic theory based on Maxwell's equations, when modelling the layer as an infinitesimally thin conducting sheet with 2D complex conductivity σ [111]. The conductivity can be derived from quantum theory: within the random-phase approximation, accounting for a weak screened Coulomb interaction, the dynamic optical response of graphene can be obtained from the Kubo formula, yielding [112]:

$$\sigma = \sigma_{\text{inter}} + \sigma_{\text{intra}}, \tag{1.12}$$

where the terms σ_{inter} and σ_{intra} describe the contributions from interband and intraband electron–photon scattering processes, respectively:

$$\sigma_{\text{intra}} = \frac{2ie^2t}{\hbar\pi\,(\Omega + i\gamma)}\ln\left[2\cdot\cosh\left(\frac{1}{2t}\right)\right],$$

$$\sigma_{\text{inter}} = \frac{e^2}{4\hbar}\left\{\frac{1}{2} + \frac{1}{\pi}\arctan\left(\frac{\Omega - 2}{2t}\right) - \frac{i}{2\pi}\ln\left[\frac{(\Omega + 2)^2}{(\Omega - 2)^2 + (2t)^2}\right]\right\}.$$

In these expressions, Ω, γ and t stand for the frequency, decay rate and temperature, respectively, normalized to the Fermi energy level E_F: $\Omega = \hbar\omega/|E_F|$, $\gamma = \hbar/\tau|E_F|$, $t = T/|E_F|$ (T is given in units of energy). Note that the mobility of the charge carriers μ_e is related to the relaxation time τ via $\tau = \mu_e|E_F|/ev_F^2$, where v_F is the Fermi velocity and e is the elementary charge. The Fermi energy level E_F determines the electrical and thermal properties of graphene, and is directly proportional to the square root of the charge carrier density n:

$$E_F = \hbar v_F\sqrt{\pi n}.$$

The real part of the complex conductivity σ characterizes the energy loss in the system, while the imaginary part of σ characterizes the phase shift between the electric field and the induced current, in a way, the *inertness* of the charge carriers to respond to changing electric field. For simplicity, we introduce the normalized conductivity α expressed as

$$\alpha = \frac{2\pi\sigma}{c}. \tag{1.13}$$

The parameters we choose for calculating $\alpha(\omega)$ are similar to those of graphene fabricated by chemical vapor deposition [113]: $T = 300$ K, $\mu = 0.3$ eV, and $\tau = 0.1$ ps. The normalized conductivity α is shown at IR and THz frequencies in Fig. 1.7a. Both Re(α) and Im(α) are positive and increase as the frequency decreases, indicating that the interband contribution to the conductivity dominates in the mid-IR frequency region [112]. Excitingly, graphene supports plasmon polaritons (GPPs) in this spectral region, which are TM polarized [87, 88, 114]. Their dispersion relation in a free-standing layer is given by [111]:

$$k_p(\omega) = \frac{\omega}{c}\sqrt{1 - \frac{1}{\alpha^2}}. \tag{1.14}$$

The resulting dispersion curve of GPPs, shown in Fig. 1.7b, at a Fermi level $E_F = 0.3$ eV, typical in experiments [87, 88, 115], is strongly separated from the light cone, revealing the extremely large momenta of GPPs. The corresponding short wavelengths of GPPs, as well as their strong confinement to the graphene sheet can also be appreciated in the field snapshots shown in Fig. 1.7c. Note that the significantly larger wavelengths of GPPs compared to the atomic distances in the graphene monolayer also justifies the validity of approximating the graphene sheet as a 2D conductivity layer.

In addition, to guarantee the existence of GPPs, the imaginary part of α must be positive. This limits the frequency range of GPPs to approximately $\omega < 2E_F$. On the other hand, the real part of α is always positive, and so is Im(k_p), guaranteeing that the energy in the system is not amplified.

Furthermore, the strong dependence of the GPP wavelengths upon the Fermi level reveals a high tunability of GPPs, for example, by applying an electrostatic potential [87, 88]. In a graphene FET, the graphene layer is typically placed on top of or below one insulating layer, or sandwiched between two insulating layers (back, top or dual gate, respectively), which are typically made of SiO_2. The insulating layers act as a gate dielectric, and by applying a voltage to the metal gate electrode, the electrical properties of the graphene layer can be tuned. Specifically, the voltage applied to the gate electrode can control the charge density of the graphene layer, and hence the electrical conductivity. Therefore, the dispersion of GPPs can be tuned by applying an external voltage to the gate electrode. Specifically, the voltage can modulate the dispersion and propagation of SPPs, as well as their confinement.

This tunability of GPPs, along with their low losses (they can propagate several wavelengths [116], as shown in Fig. 1.7c) and exceptional lateral confinement (shown in Fig. 1.7c,d), is key in many nano-opto-electronic applications requiring an active control over the electromagnetic fields, such as bio-sensors [117], lenses [115], modulators, and ultrafast electro-optical elements [118]. In addition, by converting the natural decay product of the plasmon—electronic heat—directly into a voltage through the thermoelectric effect, GPPs can be excited and detected, as shown for GPPs in the mid-IR [119] and acoustic GPPs in the THz regime [120]. This property was also central to the exploration and tunability of non-local quantum effects [121].

Out-of-Plane Hyperbolic Phonon Polaritons in Hexagonal Boron Nitride

After the advent of graphene [122], research on vdW materials expanded exponentially. One of the most widely utilized vdW materials since then has been the polar dielectric hexagonal boron nitride (hBN), initially owing to its role as the ideal substrate for graphene [123, 124] and other vdW materials. Moreover, hBN is an excellent encapsulant and gate dielectric for heterostructures of transition metal dichalcogenides (TMDs) and other 2D materials [125]. Specifically, encapsulating graphene in hBN protects it from the local environment and dramatically enhances electronic transport, with mobility values surpassed only by those of suspended graphene [116, 123, 124, 126]. This increase in mobility within hBN/graphene/hBN structures improves the lifetime of GPPs [127] and also enables their ballistic propagation at cryogenic temperatures [128].

Nevertheless, hBN is an exciting material in its own right, offering novel material properties that enable a broad range of optical [92, 129, 130], electro-optical [116, 131] and quantum-optical [132] functionalities with great potential, especially in mid-IR nanophotonics [133]. In hBN, weak vdW forces hold the layers together, akin to graphite [134]. In each of the layers, boron and nitrogen atoms form covalent bonds in a hexagonal structure (see Fig. 1.8a), exhibiting polar covalent bonds due to the notable electronegativity difference (>0.4 eV [136, 137]) between boron and nitrogen. As a result of this polar dielectric nature, hBN can support PhPs in the mid-IR spectral range.

Owing to its highly anisotropic crystalline structure, hBN displays optical uniaxial behavior, where the optical axis is perpendicular to the covalently bonded B-N layer in the out-of-plane direction. This uniaxial anisotropy results from different in-plane and out-of-plane optical phonons.[7] Therefore, the permittivity of hBN can be described by a diagonal frequency-dependent tensor $\hat{\varepsilon}(\omega)$ with two frequency-dependent components, $\varepsilon_{\perp}(\omega)$ and $\varepsilon_{\parallel}(\omega)$, that are different along the in-plane and out-of-plane directions:

[7] Analysis based on group theory reveals the existence of polar phonons, which present a net displacement of the electrical charge and, consequently, are IR active. They can be divided into modes in which the atomic motion is restricted to the hexagonal layers (in-plane vibration), and modes in which B and N atoms vibrate against each other along the out-of-plane direction (out-of-plane vibration). For more details, see Ref. [138].

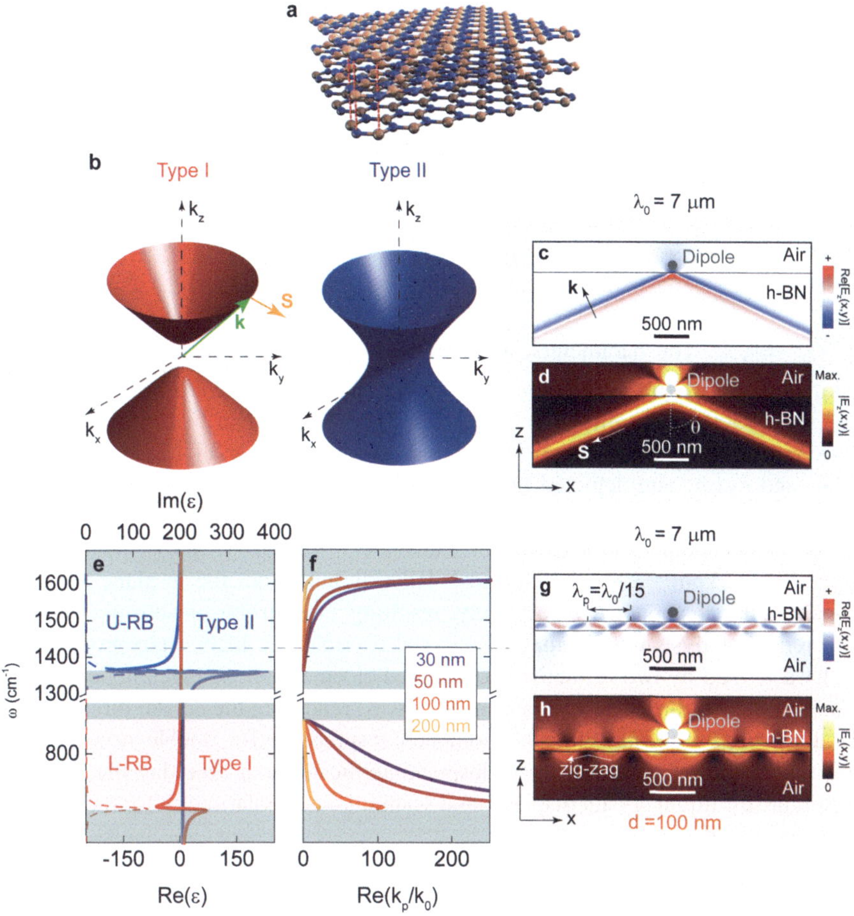

Fig. 1.8 Out-of-plane HPhPs in hBN at mid-IR frequencies. (**a**) Atomic structure of few-layer hBN. Taken from Ref. [135]. (**b**) Isofrequency surface of polaritons in hBN in the L-RB and U-RB, showing Type I and Type II hyperboloids, respectively. (**c**) Snapshot of the vertical component of the electric field of HPhPs in semi-infinite hBN, launched by a dipole source and forming a characteristic ray. The frequency is $\omega_0 = 1428\,\mathrm{cm}^{-1}$. (**d**) Field amplitude distribution of HPhPs in semi-infinite hBN, launched by a dipole source and forming a characteristic ray. The frequency is $\omega_0 = 1428\,\mathrm{cm}^{-1}$. (**e**) Permittivity functions of hBN in the mid-IR spectral range. The red line represents $\mathrm{Re}(\varepsilon_{\parallel})$ and the blue line represents $\mathrm{Re}(\varepsilon_{\perp})$. Colored spectral regions indicate the U-RB (blue) and L-RB (red). (**f**) Dispersion of HPhPs in the U-RB and L-RB for an hBN slab of thickness d surrounded by air, $\varepsilon_{1,3} = 1$, with $d = 30$, 50, 100 and 200 nm. Note that the sign of the wavevector of the modes in the L-RB is negative, but we represent the absolute value of the wavevector for compactness. (**g**) Snapshot of the vertical component of the electric field of HPhPs in a 100-nm-thick slab of hBN, launched by a dipole source and forming a characteristic *zig-zag* pattern. The frequency is $\omega_0 = 1428\,\mathrm{cm}^{-1}$. (**h**) Field distribution of HPhPs in a 100-nm-thick slab of hBN, launched by a dipole source and forming a characteristic *zig-zag* pattern. The frequency is $\omega_0 = 1428\,\mathrm{cm}^{-1}$. The frequency in (**c**,**d**,**g**,**h**) is marked by a dashed line in (**e**,**f**)

$$\hat{\varepsilon}(\omega) = \begin{pmatrix} \varepsilon_{\perp}(\omega) & 0 & 0 \\ 0 & \varepsilon_{\perp}(\omega) & 0 \\ 0 & 0 & \varepsilon_{\parallel}(\omega) \end{pmatrix}.$$

Each of the permittivity tensor components can be described by a Lorentz oscillator model [93]:

$$\varepsilon_j(\omega) = \varepsilon_j^{\infty}\left(1 + \frac{\omega_{\mathrm{LO},j}^2 - \omega_{\mathrm{TO},j}^2}{\omega_{\mathrm{TO},j}^2 - \omega^2 - i\omega\gamma_j}\right),$$

with $j = \parallel, \perp$. The parameters, as reported in Ref. [139], are given by: $\varepsilon_{\parallel}^{\infty} = 2.95$, $\omega_{\mathrm{LO},\parallel} = 760\,\mathrm{cm}^{-1}$, $\omega_{\mathrm{TO},\parallel} = 825\,\mathrm{cm}^{-1}$, $\gamma_{\parallel} = 2\,\mathrm{cm}^{-1}$, $\varepsilon_{\perp}^{\infty} = 4.9$, $\omega_{\mathrm{LO},\perp} = 1360\,\mathrm{cm}^{-1}$, $\omega_{\mathrm{TO},\perp} = 1610\,\mathrm{cm}^{-1}$, and $\gamma_{\perp} = 7\,\mathrm{cm}^{-1}$. The parallel and transversal components of the dielectric permittivity tensor of hBN, denoted as $\varepsilon_{\parallel}(\omega)$ and $\varepsilon_{\perp}(\omega)$, are depicted in Fig. 1.8e for mid-IR frequencies. In this range, hBN exhibits two distinct RBs, referred to as the upper RB (U-RB), spanning the frequency range 1360–1610 cm^{-1}, and the lower RB (L-RB), spanning the range 760–825 cm^{-1}. The U-RB and L-RB originate from the in-plane and out-of-plane optical phonons, respectively, resulting in a strong anisotropy in hBN. Theoretical analysis reveals that, in uniaxial crystals like hBN, light propagation is characterized by two refractive indices, and electromagnetic eigenmodes can be categorized as ordinary and extraordinary waves, rendering the crystal birefringent. Birefringence is an intriguing phenomenon responsible for double refraction in anisotropic crystals, and it finds widespread applications in everyday life, such as in sunglasses, liquid crystal displays, and scanning laser polarimetry for monitoring glaucoma [140].

HPHPs in Bulk hBN Remarkably, the sign of the permittivity is different along different directions within each RB of hBN: $\mathrm{Re}(\varepsilon_{\parallel}) \cdot \mathrm{Re}(\varepsilon_{\perp}) < 0$. This unique characteristic allows for the propagation of PhPs. The properties of HPhPs can be further understood by analyzing the polaritonic isofrequency surface, which represents the dispersion of polaritons in 3D wavevector **k** space at a constant frequency ω_0 [141]. In isotropic media, the isofrequency surface of free-space light or polaritons is a sphere [142], given by $k_x^2 + k_y^2 + k_z^2 = \omega^2/c^2$, where c is the speed of light. In contrast, the isofrequency surface of PhPs in hBN is described by a different equation [5, 142]:

$$\frac{k_x^2 + k_y^2}{\varepsilon_{\parallel}} + \frac{k_z^2}{\varepsilon_{\perp}} = \frac{\omega}{c}. \tag{1.15}$$

The solutions of Eq. (1.15) for $\mathrm{Re}(\varepsilon_{\parallel}) \cdot \mathrm{Re}(\varepsilon_{\perp}) < 0$ result in open hyperboloids, giving rise to hyperbolic phonon polaritons (HPhPs). Unlike surface modes confined to the interfaces of polar dielectrics, previously illustrated with the example of air/4H-SiC, HPhPs in hBN are volume modes that propagate into the bulk. In addition, the isofrequency surfaces in the two different regimes of $\varepsilon_\perp$ and $\varepsilon_\parallel$ exhibit distinct topologies, as shown in Fig. 1.8b. In hBN's L-RB, where $\mathrm{Re}(\varepsilon_\perp) > 0$ and $\mathrm{Re}(\varepsilon_\parallel) < 0$, the isofrequency surface forms a two-sheet hyperboloid (Type I), as shown in Fig. 1.8b (left). On the other hand, in hBN's U-RB, where $\mathrm{Re}(\varepsilon_\perp) < 0$ and $\mathrm{Re}(\varepsilon_\parallel) > 0$, the isofrequency surface takes the shape of a one-sheet hyperboloid (Type II), as shown in Fig. 1.8b (right). Notably, HPhPs can propagate with arbitrarily large wavevectors along the direction defined by the hyperboloid's asymptote, resulting in vanishing wavelengths. In fact, there are many different allowed values for the wavevector along the asymptotes, and thus the density of optical states can be very high along certain directions. These unique properties enable efficient canalization [143, 144] and confinement of light in extremely subwavelength volumes [145], leading to enhanced dipole-dipole interactions [146] and other intriguing effects. Moreover, the Poynting vector **S**, which indicates the direction of polaritonic energy flux and propagation, is normal to the isofrequency surface.[8] As a result, **k** and **S** are generally non-collinear in hyperbolic media (Fig. 1.8b, left). This can give rise to exotic optical phenomena, such as hyperlensing [143, 144] or anomalous reflection [149] and refraction effects, as we will see later in this work, particularly in Chaps. 5 and 6.

Additionally, experimental evidence has shown that the phase velocity of HPhPs, represented by the ratio of the angular frequency to the wavevector, is negative in hBN's L-RB [150], which means that the sign of the real part of the wavevector is negative (the absolute value of the wavevector is shown in Fig. 1.8f for compactness). However, the group velocity, represented by the derivative of the angular frequency with respect to the wavevector, is positive in both RBs, and it is significantly smaller than the speed of light in vacuum.

Localized sources generating strong near fields with large wavevectors can excite HPhPs in hBN. The color plots in Fig. 1.8c,d show the real part and the amplitude of the calculated electromagnetic field induced by a vertically oriented electric dipole near the surface of a semi-infinite hBN crystal, for a frequency inside the U-RB of hBN, $\omega_0 = 1428\,\mathrm{cm}^{-1}$. The resulting HPhPs form directional ray-like patterns, hereafter simply referred to as *hyperbolic rays*, propagating at a specific angle with respect to the z axis, given by [143, 144] $\theta = \arctan\left(\sqrt{-\varepsilon_\perp/\varepsilon_\parallel}\right)$. Notably, the direction of the hyperbolic rays is parallel to the energy flow (Poynting vector, **S**), while the direction of the phase velocity is nearly perpendicular to the hyperbolic

[8] **S** is usually assumed to be parallel to $\boldsymbol{v}_g$, and it is generally considered that both describe the waves' propagation direction. We will also assume this along this thesis. However, the collinearity between **S** and $\boldsymbol{v}_g$ has been proven only for unbounded anisotropic media [142], or bi-gyrotropic media [147]. For the in-plane Poynting vector of polaritons in strongly anisotropic layers, the proof is still an open problem. In Ref. [148], we analytically prove that, for low losses, the imaginary part of the wavevector is parallel to $\boldsymbol{v}_g$. But the analytical expression for **S** has not been derived yet. Moreover, the direction of **S** is different in different planes parallel to the layer surfaces, so to introduce the in-plane Poynting vector it is needed to at least average along the z axis. Consequently, studying the direction of **S** and the energy transfer in strongly anisotropic layers is an important problem beyond this thesis.

rays (i.e., along the direction given by the asymptote of the hyperboloid). HPhPs are extraordinary waves and can be distinguished as TM polarized, while ordinary waves (TE polarized [5]) can also propagate through hBN crystals in the L-RB (TE waves can propagate through uniaxial media if the condition $\mathrm{Re}(\varepsilon_\perp) > 0$ is fulfilled [5]). However, the wavevector magnitude of the TE waves [5], $|\mathbf{k}_0\sqrt{\varepsilon_\perp}|$, is small compared to that of HPhPs.

HPHPs in Thin Slabs of hBN Since hBN is a vdW material, due to its weak interlayer coupling, thin nanometric slabs of hBN can be easily exfoliated onto different substrates. This is a crucial advantage for the application of hBN in nanophotonics, and opens a plethora of opportunities to fabricate nanodevices. Thus, it is of great interest to analyze the polaritonic modes sustained by thin hBN films.

The propagation of HPhPs along the boundaries of hBN (and, more generally, uniaxial crystals) and inside the slabs can be analytically analyzed in the quasistatic limit $k_p \gg k_0$ [92, 151, 152]:

$$k_p(\omega) = -\frac{\psi(\omega)}{d}\left[\arctan\left(\frac{\varepsilon_1(\omega)}{\psi(\omega)\varepsilon_\perp(\omega)}\right) + \arctan\left(\frac{\varepsilon_3(\omega)}{\psi(\omega)\varepsilon_\perp(\omega)}\right) + \pi l\right], \quad l = 0, 1, 2..., \tag{1.16}$$

with $\psi(\omega) = -i\sqrt{\varepsilon_\parallel(\omega)/\varepsilon_\perp(\omega)}$. We can see that modes become quantized as the thin film forms a Fabry-Perot cavity, with being l the mode index. Importantly, the mode dispersion of HPhPs is strongly dependent on the hBN thickness, d. To illustrate the properties of Eq. (1.16), we calculate the dispersion of the first-order mode ($l = 0$) in a range of typical thickness of exfoliated thin hBN films, namely $d = 30$, 50, 100 and 200 nm (Fig. 1.8f). The film thicknesses are thus almost two orders of magnitude smaller than the wavelength of light in free space. As seen from the plot, the wavevector magnitude k_p increases with the decrease of d, for any given frequency. This dependence is consistent with Eq. (1.16), in which the wavevector k_p depends upon the hBN thickness as $k_p \alpha 1/d$.

The dispersion of the modes follows a different trend in the U-RB and the L-RB. Namely, in the U-RB, the wavevector of the modes has its minimum value near the TO phonon frequency and it increases with frequency, reaching an asymptote at the LO phonon frequency. In contrast, in the L-RB the trend is the opposite: the highest values of the wavevectors are reached near the TO phonon frequency, and its magnitude decreases with frequency, reaching its minimal value at the LO phonon frequency. Moreover, as it has been proved experimentally [150], the phase velocity of the HPhPs, $v_p = \omega/k$, in the Type I band is negative and thus, the sign of the real part of the wavevector is negative (we represent the absolute value of the wavevector in Fig. 1.8f for compactness). On the contrary, the group velocity, $v_g = \partial\omega/\partial k$, is positive in both RBs [150], and very small compared to the speed of light in free space, with measured values for the lowest-order mode of $v_g = 0.003c$ [150]. On the other hand, the propagation length of the mode $l = 0$ decreases with the decrease of

d. At the same time, the lifetime of the mode is virtually independent upon d in the shown frequency range [139].

To gain insights into the propagation features of HPhPs, we simulate the distribution of the electric field induced by a vertically oriented dipole on top of an hBN slab. Figure 1.8g,h shows the real part and the amplitude of such field, respectively. Near the dipole, the cross-section of the field distribution shows a *zig-zag* pattern. The zig-zag results from the superposition of all the excited modes [153], and can also be intuitively seen as the reflections of the hyperbolic ray from bottom and top interfaces of the slab, since the angle of the zig-zag coincides with the angle θ defining the hyperbolic ray. Because the higher order modes decay faster than the lower order ones, the zig-zag pattern fades away far from the dipole; and for long distances, only the lower modes with small wavevectors and long propagation lengths remain.

The emergence of HPhPs in hBN opened the door to achieving anisotropic optical phenomena naturally. Prominent examples are ultra-slow propagation [150], with reported values down to 0.0002c [154], anomalous reflection [149] and hyper-lensing effects [143, 144]. The long lifetimes and large quality factors of HPhPs go far beyond those of plasmon polaritons since they are associated with the low optical losses in polar dielectrics. The figures of merit of the modes in natural abundant hBN can be further improved in isotopically enriched hBN [139], enabling the fabrication of high-quality nanophotonic resonators [129, 155, 156]. HPhPs in hBN ribbons have been used for surface-enhanced infrared absorption (SEIRA) spectroscopy of small amounts of organic molecules in the strong coupling regime [157–159], demonstrating a platform for sensing [160], local control of chemical reactivity and IR quantum cavity optics experiments. HPhPs in hBN have therefore quickly attracted a large interest in the nanophotonics community due to their potential to control light propagation at the nanoscale in an unconventional way. Parallel and previous studies on hyperbolic metasurfaces—patterned structures that support guided waves with in-plane hyperbolicity—had shown potential for similar effects, including all-angle negative refraction [161], enhanced photonic density of states [162, 163], or concave wavefronts [164, 165]. Metasurface engineering was also utilized to exploit the natural out-of-plane hyperbolicity of hBN to create an in-plane hyperbolic metasurface [166], opening the door to the direct visualization of anisotropy-related phenomena, such as HPhP canalization [167]. However, the required patterning in metasurfaces typically leads to losses and limits the confinement of light. Additionally, the electromagnetic response of hyperbolic metasurfaces is governed by permittivity tensors as derived from the effective medium theory, which only holds below the long-wavelength limit where the structural periodicity is much smaller than the incident wavelength [95]. Consequently, the hyperbolic dispersion is limited to a small region in reciprocal space, constraining electromagnetic wavevectors. The discovery of HPhPs in hBN thus provided an accessible solution to the challenges and limitations associated with nanopatterning processes.

All in all, the discovery that some vdW materials can support strongly anisotropic polaritons marked a significant milestone in the field of nanophotonics. In particular, this finding opened up possibilities to create high-quality structures with desired properties, such as inherent anisotropy or low optical losses, through readily achievable fabrication processes. As commented above, this eliminated the previous need for nanostructuring to engineer such desired properties, usually introducing surface roughness or defects [104]. On top of that, vdW materials offer enhanced integration capabilities and compatibility between them and with other materials, as exemplified by the pioneering study of hybrid hyperbolic plasmon–phonon polaritons in graphene/hBN heterostructures [131]. Nevertheless, the out-of-plane character of hyperbolicity in hBN hinders a direct experimental access to phenomena associated with this pronounced anisotropy. Therefore, after the discovery of HPhPs in hBN, the prospect of an in-plane hyperbolic material held great promise for directly observing hyperbolicity-related effects.

In-Plane Hyperbolic Phonon Polaritons in Alpha-Phase Molybdenum Trioxide

A few years after the discovery of HPhPs in hBN, HPhPs with even stronger anisotropy were observed in the vdW polar insulator alpha-phase molybdenum trioxide (α-MoO_3), awakening considerable interest in the polaritonics community. This material displays fascinating optical properties, which stem from its highly anisotropic crystal structure (Fig. 1.9a,b). In it, layers formed by distorted MoO_6 octahedra (Fig. 1.9a) are weakly bound by vdW forces [168] and all three lattice constants (a, b and c) are different (Fig. 1.9b). Most importantly, α-MoO_3 has strong in-plane structural anisotropy, caused by the interlayer spacing of the (100) facet differing from that of the (001) facet by as much as 7.2%, which leads to a highly anisotropic response, different along the three main crystal axes: [100], [001], and [010] [169], which we label as x, y and z, respectively. As a result, the three principal values ε_x, ε_y and ε_z—describing the optical response along the [100], [001], and [010] crystal directions, respectively—of the dielectric tensor

$$\hat{\varepsilon}(\omega) = \begin{pmatrix} \varepsilon_x(\omega) & 0 & 0 \\ 0 & \varepsilon_y(\omega) & 0 \\ 0 & 0 & \varepsilon_z(\omega) \end{pmatrix} \tag{1.17}$$

are in general different. As in the case of 4H-SiC and hBN, each of the permittivity tensor components can be described by a Lorentz oscillator:

$$\varepsilon_j(\omega) = \varepsilon_j^{\infty} \left(1 + \frac{\omega_{\mathrm{LO},j}^2 - \omega_{\mathrm{TO},j}^2}{\omega_{\mathrm{TO},j}^2 - \omega^2 - i\omega\gamma_j} \right),$$

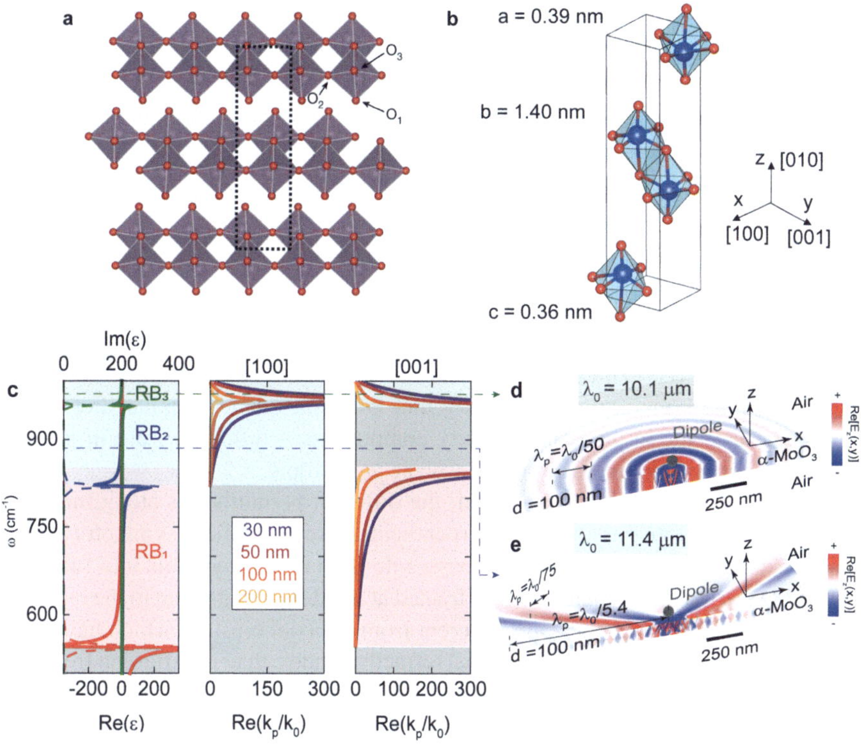

Fig. 1.9 In plane anisotropic polaritons in α-MoO_3 at mid-IR frequencies. (**a**) Illustration of the orthorhombic lattice structure of layered α-MoO_3 (red spheres represent oxygen atoms). The orthorhombic structure is based on bilayers of distorted MoO_6 octahedra stacked along the [010] direction via vdW interactions. The three possible positions of oxygen atoms are denoted O_{1-3}, and the unit cell is shown dashed. Taken from Ref. [94]. (**b**) Sketch of the unit cell of α-MoO_3 and correspondence between the crystallographic directions [100], [001], [010], and the spatial coordinates x, y, z; the lattice constants are $a = 0.396$ nm, $b = 1.385$ nm, and $c = 0.369$ nm. The blue/red spheres represent molybdenum/oxygen atoms. (**c, left panel**) Permittivity functions of α-MoO_3 in the mid-IR spectral range. The blue, red and green lines represent $\mathrm{Re}(\varepsilon_x)$, $\mathrm{Re}(\varepsilon_y)$ and $\mathrm{Re}(\varepsilon_z)$, respectively. Colored spectral regions indicate the corresponding RBs: RB1, RB2 and RB3. (**c, central and right panels**) Dispersion of HPhPs in RB1, RB2 and RB3 for an α-MoO_3 slab of thickness d surrounded by air, $\varepsilon_1 = \varepsilon_3 = 1$, with $d = 30$, 50, 100 and 200 nm, along the [100] (**center**) and [001] (**right**) crystal directions. Note that the sign of the wavevector of the modes in RB3 is negative, but we represent the absolute value of the wavevector for compactness. (**d,e**) Dipole source launching HPhPs of a 100-nm-thick slab of α-MoO_3 at two representative frequencies $\omega_0 = 880\,\mathrm{cm}^{-1}$ (**d**) and $\omega_0 = 990\,\mathrm{cm}^{-1}$ (**e**), residing into RB2 and RB3, respectively. The frequencies in (**d,e**) are marked by a dashed line in (**c**). The electric field distributions show characteristic elliptic and concave wavefronts, corresponding to in-plane elliptic and hyperbolic propagation

with $j = x, y, z$. In the mid-IR, the parameters, as reported in Ref. [95], are given by:[9] $\varepsilon_x^\infty = 4$, $\omega_{\mathrm{LO},x} = 972\,\mathrm{cm}^{-1}$, $\omega_{\mathrm{TO},x} = 820\,\mathrm{cm}^{-1}$, $\gamma_x = 4\,\mathrm{cm}^{-1}$, $\varepsilon_y^\infty = 5.2$, $\omega_{\mathrm{LO},y} = 851\,\mathrm{cm}^{-1}$, $\omega_{\mathrm{TO},y} = 545\,\mathrm{cm}^{-1}$, $\gamma_y = 4\,\mathrm{cm}^{-1}$, $\varepsilon_y^\infty = 2.4$, $\omega_{\mathrm{LO},y} = 1004\,\mathrm{cm}^{-1}$, $\omega_{\mathrm{TO},y} = 958\,\mathrm{cm}{-1}$, and $\gamma_z = 2\,\mathrm{cm}^{-1}$. The corresponding permittivity components of α-MoO_3, $\varepsilon_x(\omega)$, $\varepsilon_y(\omega)$ and $\varepsilon_z(\omega)$, are depicted in Fig. 1.9c (left panel) at mid-IR frequencies. The different directional vibrations of the α-MoO_3 crystal structure yield three RBs that locate at different frequency ranges: RB1 from 545 to $851\,\mathrm{cm}^{-1}$ originating from an in-plane phonon along the y axis (shaded in red), RB2 from 820 to $972\,\mathrm{cm}^{-1}$ originating from an in-plane phonon along the x axis (shaded in blue), and RB3 at 958 to $1010\,\mathrm{cm}^{-1}$, originating from an out-of-plane phonon along the z axis (shaded in green). More details about the specific vibrations giving rise to these resonances will be given in Chap. 4. Within these RBs, α-MoO_3 exhibits the strong reflectivity that is characteristic of polar dielectrics. However, since the in-plane permittivity components $\varepsilon_x(\omega)$, $\varepsilon_y(\omega)$ are in general different, such reflectivity is strongly in-plane anisotropic.

In fact, α-MoO_3 is a biaxial crystal: the different permittivities along the three main crystal axes give rise to two perpendicular axes of optical symmetry [142]. This means that when an optical ray passes through it, the ray splits into two, and each of those two rays which will be refracted at a different angle due to the crystal's birefringence. Biaxial crystals are different from uniaxial crystals, which have only one axis of optical symmetry and one refractive index that is different from the other two, as exemplified by hBN in the previous subsection. Biaxial crystals are less common than uniaxial crystals, and they are often more difficult to work with due to their more complex optical properties.

HPHPs in Thin Slabs of α-MoO_3 Like the previous examples of graphene and hBN, α-MoO_3 is a vdW material and thus can be easily exfoliated into nanometric-thin slabs. Therefore, studying the polaritonic modes supported by thin films of α-MoO_3 is important. As will be proven in Chap. 3, the propagation of polaritons along the main crystal axes of α-MoO_3 can be analyzed analytically in the quasistatic limit $k_p \gg k_0$ in terms of Eq. (1.16). As in the case of hBN, modes become quantized as the thin film forms a Fabry-Perot cavity. Due to the in-plane anisotropic optical response of α-MoO_3, we calculate the dispersion of the first-order mode ($l = 0$) in α-MoO_3 along the two in-plane crystal axes: $x = [100]$ (Fig. 1.9c, middle panel) and $y = [001]$ (Fig. 1.9c, right panel). We consider films of different thicknesses in the range of typical thickness of exfoliated thin α-MoO_3 films, namely $d = 30$, 50, 100 and 200 nm. The film thicknesses are thus almost two orders of magnitude smaller than the wavelength of light in free space. The plot reveals the strong confinement of polaritons in thin slabs of α-MoO_3, with wavelengths up to 300 times smaller than the corresponding photon wavelengths,

[9] Initial studies on HPhPs in α-MoO_3 focused on two [94] and three [95] RBs in the mid-IR, relying on far-field data. In Chap. 4, we will see that α-MoO_3 also hosts various RBs in the far-IR, as well as a more precise extraction of these parameters.

thus comparable to those of GPPs and HPhPs in hBN. The dependence of the mode dispersion on the thickness of α-MoO_3, represented by d, is evident. For a given frequency, the magnitude of the wavevector k_p increases as d decreases. This relationship is in line with Eq. (1.16), which states that the wavevector k_p is proportional to the inverse thickness of the slab, i.e., $k_p \alpha 1/d$.

Furthermore, the dispersion of the modes exhibits distinct propagation patterns in the various RBs. On the one hand, RB3 displays an elliptical dispersion because the allowed wavevectors differ along the [100] and [001] axes. Notably, RB1 and RB2 generate a natural hyperbolic surface on the basal plane (xy plane). In RB1, Eq. (1.16) has no physical solutions within this frequency range and along the [100] crystal axis, as illustrated by the gray shading in Fig. 1.9c (middle panel), thus prohibiting polariton propagation in this direction. Similarly, in RB2, polariton propagation is not possible along the [001] crystal axis, as shown by the gray shading in Fig. 1.9c (right panel). This shows that the propagation of PhPs along the surface of α-MoO_3 can be very directional, as not all directions are allowed. This provides a major differences with respect to all the materials previously discussed in this chapter. Additionally, the dispersion characteristics differ among RB1, RB2, and RB3. Specifically, in RB1 and RB2, the minimum value of the mode's wavevector is located near the TO phonon frequency, and it increases with frequency until it asymptotically approaches the LO phonon frequency. In contrast, in RB3, the trend is reversed: the maximum values of the wavevector occur near the TO phonon frequency, and its magnitude decreases with frequency until it reaches a minimum at the LO phonon frequency.

The initial observation of phonon polaritons in α-MoO_3 did not describe their anisotropic propagation properties [170]. However, Ma et al. [94] and then Zheng et al. [95] reported nanoimaging of polaritons with in-plane hyperbolicity in exfoliated α-MoO_3 crystals. Analogously to HPhPs in hBN, experimental evidence has also demonstrated the negative phase velocity of HPhPs in RB3 [94], which means that the sign of the real part of the wavevector is negative (again, the absolute value of the wavevector is shown in Fig. 1.9c for compactness). The group velocity is positive in all three cases, and it is significantly smaller than the speed of light in vacuum. Additionally, the mode's lifetime is very long. Polariton lifetimes of 8 picoseconds have been reported in α-MoO_3 [94], which is more than ten times larger than that of GPPs at room temperature [116]. They are also a factor of four larger than the best values reported for HPhPs in isotopically engineered hBN [139] and for GPPs at low temperatures [128].

In order to illustrate the excitation of polaritons in α-MoO_3 by a localized source, the simulated distribution of the electric field induced by a vertically oriented dipole over the slab is shown in Fig. 1.9d,e. Simulations are performed for two representative frequencies within an in-plane elliptic and an in-plane hyperbolic RB. At $\omega_0 = 990\,\text{cm}^{-1}$, residing in RB3, polaritons are allowed to propagate along every in-plane direction, forming concentric elliptic rings, as shown in Fig. 1.9d. In stark contrast, at $\omega_0 = 880\,\text{cm}^{-1}$, residing in RB2, the dispersion is hyperbolic, and the polaritonic wavefronts exhibit a concave or "V" shape (Fig. 1.9e), with extremely directional propagation, direction-dependent polaritonic wavelengths and strong misalignment between the phase and group velocities. Note that the confinement along the asymptote can reach up to $\lambda_0/75$, a larger value than that attained within the elliptic regime for the same thickness and shorter wavelengths, $\lambda_0/50$. Note that we are illustrating these values for a flake with $d = 100$ nm. The confinement factors can be considerably lower for thinner flakes.

As demonstrated in hBN and α-MoO_3, lattice anisotropy in polar vdW crystals can naturally provide in-plane optical hyperbolicity. In addition to a simplified fabrication process, naturally hyperbolic crystals offer an interatomic spacing, versus a few nanometers for the size of a man-made metamaterial unit cell, and therefore can not only complement but also surpass hyperbolic metamaterials with lower loss and higher optical confinement. This offers an unprecedented platform for controlling the flow of energy at the nanoscale. In fact, the search for ever lower symmetry materials is currently one of the main strategies in optics and nano-optics [96, 171–174], and has induced an intensive study of strongly anisotropic polaritons supported by vdW slabs and heterostructures as well as bulk crystals [175, 176].

Upon their discovery in 2018, the unique in-plane hyperbolic dispersion that HPhPs feature in α-MoO_3, together with the possibility of exfoliating thin layers of this material, opened the door to directly observe unusual optical effects at the nanoscale. This positioned α-MoO_3 as an optimal platform for uncovering the fundamentals of nano-optics in strongly anisotropic media. More specifically, the in-plane hyperbolicity of α-MoO_3 offered an exciting opportunity to explore optical phenomena that challenge our conventional intuition, such as negative reflection, anomalous refraction and natural canalization, as will be shown in Chaps. 5, 6, and 8 of this thesis. These phenomena, pertinent not only in the realm of nano-optics but also beyond, can pave the way for the development of nanoscale devices for controlling light, potentially revolutionizing how we manage light and energy at the nanometer scale.

As such, this thesis, started in 2018 shortly after the discovery of in-plane HPhPs in α-MoO_3, is dedicated to an in-depth exploration of the propagation of polaritons in hyperbolic media, aiming to unravel their fundamentals. In this regard, having a dielectric function model and an analytical dispersion is crucial for exploiting the full potential of this material and making accurate predictions about its optical response. The finding of in-plane hyperbolic polaritons in α-MoO_3 in 2018 [94, 95] highlighted the importance of these two studies. In Part II, we will derive an analytical dispersion for polaritons in biaxial crystals (Chap. 3) and will extract the accurate IR permittivity of α-MoO_3 (Chap. 4).

References

1. F.J. Alfaro Mozaz, Mid-infrared nanophotonics with hyperbolic phonon polaritons (2020). https://addi.ehu.es/handle/10810/51205
2. K.L. Kliewer, R. Fuchs, Optical modes of vibration in an ionic crystal slab including retardation. I. Nonradiative region. Phys. Rev. **144**, 495–503 (1966). https://link.aps.org/doi/10.1103/PhysRev.144.495

3. D. Marcuse, *Theory of Dielectric Optical Waveguides* (Academic Press, 1991). ISBN: 978-0-12-470951-5. https://www.sciencedirect.com/science/article/pii/B978012470951550007X
4. G. Álvarez-Pérez, K.V. Voronin, V.S. Volkov, P. Alonso-GonzáLlez, A.Y. Nikitin, Analytical approximations for the dispersion of electromagnetic modes in slabs of biaxial crystals. Phys. Rev. B **100**, 235408 (2019). https://link.aps.org/doi/10.1103/PhysRevB.100.235408
5. M. Born et al., *Principles of Optics: Electromagnetic Theory of Propagation, Interference and Diffraction of Light*, 7th edn. (Cambridge University Press, 1999)
6. D.K. Gramotnev, S.I. Bozhevolnyi, Plasmonics beyond the diffraction limit. Nat. Photon. **4**, 83–91 (2010). https://doi.org/10.1038/nphoton.2009.282
7. J. Senior, *Optical Fiber Communication Principles and Practice*, 3rd edn. (Pearson Education, Edinburgh Gate, 2009)
8. E. Abbe, Beiträge zur Theorie des Mikroskops und der mikroskopischen Wahrnehmung. Archiv für Mikroskopische Anatomie **9**, 413–468 (1873). https://doi.org/10.1007/BF02956173
9. L. Novotny, B. Hecht, *Principles of Nano-optics* (Cambridge University Press, 2006)
10. M. Von Ardenne, Das Elektronen-Rastermikroskop. Z. Phys. **109**, 553–572 (1938). https://doi.org/10.1007/BF01341584
11. I. Utke, P. Hoffmann, J. Melngailis, Gas-assisted focused electron beam and ion beam processing and fabrication. J. Vacuum Sci. Technol. B Microelectron. Nanometer Struct. Process. Meas. Phenomena **26**, 1197–1276 (2008). https://doi.org/10.1116/1.2955728
12. A. Mancini, Surface phonon polaritons in silicon carbide nanostructures revealed by near-field imaging and spectroscopy (2023). http://nbn-resolving.de/urn:nbn:de:bvb:19-315665
13. S.A. Maier, *Plasmonics: Fundamentals and Applications* (Springer New York, New York, 2007). https://link.springer.com/book/10.1007/0-387-37825-1
14. J.J. Hopfield, Theory of the contribution of excitons to the complex dielectric constant of crystals. Phys. Rev. **112**, 1555–1567 (1958). https://link.aps.org/doi/10.1103/PhysRev.112.1555
15. M. Born, K. Huang, *Dynamical Theory of Crystal Lattices* (Clarendon Press, 1954)
16. E. Galiffi et al., Extreme light confinement and control in low-symmetry phonon-polaritonic crystals. Nat. Rev. Mater. (2023). https://doi.org/10.1038/s41578-023-00620-7
17. M. Kauranen, A.V. Zayats, Nonlinear plasmonics. Nat. Photon. **6**, 737–748 (2012). https://doi.org/10.1038/nphoton.2012.244
18. C.J. Winta, S. Gewinner, W. Schöllkopf, M. Wolf, A. Paarmann, Second-harmonic phonon spectroscopy of α-quartz. Phys. Rev. B **97**, 094108 (2018). https://journals.aps.org/prb/abstract/10.1103/PhysRevB.97.094108
19. N.A. Mortensen, S. Raza, M. Wubs, T. Søndergaard, S.I. Bozhevolnyi, A generalized non-local optical response theory for plasmonic nanostructures. Nat. Commun. **5**, 3809 (2014). https://doi.org/10.1038/ncomms4809
20. C.R. Gubbin, S. De Liberato, T.G. Folland, Surface phonon polaritons for infrared optoelectronics. J. Appl. Phys. **131**. 030901 (2022). https://doi.org/10.1063/5.0064234
21. C.R. Gubbin, S. De Liberato, Quantum theory of longitudinal-transverse polaritons in nonlocal thin films. Phys. Rev. Appl. **17**, 014037 (2022). https://link.aps.org/doi/10.1103/PhysRevApplied.17.014037
22. J.D. Caldwell et al. Atomic-scale photonic hybrids for mid-infrared and terahertz nanophotonics. Nat. Nanotechnol. **11**, 9–15 (2016). https://www.nature.com/articles/nnano.2015.305
23. E.D. Palik, *Handbook of Optical Constants of Solids* (Elsevier, 1997). https://www.sciencedirect.com/book/9780125444156/handbook-of-optical-constants-of-solids#book-info
24. M. Fox, *Optical Properties of Solids* (Oxford University Press, 2010). http://site.ebrary.com/id/10838525
25. R.L. Olmon et al., Optical dielectric function of gold. Phys. Rev. B **86**, 235147 (2012). https://link.aps.org/doi/10.1103/PhysRevB.86.235147
26. M.G. Blaber, M.D. Arnold, M.J. Ford, Search for the ideal plasmonic nanoshell: the effects of surface scattering and alternatives to gold and silver. J. Phys. Chem. C **113**, 3041–3045 (2009). https://doi.org/10.1021/jp810808h

27. K.M. McPeak et al., Plasmonic films can easily be better: rules and recipes. ACS Photon. **2**, 326–333 (2015). https://doi.org/10.1021/ph5004237
28. R. Adato, J. Guo, Characteristics of ultra-long range surface plasmon waves at optical frequencies. Optics Express **15**, 5008–5017 (2007). https://opg.optica.org/oe/abstract.cfm?URI=oe-15-8-5008
29. J.A. Dionne, L.A. Sweatlock, H.A. Atwater, A. Polman, Planar metal plasmon waveguides: frequency-dependent dispersion, propagation, localization, and loss beyond the free electron model. Phys. Rev. B **72**, 075405 (2005). https://link.aps.org/doi/10.1103/PhysRevB.72.075405
30. A. Degiron, D.R. Smith, Numerical simulations of long-range plasmons. Optics Express **14**, 1611–1625 (2006). https://opg.optica.org/oe/abstract.cfm?URI=oe-14-4-1611
31. P. Berini, Long-range surface plasmon polaritons. Adv. Opt. Photon. **1**, 484–588 (2009). https://opg.optica.org/aop/abstract.cfm?URI=aop-1-3-484
32. G.P. Agrawal, *Fiber-Optic Communication Systems*, 4th edn. Wiley Series in Microwave and Optical Engineering (2010)
33. R.H. Ritchie, Plasma losses by fast electrons in thin films. Phys. Rev. **106**, 874–881 (1957). https://link.aps.org/doi/10.1103/PhysRev.106.874
34. M. Fleischmann, P. Hendra, A. McQuillan, Raman spectra of pyridine adsorbed at a silver electrode. Chem. Phys. Lett. **26**, 163–166 (1974). https://www.sciencedirect.com/science/article/pii/0009261474853881
35. D.L. Jeanmaire, R.P. Van Duyne, Surface Raman spectroelectrochemistry: Part I. Heterocyclic, aromatic, and aliphatic amines adsorbed on the anodized silver electrode. J. Electroanal. Chem. Interfacial Electrochem. **84**, 1–20 (1977). https://www.sciencedirect.com/science/article/pii/S0022072877802246
36. T.W. Ebbesen, H.J. Lezec, H.F. Ghaemi, T. Thio, P.A. Wolff, Extraordinary optical transmission through sub-wavelength hole arrays. Nature **391**, 667–669 (1998). https://doi.org/10.1038/35570
37. M.L. Brongersma, P.G. Kik, *Surface Plasmon Nanophotonics*, Springer Series in Optical Sciences (2007)
38. N. Engheta, R.W. Ziolkowski, *Metamaterials: Physics and Engineering Explorations* (Wiley and Sons, 2006)
39. S. Nie, S.R. Emory, Probing single molecules and single nanoparticles by surface-enhanced Raman scattering. Science **275**, 1102–1106 (1997). https://www.science.org/doi/abs/10.1126/science.275.5303.1102
40. K. Kneipp et al., Single molecule detection using surface-enhanced raman scattering (SERS). Phys. Rev. Lett. **78**, 1667–1670 (1997). https://link.aps.org/doi/10.1103/PhysRevLett.78.1667
41. J.D. Caldwell et al., Plasmonic nanopillar arrays for large-area, high-enhancement surface-enhanced Raman scattering sensors. ACS Nano **5**, 4046–4055 (2011). https://doi.org/10.1021/nn200636t
42. J.D. Caldwell et al., Large-area plasmonic hot-spot arrays: sub-2 nm interparticle separations with plasma-enhanced atomic layer deposition of Ag on periodic arrays of Si nanopillars. Optics Express **19**, 26056–26064 (2011). https://opg.optica.org/oe/abstract.cfm?URI=oe-19-27-26056
43. J.M. Mooney, J. Silverman, The theory of hot-electron photoemission in Schottky-barrier IR detectors. IEEE Trans. Electron Devices **32**, 33–39 (1985). https://ieeexplore.ieee.org/document/1484652
44. H.A. Atwater, A. Polman, Plasmonics for improved photovoltaic devices. Nat. Mater. **9**, 205–213 (2010). https://doi.org/10.1038/nmat2629
45. A. Aubry et al., Plasmonic light-harvesting devices over the whole visible spectrum. Nano Lett. **10**, 2574–2579 (2010). https://doi.org/10.1021/nl101235d
46. I. Goykhman, B. Desiatov, J. Khurgin, J. Shappir, U. Levy, Locally oxidized silicon surface-plasmon Schottky detector for telecom regime. Nano Lett. **11**, 2219–2224 (2011). https://doi.org/10.1021/nl200187v

47. A. Sobhani et al., Narrowband photodetection in the near-infrared with a plasmon-induced hot electron device. Nat. Commun. **4**, 1643 (2013). https://www.nature.com/articles/ncomms2642
48. C. Clavero, Plasmon-induced hot-electron generation at nanoparticle/metal-oxide interfaces for photovoltaic and photocatalytic devices. Nat. Photon. **8**, 95–103 (2014). https://doi.org/10.1038/nphoton.2013.238
49. J.B. Khurgin, G. Sun, In search of the elusive lossless metal. Appl. Phys. Lett. **96**, 181102 (2010). https://pubs.aip.org/aip/apl/article/96/18/181102/118748/In-searchof-the-elusive-lossless-metal
50. J.B. Khurgin, G. Sun, Scaling of losses with size and wavelength in nanoplasmonics and metamaterials. Appl. Phys. Lett. **99**, 211106 (2011). https://pubs.aip.org/aip/apl/article/99/21/211106/150987/Scaling-of-losses-with-size-and-wavelength-in
51. J.B. Khurgin, A. Boltasseva, Reflecting upon the losses in plasmonics and metamaterials. MRS Bull. **37**, 768–779 (2012). https://link.springer.com/article/10.1557/mrs.2012.173
52. J.D. Caldwell et al. Low-loss, infrared and terahertz nanophotonics using surface phonon polaritons. Nanophotonics **4**, 44–68 (2015). https://doi.org/10.1515/nanoph-2014-0003
53. E.N. Economou, Surface plasmons in thin films. Phys. Rev. **182**, 539–554 (1969). https://journals.aps.org/pr/abstract/10.1103/PhysRev.182.539
54. F. Kusunoki, T. Yotsuya, J. Takahara, T. Kobayashi, Propagation properties of guided waves in index-guided two-dimensional optical waveguides. Appl. Phys. Lett. **86**, 211101 (2005). https://pubs.aip.org/aip/apl/article/86/21/211101/931461/Propagation-properties-of-guided-waves-in-index
55. J.A. Dionne, L.A. Sweatlock, H.A. Atwater, A. Polman, Plasmon slot waveguides: Towards chip-scale propagation with subwavelength-scale localization. Phys. Rev. B **73**, 035407 (2006). https://journals.aps.org/prb/abstract/10.1103/PhysRevB.73.035407
56. R. Zia, M.D. Selker, P.B. Catrysse, M.L. Brongersma, Geometries and materials for subwavelength surface plasmon modes. J. Opt. Soc. Am. A **21**, 2442–2446 (2004). https://opg.optica.org/josaa/abstract.cfm?URI=josaa-21-12-2442
57. M. Schnell et al., Nanofocusing of mid-infrared energy with tapered transmission lines. Nat. Photon. **5**, 283–287 (2011). https://doi.org/10.1038/nphoton.2011.33
58. P. Sarriugarte et al., Propagation and nanofocusing of infrared surface plasmons on tapered transmission lines: Influence of the substrate. Optics Commun. **285**, 3378–3382 (2012). Special Issue: Sub-wavelength Light Localization and Focusing. https://www.sciencedirect.com/science/article/pii/S0030401812002775
59. B. Schwarz et al., Monolithically integrated mid-infrared lab-on-a-chip using plasmonics and quantum cascade structures. Nat. Commun. **5**, 4085 (2014). https://doi.org/10.1038/ncomms5085
60. S. Prucnal et al., Ultra-doped n-type germanium thin films for sensing in the mid-infrared. Sci. Rep. **6**, 27643 (2016). https://doi.org/10.1038/srep27643
61. T. Taliercio, P. Biagioni, Semiconductor infrared plasmonics. Nanophotonics **8**, 949–990 (2019). https://doi.org/10.1515/nanoph-2019-0077
62. J.R. Nolen et al., Ultraviolet to far-infrared dielectric function of n-doped cadmium oxide thin films. Phys. Rev. Mater. **4**, 025202 (2020). https://link.aps.org/doi/10.1103/PhysRevMaterials.4.025202
63. R. Soref, R.E. Peale, W. Buchwald, Longwave plasmonics on doped silicon and silicides. Optics Express **16**, 6507–6514 (2008). https://opg.optica.org/oe/abstract.cfm?URI=oe-16-9-6507
64. M. Xue, D. Pan, J. Zhao, J. Chen, Optically tunable transient plasmons in InSb nanowires. Adv. Mater. **35**, 2208952 (2023). https://onlinelibrary.wiley.com/doi/abs/10.1002/adma.202208952
65. X. Liu, M. Xue, J. Chen, Broadband plasmonic indium arsenide photonic antennas. Nanoscale **15**, 3135–3141 (2023). https://doi.org/10.1039/D2NR06590H
66. P. West et al., Searching for better plasmonic materials. Laser Photon. Rev. **4**, 795–808 (2010). https://onlinelibrary.wiley.com/doi/abs/10.1002/lpor.200900055

67. A. Boltasseva, H.A. Atwater, Low-loss plasmonic metamaterials. Science **331**, 290–291 (2011). https://www.science.org/doi/abs/10.1126/science.1198258
68. J. Frigerio et al., Tunability of the dielectric function of heavily doped germanium thin films for mid-infrared plasmonics. Phys. Rev. B **94**, 085202 (2016). https://link.aps.org/doi/10.1103/PhysRevB.94.085202
69. J.B. Khurgin, Relative merits of phononics vs. plasmonics: the energy balance approach. Nanophotonics **7**, 305–316 (2018). https://doi.org/10.1515/nanoph-2017-0048
70. N.W. Ashcroft, N.D. Mermin, *Solid State Physics* (Holt-Saunders, 1976)
71. M.A. Stroscio, M. Dutta, *Phonons in Nanostructures* (Cambridge University Press, 2001)
72. C.F. Bohren, D.R. Huffman, *Absorption and Scattering of Light by Small Particles* (Wiley and Sons, 2004). ISBN: 978-0-471-29340-8
73. R. Hillenbrand, T. Taubner, F. Keilmann, Phonon-enhanced light–matter interaction at the nanometre scale. Nature **418**, 159–162 (2002). https://www.nature.com/articles/nature00899#citeas
74. E.Y. Ma et al., The Reststrahlen effect in the optically thin limit: a framework for resonant response in thin media. Nano Lett. **22**, 8389–8393 (2022). https://doi.org/10.1021/acs.nanolett.2c02819
75. S. Adachi, *Optical Constants of Crystalline and Amorphous Semiconductors* (Springer New York, New York, 1999). https://link.springer.com/book/10.1007/978-1-4615-5247-5
76. S. Zollner et al., Dielectric functions of bulk 4H and 6H SiC and spectroscopic ellipsometry studies of thin SiC films on Si. J. Appl. Phys. **85**, 8353–8361 (1999). https://doi.org/10.1063/1.370682
77. T.E. Tiwald et al., Carrier concentration and lattice absorption in bulk and epitaxial silicon carbide determined using infrared ellipsometry. Phys. Rev. B **60**, 11464–11474 (1999). https://link.aps.org/doi/10.1103/PhysRevB.60.11464
78. J.D. Caldwell et al., Low-loss, extreme subdiffraction photon confinement via silicon carbide localized surface phonon polariton resonators. Nano Lett. **13**, 3690–3697 (2013). https://doi.org/10.1021/nl401590g
79. H.P. Boehm, A. Clauss, G.O. Fischer, U. Hofmann, Das Adsorptionsverhalten sehr dünner Kohlenstoff-Folien. Z. Anorg. Allg. Chem. **316**, 119–127 (1962). https://onlinelibrary.wiley.com/doi/abs/10.1002/zaac.19623160303
80. H. Boehm, R. Setton, E. Stumpp, Nomenclature and terminology of graphite intercalation compounds. Carbon **24**, 241–245 (1986). https://app.dimensions.ai/details/publication/pub.1000341490
81. K.S. Novoselov et al., Electric field effect in atomically thin carbon films. Science **306**, 666–669 (2004). https://www.science.org/doi/abs/10.1126/science.1102896
82. K.S. Novoselov et al., Two-dimensional gas of massless Dirac fermions in graphene. Nature **438**, 197–200 (2005). https://doi.org/10.1038/nature04233
83. Y. Zhang, Y.-W. Tan, H.L. Stormer, P. Kim, Experimental observation of the quantum Hall effect and Berry's phase in graphene. Nature **438**, 201–204 (2005). https://doi.org/10.1038/nature04235
84. K.S. Novoselov, A. Mishchenko, A. Carvalho, A.H.C. Neto, 2D materials and van der Waals heterostructures. Science **353**, aac9439 (2016). https://www.science.org/doi/abs/10.1126/science.aac9439
85. D.N. Basov, M.M. Fogler, F.J.G. de Abajo, Polaritons in van der Waals materials. Science **354**, 195 (2016). https://science.sciencemag.org/content/354/6309/aag1992
86. T. Low et al., Polaritons in layered two-dimensional materials. Nat. Mater. **16**, 182–194 (2017). https://www.nature.com/articles/nmat4792
87. J. Chen et al., Optical nano-imaging of gate-tunable graphene plasmons. Nature **487**, 77 (2012). https://www.nature.com/articles/nature11254
88. Z. Fei et al., Gate-tuning of graphene plasmons revealed by infrared nano-imaging. Nature **487**, 82 (2012). https://www.nature.com/articles/nature11253
89. T. Low et al., Plasmons and screening in monolayer and multilayer black phosphorus. Phys. Rev. Lett. **113**, 106802 (2014). https://link.aps.org/doi/10.1103/PhysRevLett.113.106802

90. E. Van Veen et al., Tuning two-dimensional hyperbolic plasmons in black phosphorus. Phys. Rev. Appl. **12**, 014011 (2019). https://link.aps.org/doi/10.1103/PhysRevApplied.12.014011
91. C. Wang et al., van der Waals thin films of WTe2 for natural hyperbolic plasmonic surfaces. Nat. Commun. **11**, 1158 (2020). https://doi.org/10.1038/s41467-020-15001-9
92. S. Dai et al., Tunable phonon polaritons in atomically thin van der Waals crystals of boron nitride. Science **343**, 1125–1129 (2014). http://science.sciencemag.org/content/343/6175/1125
93. R. Geick, C.H. Perry, G. Rupprecht, Normal modes in hexagonal boron nitride. Phys. Rev. **146**, 543–547 (1966). https://link.aps.org/doi/10.1103/PhysRev.146.543
94. W. Ma et al., In-plane anisotropic and ultra-low-loss polaritons in a natural van der Waals crystal. Nature **562**, 557 (2018). https://www.nature.com/articles/s41586-018-0618-9
95. Z. Zheng et al., A mid-infrared biaxial hyperbolic van der Waals crystal. Sci. Adv. **5**, eaav8690 (2019). https://advances.sciencemag.org/content/5/5/eaav8690
96. J. Taboada-Gutiérrez et al., Broad spectral tuning of ultra-low-loss polaritons in a van der Waals crystal by intercalation. Nat. Mater. **19**, 964–968 (2020). https://doi.org/10.1038/s41563-020-0665-0
97. A. Ramasubramaniam, Large excitonic effects in monolayers of molybdenum and tungsten dichalcogenides. Phys. Rev. B **86**, 115409 (2012). https://link.aps.org/doi/10.1103/PhysRevB.86.115409
98. A. Hartstein, E. Burstein, A.A. Maradudin, R. Brewer, R.F. Wallis, Surface polaritons on semi-infinite gyromagnetic media. J. Phys. C Solid State Phys. **6**, 1266 (1973). https://dx.doi.org/10.1088/0022-3719/6/7/016
99. F. Dirnberger et al., Magneto-optics in a van der Waals magnet tuned by self-hybridized polaritons. Nature **620**, 533–537 (2023). https://doi.org/10.1038/s41586-023-06275-2
100. M. Dapolito et al., Infrared nano-imaging of Dirac magnetoexcitons in graphene. Nat. Nanotechnol. (2023). https://doi.org/10.1038/s41565-023-01488-y
101. A. Tsiatmas, V.A. Fedotov, F.J.G. de Abajo, N.I. Zheludev, Low-loss terahertz superconducting plasmonics. New J. Phys. **14**, 115006 (2012). https://doi.org/10.1088/1367-2630/14/11/115006
102. M.E. Berkowitz et al., Hyperbolic cooper-pair polaritons in planar graphene/cuprate plasmonic cavities. Nano Lett. **21**, 308–316 (2021). https://doi.org/10.1021/acs.nanolett.0c03684
103. D.N. Basov, A. Asenjo-Garcia, P.J. Schuck, X. Zhu, A. Rubio, Polariton panorama. Nanophotonics **10**, 549–577 (2021). https://doi.org/10.1515/nanoph-2020-0449
104. M.N. Gjerding, R. Petersen, T.G. Pedersen, N.A. Mortensen, K.S. Thygesen, Layered van der Waals crystals with hyperbolic light dispersion. Nat. Commun. **8**, 320 (2017). https://doi.org/10.1038/s41467-017-00412-y
105. E.E. Narimanov, A.V. Kildishev, Naturally hyperbolic. Nat. Photon. **9**, 214–216 (2015). https://doi.org/10.1038/nphoton.2015.56
106. D. Hu et al., Probing optical anisotropy of nanometer-thin van der Waals microcrystals by near-field imaging. Nat. Commun. **8**, 1471 (2017). https://doi.org/10.1038/s41467-017-01580-7
107. R.R. Nair et al., Fine structure constant defines visual transparency of graphene. Science **320**, 1308–1308 (2008). https://www.science.org/doi/abs/10.1126/science.1156965
108. S.-E. Zhu, S. Yuan, G. Janssen, Optical transmittance of multilayer graphene. Europhys. Lett. **108**, 17007 (2014). https://doi.org/10.1209/0295-5075/108/17007
109. C. Lee, X. Wei, J.W. Kysar, J. Hone, Measurement of the elastic properties and intrinsic strength of monolayer graphene. Science **321**, 385–388 (2008). https://www.science.org/doi/abs/10.1126/science.1157996
110. K. Cao et al., Elastic straining of free-standing monolayer graphene. Nat. Commun. **11**, 284 (2020). https://doi.org/10.1038/s41467-019-14130-0
111. A.Y. Nikitin, Graphene plasmonics, in *World Scientific Handbook of Metamaterials and Plasmonics*. World Scientific Series in Nanoscience and Nanotechnology (Cambridge University Press, 2017), pp. 307–338. https://www.worldscientific.com/worldscibooks/10.1142/10642-vol4#t=toc

112. E.H. Hwang, S. Das Sarma, Dielectric function, screening, and plasmons in two-dimensional graphene. Phys. Rev. B **75**, 205418 (2007). https://link.aps.org/doi/10.1103/PhysRevB.75.205418
113. A.Y. Nikitin et al., Real-space mapping of tailored sheet and edge plasmons in graphene nanoresonators. Nat. Photon. **10**, 239–243 (2016). https://doi.org/10.1038/nphoton.2016.44
114. F.J. García de Abajo, Graphene plasmonics: challenges and opportunities. ACS Photon. **1**, 135–152 (2014). https://doi.org/10.1021/ph400147y
115. P. Alonso-González et al., Controlling graphene plasmons with resonant metal antennas and spatial conductivity patterns. Science **344**, 1369–1373 (2014). https://www.science.org/doi/abs/10.1126/science.1253202
116. A. Woessner et al., Highly confined low-loss plasmons in graphene.boron nitride heterostructures. Nat. Mater. **14**, 421–425 (2015). https://doi.org/10.1038/nmat4169
117. D. Rodrigo et al., Mid-infrared plasmonic biosensing with graphene. Science **349**, 165–168 (2015). https://www.science.org/doi/abs/10.1126/science.aab2051
118. M. Liu et al., A graphene-based broadband optical modulator. Nature **474**, 64–67 (2011). https://doi.org/10.1038/nature10067
119. M.B. Lundeberg et al., Thermoelectric detection and imaging of propagating graphene plasmons. Nat. Mater. **16**, 204–207 (2017). https://doi.org/10.1038/nmat4755
120. P. Alonso-González et al., Acoustic terahertz graphene plasmons revealed by photocurrent nanoscopy. Nat. Nanotechnol. **12**, 31–35 (2017). https://doi.org/10.1038/nnano.2016.185
121. M.B. Lundeberg et al., Tuning quantum nonlocal effects in graphene plasmonics. Science **357**, 187–191 (2017). https://www.science.org/doi/abs/10.1126/science.aan2735
122. A.K. Geim, K.S. Novoselov, The rise of graphene. Nat. Mater. **6**, 183–191 (2007). https://doi.org/10.1038/nmat1849
123. C.R. Dean et al., Boron nitride substrates for high-quality graphene electronics. Nat. Nanotechnol. **5**, 722–726 (2010). https://doi.org/10.1038/nnano.2010.172
124. A.V. Kretinin et al., Electronic properties of graphene encapsulated with different two-dimensional atomic crystals. Nano Lett. **14**, 3270–3276 (2014). https://doi.org/10.1021/nl5006542
125. G.-H. Lee et al., Flexible and transparent MoS_2 field-effect transistors on hexagonal boron nitride-graphene heterostructures. ACS Nano **7**, 7931–7936 (2013). https://doi.org/10.1021/nn402954e
126. H. Hu et al., Active control of micrometer plasmon propagation in suspended graphene. Nat. Commun. **13**, 1465 (2022). https://doi.org/10.1038/s41467-022-28786-8
127. G.X. Ni et al., Plasmons in graphene Moiré superlattices. Nat. Mater. **14**, 1217–1222 (2015). https://doi.org/10.1038/nmat4425
128. G.X. Ni et al., Fundamental limits to graphene plasmonics. Nature **557**, 530–533 (2018). https://doi.org/10.1038/s41586-018-0136-9
129. J.D. Caldwell et al., Sub-diffractional volume-confined polaritons in the natural hyperbolic material hexagonal boron nitride. Nat. Commun. **5**, 5221 (2014). https://doi.org/10.1038/ncomms6221
130. K. Watanabe, T. Taniguchi, H. Kanda, Direct-bandgap properties and evidence for ultraviolet lasing of hexagonal boron nitride single crystal. Nat. Mater. **3**, 404–409 (2004). https://doi.org/10.1038/nmat1134
131. S. Dai et al., Graphene on hexagonal boron nitride as a tunable hyperbolic metamaterial. Nat. Nanotechnol. **10**, 682–686 (2015). https://doi.org/10.1038/nnano.2015.131
132. T.T. Tran, K. Bray, M.J. Ford, M. Toth, I. Aharonovich, Quantum emission from hexagonal boron nitride monolayers. Nat. Nanotechnol. **11**, 37–41 (2016). https://doi.org/10.1038/nnano.2015.242
133. J.D. Caldwell et al., Photonics with hexagonal boron nitride. Nat. Rev. Mater. **4**, 552–567 (2019). https://www.nature.com/articles/s41578-019-0124-1
134. M. Mohr et al., Phonon dispersion of graphite by inelastic X-ray scattering. Phys. Rev. B **76**, 035439 (2007). https://link.aps.org/doi/10.1103/PhysRevB.76.035439

135. S. Ogawa, S. Fukushima, M. Shimatani, Hexagonal boron nitride for photonic device applications: a review. Materials **16** (2023). https://www.mdpi.com/1996-1944/16/5/2005
136. D. Oxtoby, H.P. Gillis, N.H. Nachtrieb, *Principles of Modern Chemistry* (Cengage Learning, 1998)
137. K. Li, D. Xue, Estimation of electronegativity values of elements in different valence states. J. Phys. Chem. A **110**, 11332–11337 (2006). https://doi.org/10.1021/jp062886k
138. B. Gil, G. Cassabois, R. Cusco, G. Fugallo, L. Artus, Nanophotonics **9**, 3483–3504 (2020). https://doi.org/10.1515/nanoph-2020-0225
139. A.J. Giles et al., Ultralow-loss polaritons in isotopically pure boron nitride. Nat. Mater. **17**, 134–139 (2018). https://doi.org/10.1038/nmat5047
140. D.S. Kliger, *Polarized Light in Optics and Spectroscopy* (Academic Press/Elsevier Science, 1990). https://www.elsevier.com/books/polarized-light-in-optics-and-spectroscopy/kliger/978-0-08-057104-1
141. A. Poddubny, I. Iorsh, P. Belov, Y. Kivshar, Hyperbolic metamaterials. Nat. Photon. **7**, 948–957 (2013). https://doi.org/10.1038/nphoton.2013.243
142. L.D. Landáu, Y.M. Lífshits, *Electrodynamics of Continuous Media* (Pergamon Press/Addison-Wesley, Oxford/Reading, MA, 1960), p. 324. https://www.elsevier.com/books/electrodynamics-of-continuous-media/landau/978-0-08-057060-0
143. S. Dai et al., Subdiffractional focusing and guiding of polaritonic rays in a natural hyperbolic material. Nat. Commun. **6**, 6963 (2015). https://www.nature.com/articles/ncomms7963
144. P. Li et al., Hyperbolic phonon-polaritons in boron nitride for near-field optical imaging and focusing. Nat. Commun. **6**, 7507 (2015). https://doi.org/10.1038/ncomms8507
145. D.A. Iranzo et al., Probing the ultimate plasmon confinement limits with a van der Waals heterostructure. Science **360**, 291–295 (2018). https://www.science.org/doi/abs/10.1126/science.aar8438
146. W.D. Newman et al., Observation of long-range dipole-dipole interactions in hyperbolic metamaterials. Sci. Adv. **4**, eaar5278 (2018). https://www.science.org/doi/abs/10.1126/sciadv.aar5278
147. E.H. Lock, A.V. Lugovskoi, S.V. Gerus, Relation between the Poynting and group velocity vectors of electromagnetic waves in a bi-gyrotropic medium. J. Commun. Technol. Electron. **66**, 834–843 (2021). https://doi.org/10.1134/S106422692107007X
148. K.V. Voronin, G. Álvarez-Pérez, C. Lanza, P. Alonso-González, A.Y. Nikitin, Fundamentals of polaritons in strongly anisotropic thin crystal layers. arXiv:2309.12935 (2023). https://arxiv.org/abs/2309.12935
149. A.J. Giles et al., Imaging of anomalous internal reflections of hyperbolic phonon-polaritons in hexagonal boron nitride. Nano Lett. **16**, 3858–3865 (2016). https://doi.org/10.1021/acs.nanolett.6b01341
150. E. Yoxall et al., Direct observation of ultraslow hyperbolic polariton propagation with negative phase velocity. Nat. Photon. **9**, 674–678 (2015). https://doi.org/10.1038/nphoton.2015.166
151. M.I. D'yakonov, New type of electromagnetic wave propagating at an interface. J. Exp. Theor. Phys. **94**, 119–123 (1988). http://www.jetp.ac.ru/cgi-bin/dn/e_067_04_0714.pdf
152. O. Takayama et al., Dyakonov surface waves: a review. Electromagnetics **28**, 126–145 (2008). https://doi.org/10.1080/02726340801921403
153. H.H. Sheinfux et al., Evolution and reflection of ray-like excitations in hyperbolic dispersion media. arXiv: 2104.00091 (2022). https://arxiv.org/abs/2104.00091
154. J. Duan et al., Launching phonon polaritons by natural boron nitride wrinkles with modifiable dispersion by dielectric environments. Adv. Mater. **29**, 1702494 (2017). https://onlinelibrary.wiley.com/doi/abs/10.1002/adma.201702494
155. I. Dolado et al., Nanoscale guiding of infrared light with hyperbolic volume and surface polaritons in van der Waals material ribbons. Adv. Mater. **32**, 1906530 (2020). https://onlinelibrary.wiley.com/doi/abs/10.1002/adma.201906530
156. J. Duan et al., Active and passive tuning of ultranarrow resonances in polaritonic nanoantennas. Adv. Mater. **34**, 2104954 (2022). https://onlinelibrary.wiley.com/doi/abs/10.1002/adma.202104954

157. M. Autore et al., Boron nitride nanoresonators for phonon-enhanced molecular vibrational spectroscopy at the strong coupling limit. Light Sci. Appl. **7**, 17172–17172 (2018). https://doi.org/10.1038/lsa.2017.172
158. A. Bylinkin et al., Real-space observation of vibrational strong coupling between propagating phonon polaritons and organic molecules. Nat. Photon. **15**, 197–202 (2021). https://doi.org/10.1038/s41566-020-00725-3
159. I. Dolado et al., Remote near-field spectroscopy of vibrational strong coupling between organic molecules and phononic nanoresonators. Nat. Commun. **13**, 6850 (2022). https://doi.org/10.1038/s41467-022-34393-4
160. N.J. Bareza et al., Phonon-enhanced mid-infrared CO_2 gas sensing using boron nitride nanoresonators. ACS Photon. **9**, 34–42 (2022). https://doi.org/10.1021/acsphotonics.1c01254
161. X. Lin et al., All-angle negative refraction of highly squeezed plasmon and phonon polaritons in graphene–boron nitride heterostructures. Proc. Natl. Acad. Sci. **114**, 6717–6721 (2017). https://www.pnas.org/doi/abs/10.1073/pnas.1701830114
162. A.V. Kildishev, A. Boltasseva, V.M. Shalaev, Planar photonics with metasurfaces. Science **339**, 1289 (2013). https://science.sciencemag.org/content/339/6125/1232009
163. Z. Jacob, I.I. Smolyaninov, E.E. Narimanov, Broadband Purcell effect: Radiative decay engineering with metamaterials. Appl. Phys. Lett. **100**, 181105 (2012). https://doi.org/10.1063/1.4710548
164. J.S. Gómez-Díaz, A. Alù, Flatland optics with hyperbolic metasurfaces. ACS Photon. **3**, 2211–2224 (2016). https://doi.org/10.1021/acsphotonics.6b00645
165. Y. Liu, X. Zhang, Metasurfaces for manipulating surface plasmons. Appl. Phys. Lett. **103**, 141101 (2013). https://doi.org/10.1063/1.4821444
166. P. Li et al., Infrared hyperbolic metasurface based on nanostructured van der Waals materials. Science **359**, 892–896 (2018). https://science.sciencemag.org/content/359/6378/892.abstract
167. P. Li et al., Collective near-field coupling and nonlocal phenomena in infrared-phononic metasurfaces for nano-light canalization. Nat. Commun. **11**, 3663 (2020). https://doi.org/10.1038/s41467-020-17425-9
168. I.A. De Castro et al., Molybdenum oxides – from fundamentals to functionality. Adv. Mater. **29**, 1701619 (2017). https://onlinelibrary.wiley.com/doi/abs/10.1002/adma.201701619
169. L. Lajaunie, F. Boucher, R. Dessapt, P. Moreau, Strong anisotropic influence of localfield effects on the dielectric response of α-MoO_3. Phys. Rev. B **88**, 115141 (2013). https://link.aps.org/doi/10.1103/PhysRevB.88.115141
170. Z. Zheng et al., Phonon polaritons: highly confined and tunable hyperbolic phonon polaritons in van der Waals semiconducting transition metal oxides. Adv. Mater. **30**, 1870091 (2018). https://onlinelibrary.wiley.com/doi/abs/10.1002/adma.201870091
171. W. Ma et al., Ghost hyperbolic surface polaritons in bulk anisotropic crystals. Nature **596**, 362–366 (2021). https://doi.org/10.1038/s41586-021-03755-1
172. N.C. Passler et al., Hyperbolic shear polaritons in low-symmetry crystals. Nature **602**, 595–600 (2022). https://doi.org/10.1038/s41586-021-04328-y
173. G. Hu et al., Real-space nanoimaging of hyperbolic shear polaritons in a monoclinic crystal. Nat. Nanotechnol. **18**, 64–70 (2023). https://doi.org/10.1038/s41565-022-01264-4
174. X. Ni et al., Observation of directional leaky polaritons at anisotropic crystal interfaces. Nat. Commun. **14**, 2845 (2023). https://doi.org/10.1038/s41467-023-38326-7
175. G. Hu, J. Shen, C.-W. Qiu, A. Alù, S. Dai, Phonon polaritons and hyperbolic response in van der Waals materials. Adv. Opt. Mater. **8**, 1901393 (2020). https://onlinelibrary.wiley.com/doi/abs/10.1002/adom.201901393
176. M. He et al., Anisotropy and modal hybridization in infrared nanophotonics using low-symmetry materials. ACS Photon. **9**, 1078–1095 (2022). https://doi.org/10.1021/acsphotonics.1c01486

Chapter 2
Experimental and Modelling Techniques

Abstract Before delving into the research results of this thesis, in this chapter we provide an overview of the primary experimental and modelling approaches employed throughout it. Firstly, we present a commonly used method for producing high-quality van der Waals (vdW) materials and heterostructures based on them: mechanical exfoliation and the dry transfer technique. Subsequently, we explore various techniques widely used for exciting and characterizing polaritons and that will be central to the development of the work presented in this thesis, with a special focus on the mid-IR and far-IR ranges. We start by introducing the fundamental tool of infrared (IR) spectroscopy, FTIR (Fourier-transform infrared spectroscopy), which enables the characterization of IR optical properties of bulk materials across extensive areas and broad spectral ranges with exceptional spectral resolution. However, traditional FTIR spectroscopy is diffraction limited, and the deeply subwavelength nature of IR polaritons necessitates characterization techniques capable of achieving subdiffraction spatial resolution. Therefore, we proceed by detailing the utilization of scattering-type near-field optical microscopy (s-SNOM), which offers spatial resolution well below the diffraction limit while also collecting topographical information about a sample, offering unprecedented capabilities for investigating subwavelength polaritonic modes. Finally, we delve into the primary strategies for modeling and interpreting experimental data resulting from s-SNOM. These strategies include numerical calculations of the momentum-dependent reflection coefficient using the transfer matrix method, as well as full-wave numerical simulations based on finite element methods.

2.1 Exfoliation and Dry-Transfer of van der Waals Materials

As discussed in Chap. 1, van der Waals (vdW) materials are composed of layers held together by weak vdW forces. When individual layers are separated from each other, they exhibit unique properties that differ from their bulk counterparts. Single or few layers of vdW materials possess extraordinary characteristics in terms of electronics, optics, mechanics, and thermodynamics, offering numerous advantages for diverse applications. For example, the bandgap of certain vdW materials, like

G. Álvarez Pérez, *Fundamentals of Nano-Optics in Hyperbolic van der Waals Materials*, Springer Theses, https://doi.org/10.1007/978-3-031-75767-9_2

molybdenum disulfide (MoS_2), is directly influenced by the number of vdW layers. Monolayers exhibit a direct bandgap, resulting in over a 104-fold enhancement in luminescence quantum emission [1]. Conversely, the bandgap becomes indirect for thicker layers. Furthermore, strain fields induce blue shifts in the Raman peaks of monolayers of hexagonal boron nitride (hBN), while bilayers experience red shifts [2]. More related to this thesis, the wavelength of phonon polaritons (PhPs), such as those found in hBN and α-MoO_3, varies according to the thickness of the slab [3, 4]. These examples emphasize the criticality of the ability to produce individual or few layers of vdW materials and control the layer count, as it allows for the prediction of their behavior and the tailored engineering of their properties [5–7].

In this regard, mechanical exfoliation represents one of the most commonly employed techniques for obtaining thin layers of materials by physically separating or peeling them from a bulk substrate. Exfoliated vdW layers possess exceptional quality since they are directly transferred from the bulk material, thereby retaining their intrinsic properties [8, 9]. Conversely, vdW layers produced through alternative methods like chemical vapor deposition (CVD) [10] may introduce defects or impurities during growth processes, potentially compromising their quality. This section provides an overview of the exfoliation process for vdW layers and their precise placement on desired locations, which are fundamental steps in exploring the unique opportunities presented by these materials [11].

Mechanical Exfoliation of van der Waals Materials

Over the past decade, a plethora of exfoliation techniques have emerged [12]. This thesis primarily employs the conventional mechanical cleavage procedure that was used for the very first isolation of graphene in 2004 [8]. It consists in a top-down method for thinning down a vdW bulk crystal using a commercially available tape. The mechanical exfoliation method usually consists of several steps (Fig. 2.1):

- (1): The vdW bulk crystals are placed on top of a commercial tape (Nitto SPV 224PRM LB, in the following, 'blue tape').
- (2–3): Part of the blue tape is folded onto the bulk crystal and retraced gently. By proceeding this way, the bulk crystal is divided into two pieces.
- (4–5): By repeating the last step several times, the bulk crystal can be progressively thinned down until an arbitrary distribution of relatively thin crystals, usually referred to as *flakes*, is observed across the tape.
- (6–9): Once thinned down, the blue tape with the crystals/flakes is put into contact with a stamp of polydimethylsiloxane (PDMS, Gelpak WF-20/1.5-X4) previously treated with ozone during 1–2 minutes to enhance the adhesion. This allows picking up a larger number of flakes presenting larger lateral dimensions. To facilitate its handling, the PDMS stamp is attached to a microscope slide by a doubled sided tape.

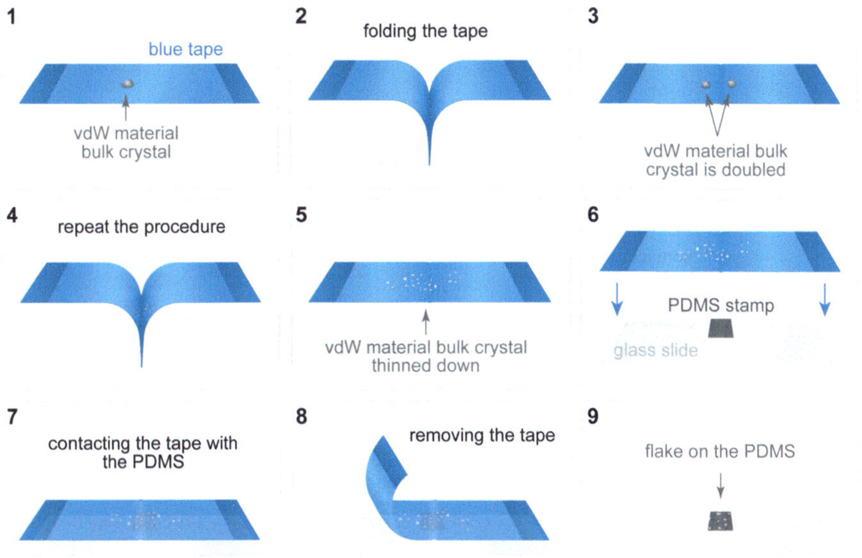

Fig. 2.1 Exfoliation of vdW materials. Schematics of the exfoliation procedure. The images show the folding-unfolding process to thin-down and exfoliate the bulk vdW crystal on the PDMS stamp in 9 steps. The PDMS stamp is inspected through the microscope after this process

Typically, one can observe a distribution of randomly dispersed flakes of varying colors on the PDMS stamp. The optical contrast between these flakes, as well as between the flakes and the PDMS material, is determined by Fresnel's coefficients [13], which themselves rely on the thickness characteristics of each layer in the air/vdW flake/PDMS system. Consequently, the color of a flake is determined by its thickness [14]. Thus, one can easily identify flakes of specific thicknesses by visually examining the PDMS substrate under an optical microscope. It is important to note that the adhesion forces between the flakes and the PDMS material are a result of the PDMS's viscoelastic properties, which are influenced by vdW forces [15].

Transfer of van der Waals Materials: The Dry-Transfer Technique

After the exfoliation process, the layers or flakes that have been successfully attached to the PDMS stamps can be released onto desired substrates with micrometer precision using the dry-transfer technique [15]. This setup involves various components, including micrometer screws employed to manipulate the spatial align-

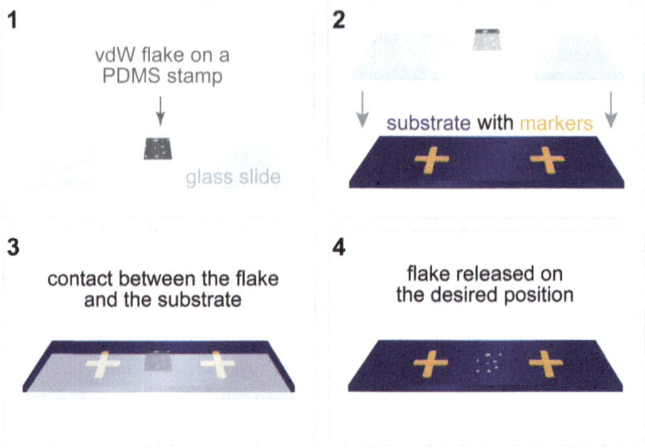

Fig. 2.2 Dry-transfer of vdW materials. Schematic of the transfer procedure in 4 steps. By increasing the temperature above 150 °C the PDMS viscoelasticity decreases strongly, enabling to release the flake with almost a 100% yield

ment between the flake and the target substrate, as well as an optical microscope utilized to monitor the transfer process in real-time (see Fig. 2.2). In the subsequent discussion, we outline the typical procedural steps associated with this technique:

1. The PDMS containing the exfoliated flakes is attached to the edge of a glass slide using double-sided tape. The glass slide is in turn attached to a micrometer XYZ stage, with the PDMS facing downwards.
2. The substrate intended for crystal transfer is fixed on a substrate holder via a vacuum pump.
3. Since PDMS is optically transparent, both the flake on the PDMS and the substrate surface are simultaneously visible in this configuration. By using an optical microscope and the XYZ micrometer screws, the relative positioning between the chosen flake and the substrate can be easily adjusted. Once aligned, the PDMS stamp with the crystal is brought closer to the substrate surface until contact is established.
4. At this point, the substrate is heated to approximately 150 °C, causing the PDMS stamps to lose their viscoelastic properties and enabling an almost 100% release rate. Subsequently, the glass slide is gradually lifted to release the crystal on the substrate. Following the transfer process, the sample is annealed during 3–5 minutes at temperatures ranging from 180 to 200 °C, ensuring enhanced adhesion between the substrate and the flake.

Sample Preparation in This Thesis

Figure 2.3 shows the dry-transfer system at the Quantum Nano-optics Lab (University of Oviedo) that was mostly utilized throughout this thesis. It is a commercial system (*HQ Graphene*) containing several micromanipulators that allow positioning the flakes with micrometer precision, while the substrate is firmly attached to the sample holder thanks to a vacuum pump. The controller allows regulating the temperature of the substrate holder between 0 and 200 °C, enabling to increase the yield with which the flake is released. Other similar systems where used in this thesis at Profs. James Hone's and Cory Dean's Labs at Columbia University (New York, USA) during a research stay undertaken in 2022.

The majority of the flakes shown throughout this thesis were exfoliated and transferred in the Quantum Nano-Optics Lab at University of Oviedo. Contributions from Dr. Jiahua Duan, Javier Taboada-Gutiérrez and Ana Isabel Tresguerres-Mata have been invaluable in this regard. Further fabrication processes, such as optical or e-beam lithographies, focused ion beam and reactive ion etching were carried out by members from different institutions. We acknowledge here contributions from Dr. Elizaveta Nikulina and Prof. Rainer Hillenbrand (nanoGUNE, Spain), Dr. Qingdong Ou and Prof. Qiaoliang Bao (Monash University, Australia), Dr. Iván Prieto (Institute of Science and Technology, Austria), and Dr. Lucía Herrer and Prof. José María de Teresa (Instituto de Nanociencia y Materiales de Aragón, Spain).

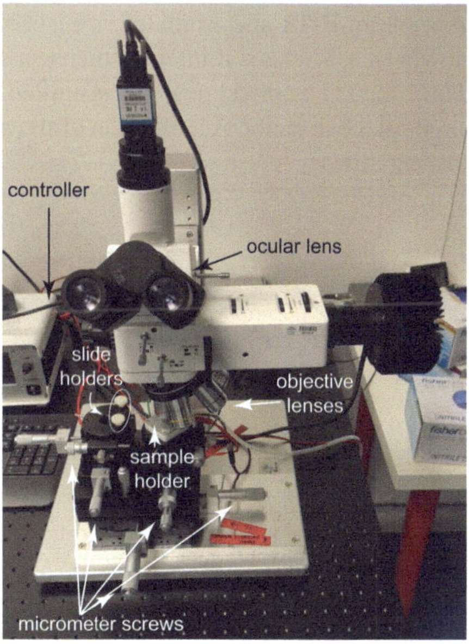

Fig. 2.3 The dry-transfer system employed in this thesis was Quantum Nano-optics Lab's (University of Oviedo) *HQ Graphene* setup

2.2 Fourier-Transform Infrared Spectroscopy (FTIR)

Given the polar dielectric nature of the materials investigated within this thesis, which possess optical resonances spanning from the mid-IR to the far-IR spectral regions, IR spectroscopy plays a crucial role in the study and characterization of their optical properties. This technique is important in the understanding of lattice vibrations within solids, as will be shown in Chap. 4, as well as the vibrational dynamics of molecules and plasmonic excitations in metals, all of which exhibit distinct resonant frequencies. Through the examination of the unique spectral patterns arising from these resonances, it becomes possible to accurately identify and characterize the material at hand, as the resulting spectrum acts as an exclusive *fingerprint*.

In particular, to fully characterize the optical response of polaritonic materials, precise measurements of the resonance frequency and linewidth are essential. This necessitates the use of appropriate IR spectroscopic techniques. The primary instrument for such measurements in the IR range is the FTIR (Fourier Transform Infrared) spectrometer. This section will provide an overview of the operation and specific techniques relevant to FTIR spectroscopy.

State-of-the-Art Setup Components

In brief, an FTIR spectrometer consists of a broadband IR source (typically a SiC glowbar), a Michelson interferometer, a sample compartment, and an IR detector (Fig. 2.4a). The optical power transmitted through the sample compartment is measured by the detector as a function of the position of the Michelson interferometer's moving mirror.

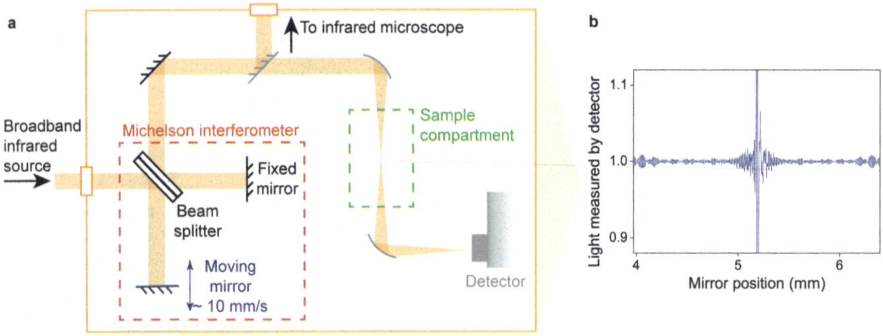

Fig. 2.4 FTIR spectroscopy. (**a**) Schematic of an FTIR spectrometer configured for transmission and reflection. (**b**) A representative interferogram obtained in an FTIR measurement, taken from [16]

The interferometer typically employs a beamsplitter, which splits the incoming IR light into two paths. One path is directed towards a fixed mirror, while the other path is directed towards a moving mirror. The light beams are then reflected back and recombined at the beamsplitter. As the moving mirror is displaced, the difference in the path length between the two beams changes. This leads to the formation of an interference pattern, known as the interferogram (Fig. 2.4b). To extracting the spectrally dependent IR signal passing through the spectrometer, a discrete Fourier Transform (FT) is then performed on the interferogram. The FT converts the interferogram from the time or mirror position domain into the frequency or wavelength domain. This transformation generally allows the extraction of the sample's reflection or transmission spectrum, which represents the intensity of IR radiation reflected or transmitted by the sample at different wavelengths. The interferogram obtained in FTIR spectroscopy is an interferometric representation of the sample's absorption spectrum. It contains both positive and negative fringes, which arise due to constructive and destructive interference between the two beams. The position and spacing of the fringes in the interferogram are directly related to the wavelengths absorbed by the sample.

The interferogram is recorded by the detector, which measures the intensity of the combined beams as a function of the mirror position. Detection is typically achieved with one of two standard detector types—pyroelectric deuterated L-alanine-doped triglycine sulfate (DLaTGS) and cryogenically cooled photovoltaic or photoconductive mercury cadmium telluride (MCT) detectors. The former requires no external cooling, is extremely broadband (covering the full IR spectrum), and has a linear response function, but is not very sensitive and has extremely slow response times (<1 kHz). MCT detectors can be extremely sensitive, are quite fast (tens of megahertz, typically), but are slightly less broadband (conventionally ≈ 20–1.3 μm, with a peak responsivity dependent on MCT alloy used), and generally exhibit nonlinearities in the output, limiting the dynamic range available for FTIR measurements. This can be alleviated with the use of superconducting bolometers, well-suited for detecting and measuring far-IR radiation in a sensitive and accurate manner, as will be shown below. Finally, we also need to address the ambient atmosphere inside the FTIR. Atmospheric water and CO_2 have large absorption bands in the IR, which can interfere with FTIR spectra. To prevent these gas-phase molecules from inducing artifacts, it is important to purge the FTIR spectrometer with either CO_2-free nitrogen or air, or to hold the FTIR under a modest vacuum.

As for the light source, conventionally, an FTIR uses an incandescent bulb, using a filament designed to operate at a temperature providing the most light at a given IR frequency. For example, while a traditional tungsten-filament incandescent bulb is appropriate for NIR applications, a lower temperature silicon carbide glowbar is more conventionally used in the MIR.

Almost all optics in the system are reflective, using gold or aluminum-coated mirrors to enable broadband operation. However, the beam-splitter is required to be made of IR transparent materials, typically KBr (transparent from 25 μm into the visible [17]). Short-wavelength measurements often use CaF_2 beam splitters, while longer wavelength measurements (FIR to terahertz) usually employ biaxially-oriented polyethylene terephthalate (BoPET) beam splitters.

The FTIR approach to optical spectroscopy has several advantages over dispersive spectroscopic techniques. First, the measurement is inherently broadband, appropriate for measurements from $>1000\,\mu\text{m}$ to approximately $1\,\mu\text{m}$ (depending on the performance of the source, beam splitter, and detector). Second, as all frequencies are collected simultaneously, there is no trade-off between spectral resolution and signal strength, generally resulting in high signal-to-noise. Finally, the spectral resolution is determined by the interferometer path length, which can be extremely large, with commercial models available with sub-GHz frequency resolution. On top of that, FTIR is a non-destructive technique, as it typically requires minimal sample preparation and does not consume or alter the sample during analysis. There are also several complications associated with FTIR spectroscopy. Some of the most significant challenges arise from the process of converting an interferogram into a spectrum. Signal processing techniques are applied in the process of creating the spectrum (notably phase correction, apodization, and zero filling [18]), which can influence the interpretation of the results if applied incorrectly. Furthermore, detector non-linearities induce significant spectral distortion into the FT process, and as such, any experiment must be carefully designed to achieve reliable results.

Reflection and Transmission

All methods of exciting surface polaritons result in resonant reflection R, transmission T, absorption A, and/or scattering S of incident electromagnetic waves (where $R + T + A + S = 100\%$). Reflection and transmission spectroscopy are the two most common and straightforward types of FTIR measurements. In both techniques, a reference spectrum is collected to (1) remove contributions from the background, such as air or water absorption, and (2) normalize the values. The reference spectrum usually consists of either reflection from a metallic mirror (near-perfect reflection at IR wavelengths), or an open path for transmission, and then the sample spectrum is collected and the two are divided. Transmission experiments are rather easy to do within an FTIR—a sample is simply mounted in the sample compartment and the transmitted light intensity as a function of frequency can be directly collected. In addition, using various commercially available tools, a sample can be rotated with respect to the incident beam path, enabling angle-dependent transmission measurements. Due to the straight beam path through the sample compartment, reflection spectra require an array of mirrors that allow the light to be incident on a sample at a fixed position, with the angle modified by controlling the mirror arrays in concert [17]. By performing correlated reflection and transmission experiments, it is possible to determine the extinction spectrum (e.g., total absorbed and scattered light), by using $E = 1 - T - R$, where $E = A$ assuming negligible scattering S (true for spectra of subdiffraction nanostructures and metamaterials).

However, both the IR reflection and transmission techniques have four major challenges. First, all common optical glasses are opaque in the IR. This means that substrates such as high resistivity silicon, germanium, or chalcogenide glasses, materials that are highly reflective or translucent in the visible, provide more appropriate IR windows [17]. Second, the long wavelength of IR light makes these experiments extremely sensitive to etalon effects arising in high-index samples and substrates that can significantly distort the spectral response. Third, conventional reflection and transmission measurements within the standard sample compartments can only be performed on samples relatively large in size, typically $>1\,\text{cm}^2$, but with defined apertures can be made on regions of interest in the millimeter range. Finally, an FTIR typically uses focusing mirrors that introduce some angular spread to the incident beam—this spread can artificially broaden the measured linewidth of resonances, including polaritonic modes due to their high sensitivity to incident angle. This will be further explored in Chap. 4, where a method to accurately measure the linewidth of resonances will be presented. In all, reflection and transmission are routine experiments for measuring relatively large (limited by diffraction) samples of materials.

Modelling of FTIR Data

FTIR reflectance and transmission spectra are commonly examined by employing algorithms such as Kramers-Kronig's relations and Fresnel's equations. Kramers-Kronig's relations are often utilized to determine the complex refractive index from the measured data. On the other hand, Fresnel's coefficients are employed to calculate the reflectance and transmission spectra, which describe the behavior of light at the interface between two media. In the particular case of dealing with thin vdW slabs, the modeling of FTIR reflectance and transmission spectra poses significant challenges associated with the small size of the structures and the consideration of the unique optical properties of these materials, such as anisotropy and interlayer coupling effects. We will not dive into these details here. Some further details regarding the modeling of experimental data will be provided in Sect. 2.5. In Chap. 4, we will demonstrate the potential of FTIR for extracting permittivity functions and elaborate on the technical intricacies and difficulties associated with analyzing this particular type of spectra.

FTIR System Mainly Used in This Thesis

The FTIR spectra shown this thesis, all presented in Chap. 4, were collected during a short research stay undertaken in December 2018 at the Nanophotonic Materials and Devices Lab led by Prof. Joshua D. Caldwell at Vanderbilt University (Nashville, USA). The spectra and their subsequent modelling were conducted in collaboration

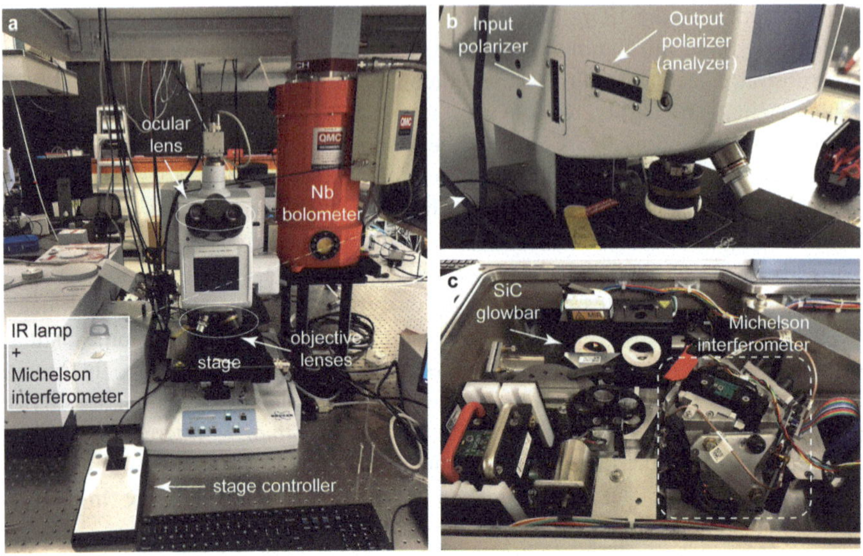

Fig. 2.5 The FTIR setup mostly employed in this thesis was a Bruker system at the Nanophotonic Materials and Devices Lab led by Prof. Joshua Caldwell at Vanderbilt University (USA)

with Prof. Thomas G. Folland and Dr. Joseph Matson. In Chap. 4, we will elaborate on the modelling of FTIR spectra, as well as on their application for the extraction of complex-valued permittivity functions. Figure 2.5a depicts the Bruker FTIR setup used, showing the objectives and the niobium (Nb) superconducting bolometer, which allows acquiring spectra down to the far-IR, reaching around $400\,\mathrm{cm}^{-1}$. In turn, the MCT detector is enclosed inside the case covering the microscope, which has two apertures for the input and output polarizers (Fig. 2.5b). The Michelson interferometer, along with the SiC glowbar, are shown in detail in Fig. 2.5c.

In addition, during the course of the thesis, the candidate also utilized the Varian FTIR setup at University of Oviedo, featuring a similar setup. However, the lack of a superconducting bolometer detector limits the acquisition capabilities, which extend only down to around $600\,\mathrm{cm}^{-1}$.

2.3 Prism Coupling and Gratings

The challenge associated with both studying and using surface polaritons is that they propagate much slower than free-space electromagnetic radiation, and thus have larger momenta than free space light. This is illustrated in Fig. 2.6a through the representative example of PhPs in 4H-SiC, which follow the dispersion already described in Fig. 1.5 in Chap. 1. Thus, due to the requirement to conserve both energy and momentum, it is impossible to directly excite surface polaritons with

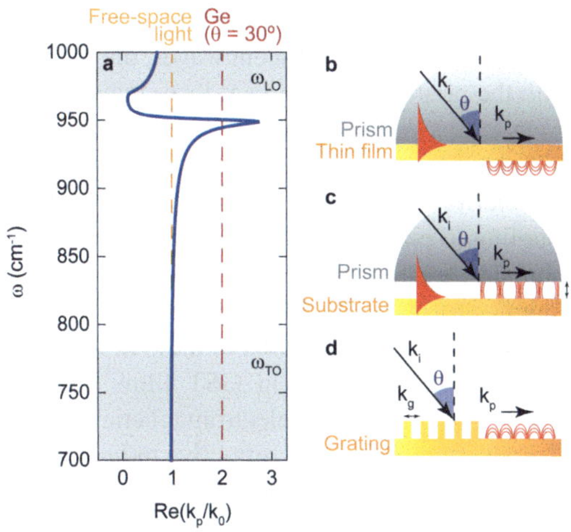

Fig. 2.6 Excitation of surface polaritons: prism coupling, gratings and nanostructuring. (**a**) Dispersion relation of SPhPs in 4H-SiC, highlighting the region between ω_{TO} and ω_{LO}. (**b**)–(**d**) Due to the momentum mismatch between free-space light and polaritons, methods to excite and probe them require this mismatch to be overcome. This can be realized by coupling light to the polaritonic medium through a high-index prism in either the (**b**) Kretschmann or (**c**) Otto configurations, or by (**d**) imparting higher momentum through grating coupling or through nanostructuring of the polaritonic medium, resulting in sub-diffractional resonant cavities

free-space light, whose dispersion is given by the so-called light line (yellow dashed line in Fig. 2.6a). This *momentum mismatch* between free-space light and polaritons can be overcome by slowing down the incident light, providing it with the necessary *extra* momentum. This can be done using a range of techniques, such as prism coupling, shown in Fig. 2.6b,c [19–21] or gratings Fig. 2.6d.

Prism coupling exploits the evanescent wave that results at the boundary of a high refractive index prism, either in contact with (Fig. 2.6b) or close to (Fig. 2.6c) the sample, the so-called Kretschmann and Otto configurations, respectively. This allows coupling into polaritons with $k_p = n_{\mathrm{prism}}\sin(\theta)k_0$, where θ is the incident angle. As a representative example, we show k_p for a germanium (Ge) prism under an incidence angle $\theta = 30°$ (dark red line in Fig. 2.6a). This angle has to be larger than the angle of total internal reflection, given by $\arcsin(1/n_{\mathrm{prism}})$. The refractive index of Ge can reach a value of about 4 in this frequency range [22], thereby enabling coupling to polaritons with momenta larger than that of free space light at $\theta > 14°$. On top of that, in the Otto configuration the decay length L of the evanescent waves into the air gap for both the totally reflected incoming light and the SPhP strongly varies with the in-plane momentum k_p [19, 23]:

$$L = \frac{\lambda}{2\pi\sqrt{\frac{k_p^2 c^2}{\omega^2} - 1}},$$

where λ is the wavelength and c the speed of light in a vacuum. For small gaps $d \ll L$ with large overlap of the two evanescent waves, the strong radiative coupling of polaritons back into the prism is a significant loss channel and prevents efficient excitation, while for large gaps $d \gg L$ the small overlap between the two evanescent waves inhibits an efficient energy transfer. There is, however, a critical coupling gap width d_{crit} where the polaritonic radiative and intrinsic losses exactly balance each other, and critical coupling to polaritons is achieved.

Both the Kretschmann and Otto configurations offer effective methods for exciting and observing polaritons, including strongly anisotropic shear polaritons in low-symmetry crystals [24], coupling of PhPs in thin AlN slabs and SiC substrates [25], and s-polarized waveguide modes in GST films [26]. However, when it comes to probing deeply sub-diffractional electromagnetic modes, they have certain limitations, as prisms with extremely large refractive indices would be needed.

A second route is through exploiting diffraction gratings or nanostructuring of the polaritonic medium, where additional momenta are provided by the periodicity of the grating or the nanostructure $k_p = k_0 \sin(\theta) + 2m\pi/k_g$, where k_g is the grating or nanostructure period and m is an integer. This is essentially a form of Bragg scattering—light interacts with the grating, coupling into diffractive modes, which results in a slowing of the light propagation. In all of these approaches by changing the angle of incidence θ or the geometry of the structures, the momentum at a given incident frequency can be tuned to match the momentum offset between free-space light and the polariton mode. This approach has been successfully exploited to explore sub-diffractional electromagnetic modes in metal surfaces [27, 28], arrays of nanoresonators [29] and polaritonic crystals in hBN [30] or, more recently, gold gratings underneath α-MoO_3 slabs [31].

However, even if polaritons can be excited with these structures, the conventional optical microscopes used to image their propagation are inherently constrained by the diffraction limit of light, which limits the smallest distance between two resolvable points to about half the wavelength of the illuminating light. Consequently, even the best optical microscopes cannot resolve features smaller than about 200 nm, while for IR the limit is almost always above a micrometer.

2.4 Scattering-Type Near-Field Optical Microscopy (s-SNOM)

To overcome these limitations, alternative methods have been developed. Scanning Probe Microscopy (SPM) is a family of high-resolution imaging techniques that routinely achieve resolutions in the nanometer range, with some of them even capable of resolving atomic-scale details. The key feature shared by all SPM variants is the utilization of sharp probing tips to scan the surface of the sample. The scanning process involves mechanical movement of either the probe or the sample, pixel by pixel and line by line, resulting in a raster image of the sample by recording its interaction with the probe as their relative positions vary. However, certain drawbacks arise from this type of image acquisition in SPM techniques. Notably, the probing is confined to the sample surface, and the image acquisition times tend to be relatively long to complete the scanning process.

SPM was triggered by the groundbreaking invention of the Scanning Tunneling Microscope (STM) by Binnig et al. Their pioneering work [32–34] addressed the challenge of achieving precise control over the distance between the probing tip and the sample surface despite external disturbances such as noise and vibration. STM measures electron current tunneling between probe and sample surface, and thus is limited to conductors and semiconductors. Shortly thereafter the invention of STM, Binnig et al. developed the Atomic Force Microscope (AFM) [35], which revolutionized imaging by enabling the examination of various samples regardless of their conductivity. In the AFM, a small cantilever holds the probing tip, which responds to changes in the interaction force between the tip and the sample's surface. By utilizing extremely sharp probing tips, even subatomic resolution can be achieved [36]. Moreover, depending on the coating or functionalization of the tip, different types of forces can be probed.

The principles of SPM can also be applied to optical probing. The concept itself predates the development of STM and is credited to Synge, who proposed a method for high-resolution optical imaging back in 1928. His approach involved creating a small aperture in an opaque screen [37]. By placing the aperture close to the sample, the detrimental effects of diffraction were minimized, allowing the transmitted light to reach the sample without significant degradation. Nowadays, the most commonly utilized aperture probes for optical imaging are metal-coated single-mode optical fibers that possess a small opening at one end [38], in the so-called aperture-based scanning near-field optical microscopy (a-SNOM). As the size of the aperture decreases, the resolution improves, but this improvement is accompanied by a decrease in the transmission efficiency due to the waveguide cut-off effect [39]. In practice, the resolution of a-SNOM is typically limited to approximately $\approx \lambda_0/10$. Consequently, for mid-IR light, the achievable resolution ranges from a few hundred nanometers to 1 μm, which often falls short of meeting the desired nanoscale criteria. When employing THz illumination, the resolution is typically in the range of tens of micrometers [40], making practical applications in nanoscience challenging to accomplish.

An alternative approach that offers resolution independent of the wavelength is to leverage the enhanced field near a small particle for illuminating the sample [41]. Early near-field optical microscopes based on this principle utilized metalized AFM probes instead of the initially proposed small particles to create an optical nano-focus [42, 43]. In contrast to a-SNOM, this approach employs a metalized AFM probe that is illuminated by a focused laser beam at an oblique angle of incidence. A metal coating is applied to enhance the scattering efficiency [44]. The metal-coated tips are commonly based on commercially available and cost-effective Si cantilevers. The nano-focus size solely depends on the tip apex radius, ensuring consistent resolution with visible and IR light, typically around 10 nm [45–47] and below [48].

The imaging mechanism in SNOM relies on the optical interaction between the probe and the sample mediated by evanescent fields. This interaction alters the amplitude and/or phase of the light waves scattered by the probe, which are then detected through far-field methods (see the subsection below). Since the scattered light is the measured quantity in apertureless SNOM, this variant is also referred to as scattering-type SNOM (s-SNOM). Elastic scattering is predominantly the main scattering channel, although Raman scattering can also occur [49], which can be separated from the elastic scattering background through filtering. Additionally, by generating light with double the frequency via second harmonic generation, it is possible to extract this harmonic signal if the field intensity at the tip apex is sufficiently strong [50, 51]. Furthermore, s-SNOM probes have been reported to enable single- and two-photon fluorescence excitation [52]. It is worth noting that unlike most other scanning probe techniques, near-field optical microscopy can detect objects located below the sample surface [53, 54]. Lastly, since s-SNOM is based on AFM, it can simultaneously provide the topography of the sample surface along with the optical image.

Probing the Near Field

The near fields serve as the key means for probing and extracting detailed information about a sample in SNOM. Near fields refer to the electromagnetic fields present in extremely close proximity to an object's surface. These fields exhibit strong localization and carry the spatial variations within the optical signal under investigation, therefore containing valuable information regarding the object's optical properties on a subwavelength scale. When a sharp probe tip is scanned in close proximity to the sample surface, it interacts with the near fields. This interaction induces alterations in the scattering of light. By detecting and analyzing these changes, high-resolution images or spectroscopic data can be generated. This process enables the characterization of the sample's optical properties with a spatial resolution that is only limited by the size of the tip.

In order to grasp the fundamental distinction between conventional far-field optical measurements and near-field measurements, one can envision a basic scattering experiment involving a macroscopic object, as depicted in Fig. 2.7. In general, the optical properties of this object can be described by its optical dielectric function $\varepsilon(\mathbf{r}, \omega, \mathbf{q})$, which depends on its position $\mathbf{r}$, frequency ω, and momentum $\mathbf{q}$. When the object is illuminated with propagating electromagnetic radiation (depicted in blue), it induces both a propagating field (blue) and a non-radiating field (yellow) [55], where the latter decays exponentially perpendicular to the surface of the sample within the so-called near-field zone. Non-radiating Coulombic near-field interactions depart from the far-field scaling of interaction with distance ($\propto 1/r$) and scale considerably faster ($\propto 1/r^3$), and thereby are conventionally limited to the very close distances from the object [56]. In the case of IR illumination, the non-radiating field typically diminishes within a range of tens to hundreds of nanometers

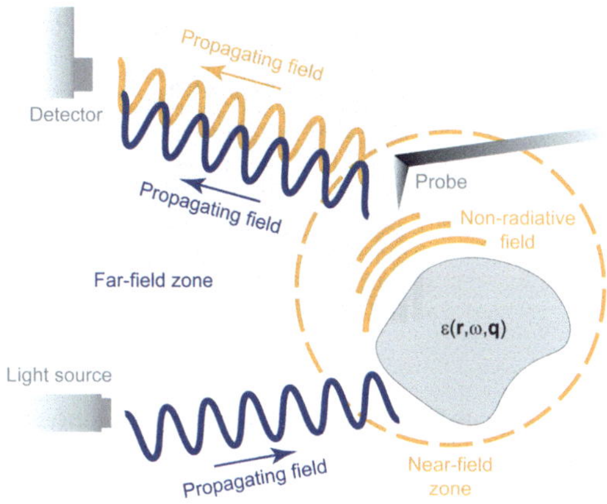

Fig. 2.7 Probing the near field with a s-SNOM. Far-field (blue) and near-field (yellow) measurements, accessing the propagating field and the evanescent field, respectively

[57]. Conversely, the propagating field can radiate into free space and be captured by a detector situated far from the object. By analyzing the detected far-field signal, classical electromagnetic theory enables us to deduce the material properties of the object, such as the complex dielectric permittivity (or equivalently, complex optical conductivity or refractive index) [58]. However, due to the limitations imposed by diffraction, the focused spot size of the illumination cannot be smaller than the wavelength. Consequently, the detected signal cannot perceive local variations in the object at a subwavelength scale. Furthermore, the non-radiating near field, which potentially carries crucial information about the object's high spatial frequency (momentum) properties, is completely lost. The primary objective of near-field measurements is to recover this lost information with a spatial resolution that is not constrained by conventional free space optics. This is done in s-SNOM by scanning a sharp tip in close proximity to the object's surface.

Nevertheless, to achieve this objective it is necessary to overcome a series of trade-offs. Firstly, a complex sample-probe arrangement is needed to position the probe close to the sample surface. Additionally, comprehensively understanding the optical properties of the probe can be challenging, as the probe can act as both a light confiner (due to its extremely small apex curvature) and a scatterer (partly due to its elongated shape, resembling an antenna), as will be shown below. Furthermore, the non-radiating field directly affects the probe's polarizability, which subsequently manifests itself in a nontrivial manner in the far-field-detectable wave (depicted in yellow). By meticulously conducting experiments and employing theoretical analysis, it becomes possible to access surface information of the object at a small length scale.

Therefore, the essential ingredient to nanoscale-resolved near-field microscopy is a strongly confined near field for generating highly localized scattering. To understand the physics of the interaction between the near field and the tip, we need to consider that the tip, when illuminated by the incident light, gives rise to evanescent fields with a wide range of in-plane momenta k. This is essentially a consequence of its sharp geometry and the lightning-rod effect [44]. When the

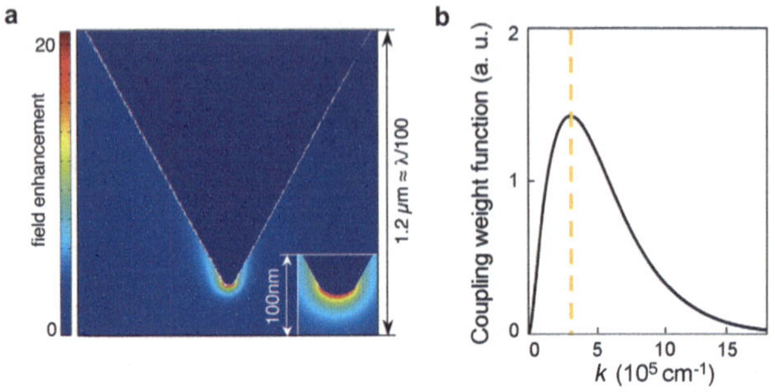

Fig. 2.8 Near-field microscopy based on field concentration at sharp metal tips. (**a**) Calculated field distribution at the tip apex of a metal cone with length $L = 1\ \mu$m, at $\omega \approx 84.7\,\text{cm}^{-1}$ ($\lambda = 118\,\mu$m), showing nanoscale field confinement and 25-fold field enhancement at the tip apex. Taken from Ref. [59]. (**b**) Momentum dependence of the time-averaged near-field coupling weight function $\langle k^2 \exp(-2kz_d)\rangle_t$ at $\omega \approx 1000\,\text{cm}^{-1}$ ($\lambda \approx 10\,\mu$m), which peaks at $k = 3.4 \times 10^5\,\text{cm}^{-1}$ for a tip radius $a = 30$ nm. Taken from Ref. [60]

tip approaches a polar and/or conducting surface, the evanescent fields are altered, which in turn affects the tip polarization. To quantify this interaction, several analytical models can be worked out (see Sect. 2.5). Here we only briefly comment on the essential aspects of the modeling procedure to show that nanoscale near-field confinement can be achieved by plane-wave illumination of a conical metal tip, (see Fig.2.8a, which is taken from Ref. [59]). The field distribution is calculated using one of the simplest models, the point dipole model, in which the apex of the tip is modeled as a point dipole. This will be further elaborated in Sect. 2.5. The numerically calculated field distribution at the apex, with curvature radius $a =$ 30 nm, of a short metal tip, with length $L = 1\ \mu$m, clearly shows field confinement to about 30 nm in all three directions. This result, adapted from Ref. [59], was obtained for 2.54 THz radiation ($\lambda = 118\ \mu$m) incident at 60° from the tip axis, but analogous field confinement values can be obtained in other frequency ranges, such as the visible and IR regimes. The strong confinement as well as an about 25-fold field enhancement, compared to the incident field.

The function that determines the electromagnetic coupling between the tip and the sample, and which thus defines the range of momenta where the s-SNOM can probe the surface excitations of the system most effectively, can also be computed via the dipole-model analysis. Such coupling function contains a weight term of the form $k^2\exp(-2kz_d)$, where z_d is the distance between the tip dipole and the sample surface [60]. As will be shown below, usually the tip is tapping and thus the magnitude of z_d is varying with time. The plot of the time-averaged weight function $\langle k^2\exp(-2kz_d)\rangle_t$ at $\omega \approx 1000\,\text{cm}^{-1}$ ($\lambda = 10\,\mu$m) reveals a bell-shaped momentum dependence that peaks around $k = 1/a$ (Fig 2.8b). Thus the dominant in-plane momenta contributing to near-field coupling are distributed around $k = 1/a$ (dashed line in Fig 2.8b). For that reason, the near-field spectra show resonances if and only if the dispersion curve of a mode intersects the dashed line that marks the dominant near-field momentum. For a typical value of the tip radius, $a = 30\,\text{nm}$, the probing in-plane momentum exceeds that of the incident light at $\omega \approx 1000\,\text{cm}^{-1}$ ($\lambda \approx 10\,\mu\text{m}$) by about 2 orders of magnitude. These virtues of tip-enhanced near-field coupling enable the exploration of a variety of polaritonic excitations, in particular the ones that will be shown all along this thesis, which are fundamentally finite-momenta effects.

State-of-the-Art Setup Components

The most commonly used contemporary s-SNOM systems are those applicable to the visible and IR frequency ranges. A typical visible/IR s-SNOM setup is illustrated in Fig. 2.9.

s-SNOM imaging is best when using monochromatic continuous wave (CW) light sources. Historically, the coherent sources available in the IR were limited to a number of gas lasers, specifically HeNe (632 nm, and 1.15, 1.52 and 3.39 μm) and CO_2 (9.5–11 μm) lasers [61]. The advent of the quantum cascade laser (QCL) [62, 63] and its counterpart the interband cascade laser (ICL) [64] have provided compact, continuously frequency tunable, high-power, commercially available coherent sources across a wide range of IR frequencies, with which propagating polaritons can be probed across various frequency ranges.

s-SNOM measurements commonly require laser power in the mW regime, typically between 0.5 and 5 mW [65], which has become easily reachable with the

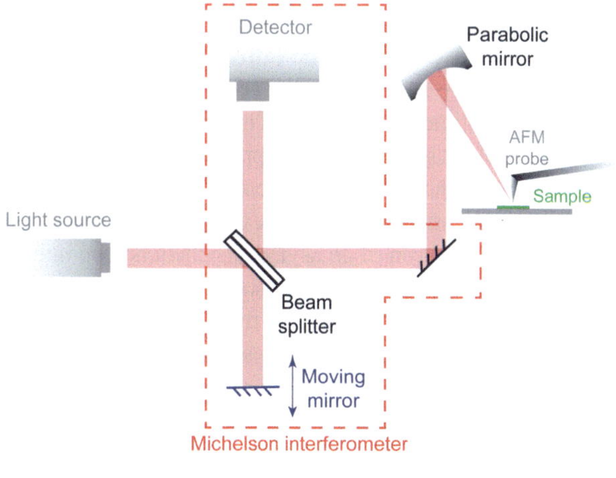

Fig. 2.9 A typical s-SNOM setup with visible or IR light sources. The incident light is split into two beams by a beam-splitter. The reflected beam by the reference moving mirror and the beam scattered by the AFM tip are paired again at the beam splitter which, in turn, brings them to the detector. See more details in the next subsection about the need for a reference mirror and different s-SNOM signal detection schemes

development of highly stable commercial lasers. Additionally, having a low noise in the optical signal is crucial for the acquisition of high-quality s-SNOM images. The noise in the context of electrical signals in often expressed in terms of voltage. More specifically, in terms of the root mean square (RMS) voltage spectral density, often expressed in units of $V/\sqrt{Hz}$. The RMS voltage spectral density is a measure of the noise power per unit bandwidth. It indicates how much noise power is present in a specific frequency range. The square root reflects the fact that noise power is proportional to the square root of the bandwidth. In s-SNOM, the RMS can reach below 100 $nV/\sqrt{Hz}$ at $\approx$100 kHz. The result is a promisingly good signal to noise ratio (SNR) for near-field detection, typically above 50–1, calculated by comparing the strength of the desired signal to the level of background noise. This SNR is critical for s-SNOM because laser noise is usually the largest limiting factor as the noise of state-of-the-art IR detectors is typically below 30 $nV/\sqrt{Hz}$. Most CW lasers in the IR range are capable of providing high output power of up to hundreds of mW and maintain a high level of stability over long periods of time. The high power stability and low RMS noise are crucial for performing long raster scans for high-resolution images.

In addition to CW lasers, ultrabroadband or tunable light sources, including free-electron lasers (FELs) [66], synchrotron IR and THz [67] radiation and ultrafast fiber or solid-state lasers have been successfully used for s-SNOM excitation and characterization of IR polaritonic modes. These sources offer improved temporal and spectral resolutions compared to other source types. Recently, optical parametric amplifiers (OPAs) have extended the tuning range up to a wavelength of around 20 μm, with several tens of mW average power, excellent long-term stability and high brilliance.

In common visible/IR s-SNOM setups (as the one sketched in Fig. 2.9), the incident light is divided by a beamsplitter into a reference arm and a probing arm. The probe beam is focused (within the diffraction limit) with focusing optics, typically a parabolic mirror or a lens, onto the apex of the AFM tip. Light interacts with the tip-sample system and is scattered. The backscattered (or in some cases forward scattered) light follows the same (or opposite) beam path back to the beam splitter and is eventually collected by the detector. The reference beam is reflected from a flat mirror on a delay stage and interferes with the backscattered light at the detector. The result is the formation of an asymmetric Michelson interferometer (see more details on the detection schemes in the subsection below). In this approach, both the amplitude and phase information of the scattered light can be obtained, yielding the possibility of extracting local complex dielectric functions of the sample volume under the tip without the need for applying Kramers–Kronig's relations.

Like in any other tip-based scanning technique, the collected information is difficult to interpret due to nontrivial tip-sample interactions. In s-SNOM, these difficulties result from at least three factors. First, the well-known antenna effect causes light to be highly confined between the probe apex and the sample surface, and thus the specific geometry of the tip apex plays a significant role in determining the intensity of the scattered signal [68]. Second, in addition to the local optical information, a strong but undesired background signal can be detected. This background signal can be mainly attributed to the light scattering from the tip shank, cantilever, and sample surface. Third, due to the broad momentum distribution of the localized radiation, non-trivial phenomena such as the non-locality of electrons and phonons must be carefully addressed in some cases [69, 70]. Relevant to this thesis, these complications can occur in layered structures, thin films, plasmonic structures, and low-dimensional materials, among others [71, 72].

Near-Field Signal Detection Schemes

In addition to the numerous benefits offered by s-SNOM, its application in optical imaging encounters a significant hurdle stemming from the considerable background scattering mentioned earlier. As has been discussed, s-SNOM experiments involve the detection of a far-field signal comprising both the desired near-field information and an unwanted background signal resulting from far-field scattering caused by the tip, cantilever, and sample. This background signal is unrelated to the near-field interaction between the probe and sample, yet it can introduce undesired artifacts or contrasts in s-SNOM images, thereby distorting them and complicating the extraction of optical properties from the samples. Consequently, the inclusion of a detection scheme capable of eliminating the background is imperative in s-SNOM experiments. Since the initial experimental realizations of s-SNOM, a substantial portion of research efforts has been dedicated to suppressing the background contribution [48, 57, 73–76].

To mitigate the background, the near-field signal is commonly modulated using an oscillating AFM probe. In tapping mode, the AFM tip vibrates at a mechanical resonance frequency Ω, typically ranging from 20 to 300 kHz depending on the specific cantilever and driving voltage employed. The near-field signal exhibits a nonlinear variation with respect to the tip-sample distance, while the background signal exhibits a linear variation. Consequently, the demodulated scattering signal at higher harmonics of the tip tapping frequency (2Ω, 3Ω, etc.) predominantly contains the near-field signal, effectively suppressing the background.

To optimally preserve the near-field signal and minimize the background signal, several detection schemes have been developed. In this context, we will primarily focus on intensity detection techniques, among which, the most widely employed are the self-homodyne, homodyne, heterodyne, and pseudo-heterodyne schemes. We will discuss these four techniques, as well as their respective advantages and disadvantages, with the aim of pedagogically introducing the pseudo-heterodyne method, which has been extensively employed in this thesis. Other detection schemes, such as electro-optic sampling [77], phase-shifting interferometry [78] and the synthetic optical holography technique [79, 80], have demonstrated successful results. However, they are less common, and they have not been used in this thesis. Therefore, a discussion of these schemes is beyond the scope here.

Self-Homodyne Detection (SHD) In the early stages of s-SNOM, the most common and straightforward detection scheme was SHD, illustrated in Fig. 2.10a. The detector only records the signal intensity, which is the squared amplitude of the incident electric field:

$$I = (\mathbf{E}_n + \mathbf{E}_b)(\mathbf{E}_n^* + \mathbf{E}_b^*),$$

where $\mathbf{E}_n$ is the complex near-field scattered electric field and $\mathbf{E}_b$ is the complex undesired background. "*" denotes the complex conjugate. By multiplying the terms and expanding, the following equation can be obtained:

$$I = |\mathbf{E}_n|^2 + |\mathbf{E}_b|^2 + 2|\mathbf{E}_n||\mathbf{E}_b|\cos(\varphi_n - \varphi_b).$$

The lock-in amplifier demodulates the signal at higher harmonics of the tip tapping frequency. In practice, each higher harmonic improves this filtering, but also comes with weaker signal strength. Typically for polaritonic measurements, the second to fourth harmonics are plotted. Since $\mathbf{E}_b$ varies little with the tip-sample distance, it is mainly removed by filtering after demodulation. Thus, the modulated signal takes the following form:

$$I = |\mathbf{E}_n|^2 + 2|\mathbf{E}_n||\mathbf{E}_b|\cos(\varphi_n - \varphi_b).$$

In this scenario, the detected signal comprises a subtle combination of both the near-field and background components. Fortunately, in most situations, the second term in the equation significantly outweighs the first term. This is because background scattering tends to be orders of magnitude larger than near-field scattering. Consequently, the detected signal can be structured based on the near-field signal while being augmented by the background signal. However, it is important to note that the SHD scheme has a major drawback. It only provides access to intensity information, completely losing the phase information. This limitation arises because the parameter φ_b is not controllable in the same way as in more sophisticated detection schemes introduced below. Furthermore, variations in both the spatial distribution and phase of the background component, denoted as $\mathbf{E}_b$ and φ_b respectively, can introduce a certain level of randomness or artifacts in the resulting near-field images. This randomness makes it challenging to interpret the acquired data quantitatively.

Homodyne Detection Currently, SHD is rarely used. Instead, a more advanced detection scheme, HD, is often utilized in s-SNOM setups. Essentially, the HD scheme is like the SHD scheme with an added reference arm, such that an asymmetric Michelson interferometer is formed (Fig. 2.10b). With this coherent detection setup, both the amplitude and phase information can be obtained. Based on the detected signal intensity, we have the following equation in this case:

$$I = (\mathbf{E}_n + \mathbf{E}_b + \mathbf{E}_r)(\mathbf{E}_n^* + \mathbf{E}_b^* + \mathbf{E}_r^*),$$

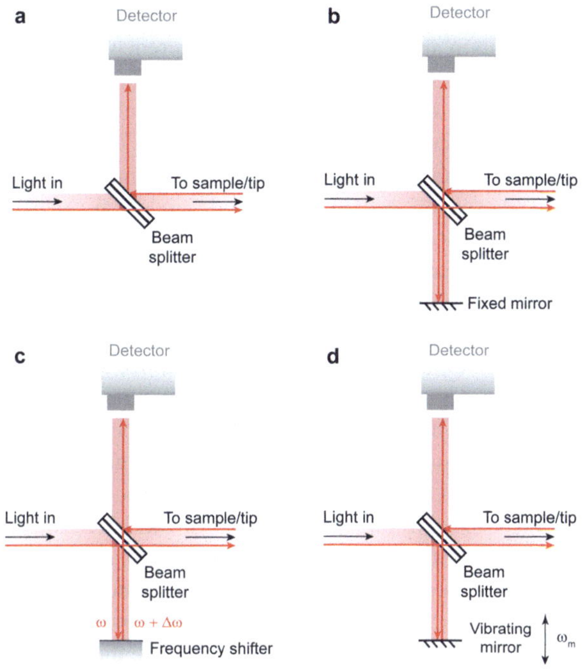

Fig. 2.10 Most common s-SNOM intensity detection schemes. (**a**) Self-homodyne detection. (**b**) Homodyne detection. (**c**) Heterodyne detection. (**d**) Pseudo-heterodyne detection

where $\mathbf{E}_r$ is the reference electric field. By multiplying the terms and expanding, we obtain the following formula:

$$I = |\mathbf{E}_n|^2 + |\mathbf{E}_b|^2 + |\mathbf{E}_r|^2 + 2|\mathbf{E}_n||\mathbf{E}_b|\cos(\varphi_n - \varphi_b) + 2|\mathbf{E}_n||\mathbf{E}_r|\cos(\varphi_n - \varphi_r) \\ + 2|\mathbf{E}_b||\mathbf{E}_r|\cos(\varphi_b - \varphi_r).$$

After demodulation by the lock-in amplifier, the $|\mathbf{E}_b|^2$, $|\mathbf{E}_r|^2$ and $2|\mathbf{E}_b||\mathbf{E}_r|\cos(\varphi_b - \varphi_r)$ terms are filtered since both $\mathbf{E}_b$ and $\mathbf{E}_r$ are not modulated by the tip tapping frequency. Therefore, the detected signal becomes:

$$I = |\mathbf{E}_n|^2 + 2|\mathbf{E}_n||\mathbf{E}_b|\cos(\varphi_n - \varphi_b) + 2|\mathbf{E}_n||\mathbf{E}_r|\cos(\varphi_n - \varphi_r).$$

By intentionally increasing the intensity of the reference beam to a significantly higher level than that of the probing beam, the first two terms in the equation become relatively small compared to the third term. As a result, the near-field term, which is scaled by the reference signal, can be obtained.

It should be noted that, in the HD scheme, the second term containing the background cannot be completely ignored, particularly for reflective samples where the electric field vector $\mathbf{E}_b$ is large. Even with the implementation of the demodulation procedure, background artifacts can still be present. To minimize these artifacts, a common approach is to place a reference material (such as a thick metal pad) adjacent to the sample. Dividing the signal detected from the sample by the signal detected from the reference can effectively eliminate spatial variations caused by $\mathbf{E}_b$ [81]. Furthermore, since the optical path of the reference beam can be easily adjusted, the parameter φ_r becomes the controlling factor. Therefore, coherent detection based on both the amplitude and phase can be achieved. To acquire complete information about both the amplitude and phase of the near-field signal, two consecutive scans are required with the reference mirror set at different optical path lengths. The difference between these lengths should be precisely one-eighth of the wavelength, which corresponds to a 90° phase delay [46]. The real part of the signal can be determined when the reference optical path matches the sampling optical path, while the imaginary part is identified when the optical path difference is one-quarter of the wavelength. This corresponds to acquiring both the sine and the cosine components from a complex signal, and then reconstructing its amplitude and phase. This technique is relatively time-consuming and necessitates high signal stability. In addition, the influence of the background scattering can still have a noticeable impact on the amplitude and phase measurements by the HD method. The complete elimination of the background therefore requires more sophisticated detection methods, presented below. Nevertheless, the HD scheme is relatively straightforward to implement compared to the more advanced techniques discussed next. As a result, it remains one of the most commonly utilized detection methods [82].

Heterodyne Detection (HTD) In principle, the HTD detection scheme [83–85] can completely eliminate the background signal. The key difference between HTD and HD schemes lies in the fact that, in the former, the reference beam undergoes a frequency shift relative to the sampling beam, while it remains unchanged in HD detection (see Fig. 2.10c). The HTD detection uses a reference wave of a slightly different frequency from the one used for probe illumination. The reference must nevertheless be coherent with the illumination, and the relative difference in their frequencies must be very small [81]. Such frequency shift may be produced by an acousto-optic modulator (AOM), exploiting the Doppler shift of light upon its diffraction by traveling acoustic waves [81]. Mathematically, the detected signal is expressed as:

$$I = (\mathbf{E}_n + \mathbf{E}_b + \mathbf{E}_r)(\mathbf{E}_n^* + \mathbf{E}_b^* + \mathbf{E}_r^*).$$

However, in the case of HTD detection, $\mathbf{E}_r$ represents the electric field of the frequency-shifted reference. Once again, by multiplying and expanding the terms, we arrive at the following equation:

$$\begin{aligned} I =& |\mathbf{E}_n|^2 + |\mathbf{E}_b|^2 + |\mathbf{E}_r|^2 + 2|\mathbf{E}_n||\mathbf{E}_b|\cos(\varphi_n - \varphi_b) \\ &+ 2|\mathbf{E}_n||\mathbf{E}_r|\cos(\Delta\omega t + \varphi_n - \varphi_r) + 2|\mathbf{E}_b||\mathbf{E}_r|\cos(\Delta\omega t + \varphi_b - \varphi_r). \end{aligned}$$

After demodulation, $|\mathbf{E}_b|^2$ and $|\mathbf{E}_r|^2$ are effectively removed by filtering due to their lack of time dependence. It is worth noting that $2|\mathbf{E}_b||\mathbf{E}_r|\cos(\Delta\omega t + \varphi_b - \varphi_r)$ is associated with the frequency $\Delta\omega$, providing a convenient reference for aligning the interferometer optically. The term $2|\mathbf{E}_b||\mathbf{E}_r|\cos(\varphi_b - \varphi_r)$ is primarily modulated at the tip-tapping frequency, Ω. To extract the desired signal, namely, the $2|\mathbf{E}_n||\mathbf{E}_r|\cos(\Delta\omega t + \varphi_n - \varphi_r)$ term, we perform demodulation at $n\Omega \pm \Delta\omega$ instead of $n\Omega$. To clarify the rationale for demodulating at $n\Omega \pm \Delta\omega$, we represent $2|\mathbf{E}_n||\mathbf{E}_r|\cos(\Delta\omega t + \varphi_n - \varphi_r)$ as a Fourier series:

$$\sum_k A_k |\mathbf{E}_n| \cos(k\Omega t) |\mathbf{E}_r| \cos(\Delta\omega t + \varphi_n - \varphi_r).$$

Hence, when demodulated at $n\Omega \pm \Delta\Omega$, the amplitude information predominantly encompasses the near-field amplitude amplified by the reference amplitude. The detected phase information solely pertains to the near-field characteristics because the reference phase and frequency shift are controllable parameters. There is no presence of uncontrollable background noise.

Nevertheless, the HTD method has some practical disadvantages. Firstly, it typically involves the utilization of an AOM to induce a frequency shift in the reference beam. However, the applicability of this method is limited due to the fact that commercially available AOMs have a restricted spectral range in which they can operate. In particular, commercially available AOMs cover only a limited portion of the IR spectral range. Moreover, IR AOMs block visible light, complicating alignment. Additionally, AOMs operate in a high-frequency range, resulting in lower signal quality for light detectors compared to the near-field signal. The setup requires additional components like a beam expander, since frequency-shifted beam from the AOM is only slightly separated from the undiffracted beam, making it less compact. Further adjustments are needed for different spectral lines, as the output angle of the frequency-shifted beam changes with wavelength. Finally, the HTD scheme relies on the phase of the probe vibration, which can change due to mechanical interactions with the sample.

Pseudo-Heterodyne Detection (PHD) To overcome these limitation, the PHD scheme was proposed and has been widely adopted in many studies [76, 86, 87], including those carried out in this thesis. The PHD setup, schematized in Fig. 2.10d, bears resemblance to the HD setup. The key distinction lies in the fact that, while in the HTD scheme the reference beam is shifted in frequency, in the PHD method the reference beam undergoes a sinusoidal phase modulation. Specifically, the reference mirror undergoes harmonic vibrations characterized by a small spatial amplitude a at a much lower frequency, denoted as ω_m, compared to the vibration frequency of the AFM probe, denoted as Ω. These vibrations alter the optical path of the reference arm, resulting in phase modulation of the electric field of the reference beam.

This is achieved with a Michelson interferometer configuration, using only one beam splitter and a vibrating mirror. The PHD scheme is therefore much easier to implement than a HTD setup. Also, the required components are readily available

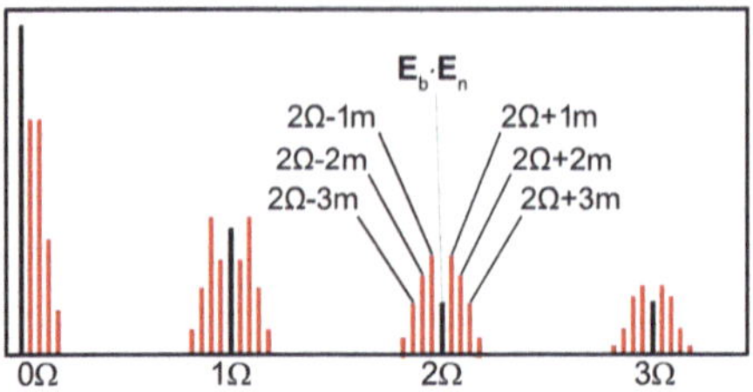

Fig. 2.11 Pseudo-heterodyne detection. Schematic of the side-band maxima in the vicinity of the harmonic frequencies of the resonantly oscillating AFM tip. The black maxima $m = 0$ still present background contribution. Reproduced from [81]

over the entire near-UV to far-IR spectral range. Furthermore, the vibrating mirror can be driven by piezoelectric actuators with kHz frequencies and thus leave the useful signal in the sub-MHz range where light detectors offer a higher signal-to-noise ratio compared to those required by the HTD technique.

The modulated electric field can be expressed in the form of a Fourier series, as follows:

$$\mathbf{E}_r = \sum_k A_k e^{ik\omega mt}.$$

For this equation, the Fourier coefficients are given by the following formula

$$A_k = AJ_k(a)e^{i\left(\frac{k\pi}{2}+\delta\varphi\right)},$$

where J_k is the k-th Bessel function of the first kind and $\Delta\varphi$ is the phase difference between the reference beam and sampling beam. As a result of phase modulation occurring in the reference arm, the detected signal's nth harmonics are effectively split into sidebands positioned at frequencies of $n\Omega \pm m\omega_m$ (Fig. 2.11). It should be noted that within these sidebands, the signal is directly proportional to both the real and imaginary components of the complex signal. In particular, the real part is obtained for even, and the imaginary part for odd sidebands [81]. By selecting a specific value for the modulation amplitude a, the proportionality between the real and imaginary components can be adjusted to be equal. Consequently, the PHD scheme enables the simultaneous measurement of both the signal's amplitude and phase during a single raster scan, resulting in highly efficient time usage. This advantageous time efficiency, combined with the absence of background interference and its easy implementation, makes PHD a widely used technique. Indeed, all the s-SNOM nanoimaging experiments performed in this thesis have consistently relied on the PHD demodulating technique (see Fig. 2.15 below, which illustrates a typical utilized setup).

Nanoimaging Polaritons in van der Waals Materials at Mid-IR Frequencies Using s-SNOM

The advancements enabled by s-SNOM in the realm of IR nanophotonics have been truly revolutionary. It introduced an experimental technique that, for the first time, enabled the quantification of the frequency-dependent optical behavior of materials or structures with spatial resolution comparable to the length-scale of polaritonic effects. Presently, s-SNOM offers a deterministic characterization of optical properties at the nanoscale, providing an incredibly fine spatial resolution (less than 10 nm) across a wide spectral range, as well as high temporal (less than 10 fs) and spectral (less than 1 cm^{-1}) resolution. Furthermore, it enables ultra-broadband optical (0.5–3000 μm) nanoimaging and nanospectroscopy. It is important to note that the AFM tip used in s-SNOM also serves as a light momentum matcher (as was shown in Fig. 2.8), facilitating momentum-resolved optical characterization typically in the range of 10–100 nm^{-1} based on the dimensions of the tip.

Perhaps, one of the most powerful features of s-SNOM methods for characterizing polaritons is in the ability to directly image the wavelength of a propagating mode [3, 88–90]. Initially this was realized with metallic elements fabricated on the top of SiC to focus SPhPs launched by the s-SNOM tip [91]. Over time, this technique has evolved into a method for extracting the polariton dispersion relation [3, 88–90]. The original demonstration of this method involved the investigation of plasmon polaritons propagating on graphene gated structures [88, 89]. As discussed in section "Plasmon Polaritons in Graphene" in Chap. 1, plasmon polaritons in graphene (GPPs) feature an exceptionally high confinement factor and can actively tuned by applying an external voltage. This, along with their relatively long propagation lengths makes them a viable platform to control IR light at the nanometric scale.

The experimental demonstration of these phenomena requires the implementation of a sharp edge, defect, or step, that can reflect the wave launched by the tip and, in certain cases, directly launch the polaritonic waves. Specifically, the scattering of the incident light from the tip couples to high-momentum polaritons, thereby launching a mode that propagates radially. This phenomenon is illustrated in Fig. 2.12a (top panel), which portrays the experimental setup employed to launch and detect the propagation of surface waves in graphene (represented as blue rings). The metallized AFM tip (depicted in yellow) is illuminated by an IR laser beam with a wavelength denoted as λ_0. The aforementioned mode continues to propagate radially until it encounters a sharp boundary, such as a defect or the edges of the flake shown in Fig. 2.12a. At this point, the wave is reflected, generating a standing wave as a result of the interference between the forward- and backward-scattered waves (Fig. 2.12a). Moreover, the polariton can also be directly launched due to scattering off the flake edge or boundary, leading to the formation of two distinct interference patterns characterized by periodicities of $\lambda_p/2$ and λ_p for tip- and edge-launched polaritons, respectively.

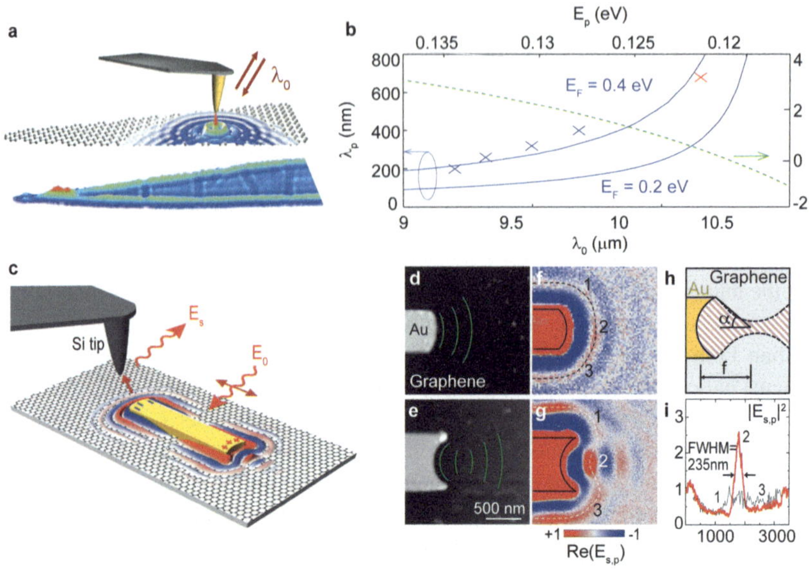

Fig. 2.12 **IR real-space nanoimaging of GPPs using s-SNOM**. (**a, top panel**) Diagram of the experimental configuration used to launch and detect propagating GPPs (represented as blue fringes). The metallized AFM tip (shown in yellow) is illuminated by an IR laser beam with wavelength λ_0. (**Bottom panel**) Near-field amplitude image acquired for a tapered graphene ribbon on top of 6H-SiC. The imaging wavelength is $\lambda_0 = 9.7\ \mu m$. The tapered ribbon is 12 μm long and up to 1 μm wide. (**b**) Experimentally extracted GPP wavelength λ_p as a function of incident wavelength λ_0. Values for λ_p are obtained from interference fringes (blue crosses) and localized modes (red cross), compared to the calculated GPP dispersion (blue curves). The green dashed line represents the SiC substrate permittivity. (**c**) Illustration of a s-SNOM experiment used to image graphene nanoantennas. (**d–g**) Topography and experimental near-field images $\text{Re}(E_{s,p})$, of a convex and a concave antenna extremity, respectively ($\lambda_0 = 11.06\,\mu m$). By employing a Si dielectric tip, the polaritonic fields can be directly visualized, avoiding detection via polaritonic interferometry. The green lines in the topography images illustrate the GPP wavefronts. (**h**) Illustration of GPP focusing with a concave antenna extremity. (**i**) Experimental near-field intensity (red curves) along the white dashed lines in (**e**,**g**). The dark gray curves show the experimental near-field intensity along the black dashed lines in (**e**). All curves and near-field images are normalized to their maximum in the lower left corner of (**g**). (**a**,**b**) are adapted from [88] and (**c–i**) from [92]

Importantly, by extracting a series of linescans normal to the flake edge, the periodicity of the fringes can be quantitatively determined. This is typically performed by fitting a damped oscillatory function to the data, yielding valuable information such as the polaritonic wavelength and other relevant figures like the propagation length. By varying the incident frequency during this process, the polariton dispersion can be experimentally measured (Fig. 2.12b), from which quantities such as the group velocity or the lifetime can be calculated. This approach holds several notable implications beyond the potential for manipulating light at the nanoscale. For instance, studying the polariton dispersion offers a pathway to investigate various material-specific properties of interest in IR nanophotonics.

These properties include the optical conductivity, Fermi energy, scattering rate of free carriers, as well as phonon scattering and energies in polar materials. Remarkably, this technique can achieve spatial precision that surpasses the curvature radius of the scanning near-field optical microscopy (s-SNOM) tip, allowing for probing subsurface features [93] and characterizing defects in semiconducting materials and devices [94]. In Chap. 4 we will demonstrate, as an original result of this thesis, how this technique can allow for a very accurate extraction of dielectric functions, even of strongly anisotropic media.

On top of that, s-SNOM also provides a means to directly map localized electromagnetic field profiles when combined with nanostructured polaritonic antennas. Nevertheless, it is important to consider that the utilization of a metallized AFM tip can induce alterations to the fields. This is primarily due to the substantial dipolar field associated with the additional antenna and its intricate interaction with the polaritonic dipole(s). In order to accurately map the resonant modes, it is necessary to determine the spectral position of the antenna resonances. Typically, this information is extracted through FTIR measurements conducted on periodic arrays of the antennas under investigation. However, it should be noted that a spectral shift exists between the resonant conditions in the far-field and near-field due to the presence of the s-SNOM tip [95]. Consequently, far-field measurements only provide a general estimation of the spectral position. Therefore, ideally one should measure the near-field spectra in the presence of the s-SNOM tip. This can be accomplished by employing a broadband light source in nano-FTIR [96], a technique that combines FTIR and s-SNOM, and which allows acquiring FTIR spectra with nanoscale resolution, thereby enabling direct measurements of the local spectral response. A detailed description of this technique falls beyond the scope of this thesis; interested readers are referred to [96].

In order to mitigate these effects, it is possible to employ a dielectric tip, typically made of silicon (Si), to extract the topographic and optical fields. This tip is designed to minimize the launching of polaritons, thus ensuring that the local near-field distributions remain unaffected by tip-induced artifacts. The use of resonant optical antennas and a Si tip was successfully implemented to directly visualize the wavefronts and propagation direction of GPPs [92]. Figure 2.12c illustrates the controlled and engineered excitation of GPPs using a tailored gold antenna. To efficiently excite the GPPs launched by the antenna, the dielectric tip and the sample are illuminated with s-polarized light parallel to the antenna axis, and the real part of the p-polarized tip-scattered light is recorded along with the topography during the scanning process. In the case of metal plasmonic antennas, this imaging approach was used in Ref. [92] to extract the vertical near-field component $\mathrm{Re}(E_z)$. However, we note here that the near-fields detected by polarization-resolved interferometric s-SNOM do not correlate with a single component of the local near field, but rather with a complex combination of the different local near-field components at each point (E_x, E_y and E_z). Furthermore, depending on the detection scheme (s- or p-polarization), a different scaling of the scattered fields as a function of the local near-field enhancement is obtained [97]. Typically, the measurement focuses on the z component due to its easier interpretability. This is primarily because it dominates under p-polarization detection schemes, aligning parallel to the tip's long symmetry axis [98]. As such, this component will also be the one measured throughout this thesis.

The utilization of a Si tip thus allows mapping the near field of the GPP wavefronts excited by antennas, in this case resonant concave and convex antenna extremities. A convex extremity launches diverging GPPs (Fig. 2.12d,f), while a concave extremity launches converging GPPs (Fig. 2.12e,g), leading to a focus at a distance of approximately 400 nm, as evidenced by the change in wavefront curvature (green lines in Fig. 2.12e). The concave extremity effectively acts as a 2D planar lens for focusing GPPs, achieving a focus spot size of 235 nm, which is 47 times smaller than the incident wavelength $\lambda_0 = 11.06\ \mu m$ (Fig. 2.12h,i). A significant advancement of this approach over earlier experiments [88, 89] is the separation of the excitation and detection processes for GPPs. In those previous experiments, a metal tip was used for both exciting and detecting GPPs, resulting in complex images of polariton interference. In contrast, by adopting this approach, local near-field distributions can be obtained, which are nominally unaffected by tip-induced artifacts. A more powerful focusing device, beating this ratio between the focusing spot and the incident wavelength, will be developed in Chap. 6 by taking advantage of the properties of in-plane hyperbolic polaritons.

The capabilities of s-SNOM have also been employed to nanoimage out-of-plane hyperbolic phonon polaritons (HPhPs) within nanoscale thickness slabs of hBN [3], as shown in Fig. 2.13a. As described in section "Out-of-Plane Hyperbolic Phonon Polaritons in Hexagonal Boron Nitride" in Chap. 1, HPhPs exhibit remarkable confinement and their wavelengths are strongly dependent on the thickness of the material. This phenomenon was initially demonstrated through nanoimaging data presented in Fig. 2.13b–g, wherein the normalized near-field amplitude, with respect to the near-field signal of gold, is plotted for various IR frequencies ranging from 1550 to 1580 cm^{-1}. Figure 2.13g showcases HPhPs in three-layer (left) and four-layer (right) hBN, while the hBN slab thickness in (b) to (f) is d = 256 nm. These images reveal a periodic pattern of maximum intensity regions (fringes), denoted as $s(\omega)$, running parallel to the edges, with *hot spots* occurring at the intersections of two or more fringes. By doubling the fringe period observed in Fig. 2.13, the wavelength of the polaritons, λ_p, can be determined, and the corresponding momentum can be calculated as $k_p = 2\pi/\lambda_p$.

s-SNOM has also played a crucial role in demonstrating the practical applications of hBN HPhPs in the field of nano-optics, arising from the exceptional directional properties and extremely tight confinement of HPhPs at sub-wavelength scales. For instance, super-resolution focusing in the type II hyperbolic band of hBN has been successfully demonstrated by performing nanoimaging of an array of gold (Au) nanodisks positioned underneath an hBN slab (see Fig.2.13h–k) [99, 100]. The gold nanodisks can only be resolved when the incident light's frequency falls within the range of *Reststrahlen* bands (RBs) (see Fig. 2.13h–j). This is due to the fact that HPhPs possess the requisite high momentum components necessary for experimentally resolving objects at sub-diffractional scales (see Fig. 2.13k) [99, 100]. Essentially, the hyperbolic volume-polaritonic rays function as a magnifying lens,

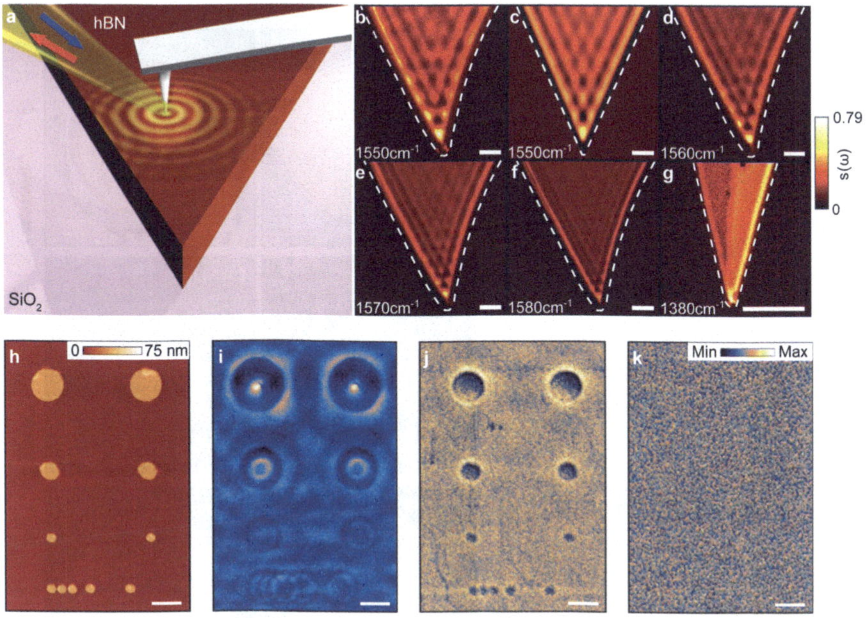

Fig. 2.13 IR real-space nanoimaging of out-of-plane HPhPs in hBN using s-SNOM. (**a**) Schematic of the s-SNOM experimental configuration used to image hBN flakes. Arrows denote the incident and back-scattered IR light. Concentric yellow circles illustrate out-of-plane HPhPs launched by the AFM tip and reflected by the two edges of a tapered hBN crystal. (**b**, and **d–f**) IR near-field images of the normalized amplitude $s(\omega)$ defined in the text and taken at different IR frequencies [hBN thickness in (**b–f**): $d = 256$ nm]. (**c**) Simulation of the HPhP interference pattern. (**g**) Out-of-plane HPhPs probed in three-layer **(left)** and four-layer **(right)** hBN crystals. White dashed line tracks the hBN edges according to the AFM topography. Scale bars indicate 800 nm. (**h**) An AFM image of Au disks defined lithographically on SiO_2/Si substrate before hBN transfer. (**i**) Near-field amplitude image of the top surface of a 395-nm-thick hBN at IR laser frequency $\omega = 1515\,\text{cm}^{-1}$ ($\lambda = 6.6\,\mu\text{m}$). The observed *hot rings* are concentric with the Au disks. (**j**) Near-field image of the same sample as in (**i**) at $\omega = 1610\,\text{cm}^{-1}$ ($\lambda = 6.2\,\mu\text{m}$) where HPhPs propagate almost vertically. **(k)** Near-field image of the same sample at $\omega = 1740\,\text{cm}^{-1}$ ($\lambda = 5.7\,\mu\text{m}$) showing complete homogeneity and lack of any distinct features. The colour scales for (**i–k**) are indicated in (**k**). The scale bars in all panels are 1 μm long. (**a–g**) are taken from [3], (**h–k**) from [99]

as illustrated in Fig.2.13h–k. The concentric rings observed on the top surface of the hBN (depicted in Fig. 2.13i) emerge as a result of multiple reflections occurring at the interfaces of the slab. These rings, manifesting in real-space, are an outcome of the superposition of various discrete modes simultaneously launched in **k**-space, as described by Eq. (1.8).

As a key result that triggered most of the work carried out in this thesis, the utilization of s-SNOM has played a critical role in visualizing in-plane HPhPs in α-MoO_3. This crystal exhibits surface anisotropy, as discussed in section "In-Plane Hyperbolic Phonon Polaritons in Alpha-Phase Molybdenum Trioxide" in Chap. 1, making all the exotic phenomena stemming from this anisotropy directly accessible

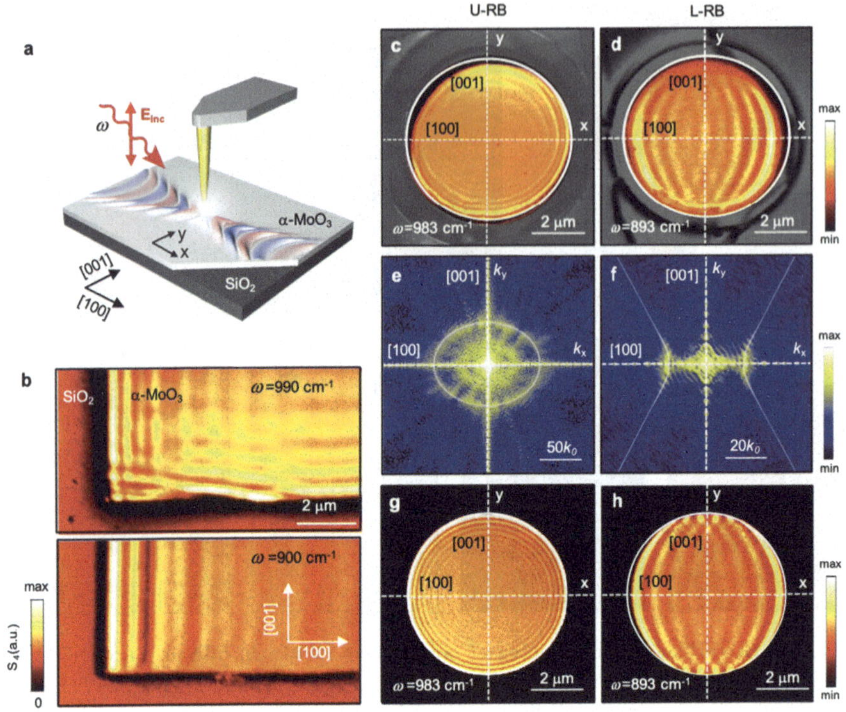

Fig. 2.14 **IR real-space nanoimaging of in-plane HPhPs in $\alpha-MoO_3$ using s-SNOM.** (**a**) Schematic of the s-SNOM experimental configuration used to image an α-MoO_3 flake. (**b**) Near-field amplitude images $|s_4(x, y)|$ of an α-MoO_3 flake with thickness $d = 250$ nm at an illuminating frequencies $\omega = 990\,\text{cm}^{-1}$ (**top panel**) and $\omega = 900\,\text{cm}^{-1}$ (**bottom panel**). Scale bars, 2 μm. (**c**, **d**) Near-field amplitude images $|s_4(x, y)|$ of an α-MoO_3 disk (colour key at right) with $d = 144$ nm. The imaging frequencies are $\omega = 983\,\text{cm}^{-1}$ (RB3) (**c**), and $893\,\text{cm}^{-1}$ (RB2) (**d**). Dashed white lines indicate the [100] and [001] surface directions. Scale bars, 2 μm. (**e**,**f**) Absolute value of the Fourier transform $|s_4(k_x, k_y)|$ of the near-field images shown in (**a**) and (**b**), respectively (colour key at right), revealing the IFCs for each RB. Solid lines show analytically calculated IFCs (note that they correspond to $2k$). The scale bars are $50k_0$ and $20k_0$, for the RB3 and RB2 respectively. (**g**,**h**) Calculated near-field amplitude images $|E_z(x, y)|$ for an α-MoO_3 disk (colour key at right) at $\omega = 983\,\text{cm}^{-1}$ (RB3) (**e**) and $893\,\text{cm}^{-1}$ (RB2) (**f**). Scale bars, 2 μm. Taken from [4]

for investigation using s-SNOM, as schematized in Fig. 2.14a. In Fig. 2.14b, near-field amplitude images obtained via s-SNOM depict an α-MoO_3 flake with a thickness of $d = 250$ nm, taken at frequencies of $\omega = 990\,\text{cm}^{-1}$ and $\omega = 900\,\text{cm}^{-1}$, both falling within two of the RBs of α-MoO_3 [101]. At $\omega = 990\,\text{cm}^{-1}$ (top image), bright fringes parallel to all the flake edges are observed, bearing a strong resemblance to out-of-plane HPhP fringes observed in hBN, as shown in Fig. 2.13. However, in the case of α-MoO_3, the periodicity of these fringes is highly dependent on the propagation direction. Specifically, the fringe periodicities are $\lambda_x = 950$ nm and $\lambda_y = 1200$ nm along the [100] and [001] crystal directions, respectively. This discovery not only revealed the deep subwavelength confinement of polaritons but also demonstrated a pronounced anisotropic in-plane propagation along the flake. The extent of this anisotropy becomes even more pronounced at $\omega = 900\,\text{cm}^{-1}$ (lower image in Fig. 2.14b), where the bright fringes are observed only parallel to the [001] crystal direction, as depicted in the lower panel of Fig. 2.14b.

Pivotal to this thesis, s-SNOM enables the quantification of polaritonic anisotropy and the measurement of isofrequency curves (IFCs) in wavevector space. To achieve this, thorough analysis of polariton propagation along various directions on the flake is necessary. To accomplish this analysis, one common solution is to fabricate material disks and conduct polariton interferometry experiments. Figure 2.14c and d illustrates representative near-field amplitude images captured at frequencies corresponding to *Reststrahlen band* 3 (RB3) and *Reststrahlen band 2* (RB2), respectively. In RB3, the interference pattern exhibits an elliptic shape, with the largest wavelength of PhPs aligning along the [001] surface direction, gradually diminishing to its smallest value in the orthogonal [100] surface direction. Conversely, in RB2, the interference pattern takes on an almond shape, where PhPs possess their longest wavelength along the [100] direction and gradually vanish until no observable polariton propagation occurs along the perpendicular [001] direction. By performing Fourier transformations of these near-field patterns, the IFCs can be directly obtained (Fig. 2.14f, lower panels). In RB3, an ellipse is observed, while in RB2, a hyperbola is manifested, indicating that PhPs exhibit elliptic and hyperbolic dispersions, respectively. The presence of two ellipses in Fig. 2.14c, differing by a factor of 2 in their semi-axes, is attributed to the simultaneous presence of tip-launched and edge-launched PhPs. On the other hand, the hyperbola in Fig. 2.14f opens along the [001] direction, signifying the absence of PhPs along this crystal direction, which also elucidates the observation depicted in Fig. 2.14b (lower panel).

Nevertheless, the acquisition of s-sNOM images and data is of little use without a proper and quantitative interpretation of the results. Developing models and simulations that enable clear interpretation of the results is not only imperative to comprehend and explain the underlying phenomena, but also to predict novel effects. This has been a central aspect in this thesis, with the visualization and modelling of effects such as hyperbolic propagation in biaxial crystals, negative reflection, anomalous refraction or polariton canalization. In the last section of this chapter, some of the most common models in the interpretation of s-SNOM data will be introduced. In the two subsequent chapters, key tools will be developed to predict the polaritonic behavior, particularly in α-MoO_3 and other biaxial materials. More specifically, when the results shown in Fig. 2.14 were published, there was not an available dispersion relation for polaritons in biaxial layers in the literature. Furthermore, the permittivity of α-MoO_3 in the IR range was not precisely measured. There was also a lack of a clear understanding of how polaritons would behave when reflecting and refracting within highly anisotropic media. Therefore, the analytic dispersion and the permittivity of α-MoO_3, as were as studies on reflection and refraction of hyperbolic polaritons, were instrumental in carrying out proper prediction and interpretation of the experimental data in the study of strongly anisotropic polaritons.

s-SNOM in This Thesis

A typical s-SNOM setup, and the one widely utilized throughout this thesis, is shown in Fig. 2.15a. It consists of a commercial system (NeaSNOM) coupled to an MCT detector, a QCL (Daylight) with four chips covering the frequency ranges 853–980 cm^{-1}, 920–1210 cm^{-1}, 1280–1465 cm^{-1}, 1445–1690 cm^{-1}, two CO_2 lasers discretely tunable between 9.25 μm and 11.29 μm, and one visible HeNe laser at 633 nm. Two similar s-SNOM systems were also used during a nine-month Fulbright research stay in 2022 at the Infrared Research Lab led by Prof. Dmitri Basov at Columbia University (New York, USA).

Figure 2.15b provides a closer view of the setup, showing the parabolic mirror, sample holder, and Michelson interferometer. This configuration is further mag-

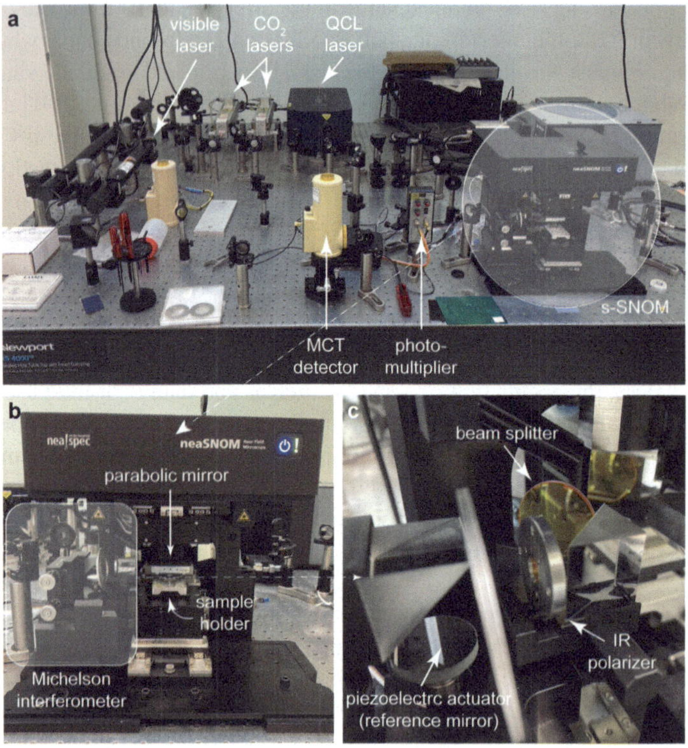

Fig. 2.15 s-SNOM system mostly employed in this thesis at the Quantum Nano-optics Lab (University of Oviedo). (**a**) S-SNOM setup showing the optical table, the laser sources, detector. The s-SNOM is marked by the dashed shadowed circle. (**b**) Zoom in of the s-SNOM system, showing the parabolic mirror, the sample holder, and the Michelson interferometer, marked by the dashed rectangle. (**c**) Zoom in of the Michelson interferometer, showing the reference mirror and the beam splitter for the PHD detection scheme, as well as the IR polarizer

nified in Fig. 2.15c, highlighting the reference mirror and beam splitter for PHD detection.

The polarization of the incident light is a critical factor in s-SNOM. Traditionally, p-polarized light is used because it aligns parallel to the tip and thus tends to dominate. In that case, an iris is commonly used instead of the IR polarizer in the experimental setup shown in Fig. 2.15c to regulate the signal entering the reference arm. Conversely, when working with incident s-polarized light, the iris is substituted by an IR polarizer, as the one shown in Fig. 2.15c. Without a polarizer in front of the reference arm, the amplified signal will mainly reflect the laser's polarization. Adjusting the IR polarizer to 45° enhances the signal from the out-of-plane components. These components are typically more easily interpretable in s-SNOM measurements, as demonstrated in this thesis through comparisons with simulations of the out-of-plane fields. Therefore, for a comprehensive measurement of both components using near-field signals, it is recommended to incorporate both s and p polarizations in the interferometer. Conversely, if the objective is to exclusively measure in-plane components, the polarizer can be omitted. When exclusively measuring s polarization, both in-plane components can be obtained, with the y component usually being smaller in magnitude. To isolate a signal along the x axis, a polarizer should be placed just before the detector.

The s-SNOM technique played a pivotal role in the development of this thesis. Every near-field image shown, with the exception of those presented in Chap. 10, was acquired at the Quantum Nano-optics Lab at University of Oviedo. This work was accomplished through close collaboration with various members of the research group, and specially with Dr. Jiahua Duan, whose contributions were key to the overall development of this thesis, particularly in Chaps. 5, 6, 7, and 8. Additionally, contributions from Ana Isabel Fernández-Treguerres Mata were very helpful in the acquisition of the s-SNOM nanoimaging results shown in Chap. 4.

The near-field images shown in Chap. 10 were acquired in collaboration with Francesco L. Ruta and Samuel L. Moore in Prof. Dmitri Basov's Infrared Research Lab at Columbia University. The AFM images shown in Chap. 10 were acquired in collaboration with Dr. Maëlle Kapfer from Cory Dean's Lab during the same stay at Columbia University.

All these measurements have been crucial in validating the prediction and validation of novel optical phenomena, analytical frameworks, and simulations that were developed throughout the course of this thesis.

2.5 Analytical and Numerical Models of Far- and Near-Field Signals

The intricate nature of propagating and non-propagating waves in near-field detection poses some difficulties to data interpretation. Understanding the scattering pattern of the tip—the scatterer that initiates both the near-field and free-space far-field waves—is the first step in obtaining the sample properties encoded in the near-field signal [55] (Fig. 2.16a). In addition, due to the strong interaction between the tip and the sample, they lose their distinct identity, and become a single entity. Thus, scattering problems involve tip-sample ensembles, the intrinsic length scale of which is on the deep subwavelength scale. This can pose a challenge for calculations based on classical electrodynamic theory, as well as calculation divergence issues and large mesh cell requirements. In addition, the tip geometry approximates a cone, which creates a non-integrable singularity point and length scale mismatch near the tip apex. As a result, analytical modeling or numerical simulation is difficult to perform. Nevertheless, the complexity and predictive ability of theoretical and simulation methods have significantly advanced over the last decades. In this section, we will review some of the simplest modelling approaches, as well as those that have been mostly utilized throughout this thesis.

Momentum-Dependent Reflection Coefficient: Transfer-Matrix Method Calculations

A straightforward but crude approximation for understanding the s-SNOM signal is to calculate the momentum-dependent reflection coefficient of the polaritonic medium. In this approximation we completely ignore the geometry of the tip and only account for its presence by considering the momentum-dependent sample response.

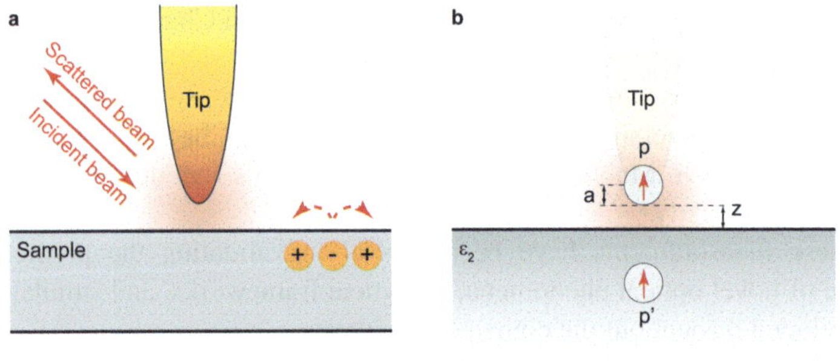

Fig. 2.16 Modelling the interaction between the s-SNOM tip and the sample. (**a**) Schematic of a metal-coated AFM tip being illuminated. The impinging beam produces a concentration of light (hot spot) at the tip apex enhancing the electric near fields in the vicinity of the sample. The tip acts also as a detector scattering the near fields into the far field. The induced charges in the sample are depicted by yellow circles with + and − symbols. Red dashed lines represent the electric field induced. (**b**) In the point dipole model, the tip is modelled as a small sphere at its apex with an electric dipole moment **p**. A mirror dipole is induced in the sample, **p**′ acting back on the tip dipole, which in turn acts again on the sample, and so on

Since surface polaritons at interfaces between isotropic media are TM polarized [102] (see also Chap. 1), this method usually reduces to calculate the momentum-dependent p-polarized reflection coefficient, the so-called r_p approximation, with p denoting the p-polarized electric field. In the simplest case, an interface between two semi-infinite isotropic media with permittivities ε_1 and ε_2, the momentum-dependent p-polarized reflection coefficient is given by [103]:

$$r_p(\omega, k) = \frac{\varepsilon_1 k_2 - \varepsilon_2 k_1}{\varepsilon_1 k_2 + \varepsilon_2 k_1},$$

where ε_1 denotes the superstrate permittivity and ε_2 denotes the sample permittivity. k_1 and k_2 denote the z-components of the momenta for superstrate and sample. They are given by

$$k_{1,2} = \sqrt{\varepsilon_{1,2}\frac{\omega^2}{c^2} - k^2},$$

where ω is the light frequency and k is the in-plane component of the momentum. Since the confined electric field is evanescent, the imaginary part of $k_{1,2}$ must be a positive nonzero number. Due to tip scattering, the near-field electromagnetic wave has a wide in-plane momentum distribution, with most values being larger than the free space momentum. As shown in Fig. 2.8, the k distribution follows a weight function that peaks at roughly $\approx 1/a$, where a is the tip apex curvature radius, which typically varies from 10 to 100 nm. Since $k \gg \omega/c$ in most cases, r_p can be further approximated by a k-independent function obtained found by taking the quasistatic limit as $k \to \infty$:

$$r_p(\omega) = \frac{\varepsilon_2 - \varepsilon_1}{\varepsilon_1 + \varepsilon_2} = \frac{\varepsilon_2 - 1}{\varepsilon_2 + 1}.$$

where in the last step we have considered that the superstrate is air, thus $\varepsilon_1 = 1$. We must stress that the r_p approximation only provides a simple understanding of the physical process in a near-field experiment and is not expected to provide precise quantitative predictability of the tip-sample interaction. However, the importance of the approximation should not be overlooked as it functions as a stepping stone for future developments of more complicated models. Many studies have been successfully interpreted based on this simple approach, also involving 2D materials, and it has as well as yielded several results within this thesis, as shown in Chaps. 4, 7, 8, and 9.

As will be elucidated in Chap. 3, in the case of anisotropic media, particularly those exhibiting biaxial characteristics, electromagnetic modeling becomes more intricate. A specific method for computing the characteristics of light propagation in strongly anisotropic layered structures is needed. In this thesis, we employed the approach outlined in [104]. By employing this algorithm, the reflection and transmission coefficients for s- and p-polarized light, i.e., r_{ss}, r_{pp}, t_{ss} and t_{pp}, and the mode coupling reflection and transmission coefficients r_{sp}, r_{ps}, t_{sp} and t_{ps} can be calculated, where the subscripts refer to incoming and outgoing polarization, respectively. We can effectively compute the reflection and transmission coefficients for various layer configurations and orientations. This algorithm specifically caters to non-magnetic materials and assumes a linear response from the structure, which will be fulfilled by the systems that are studied in this thesis. Due to limitations in space, we refrain from presenting the formalism here. For more comprehensive information, interested readers are directed to [104].

Point-Dipole Model

To account for tip-sample interactions, several models have been developed. The simplest one is the point-dipole model, derived in the late 1990s [48]. In this model, the tip is approximated as a polarized sphere and then as a point dipole with the same polarizability [39, 75, 105] (Fig. 2.16b). This model has successfully produced results that agree with the experimental data. With this approximation, the problem at hand essentially becomes a light scattering problem involving a small particle (the AFM tip) above an infinite half space (the sample). A rigorous mathematical treatment of this problem can be found in references [103, 106]. Here, we only briefly discuss the concept of the model without providing the details. Additionally, we only present the simplest version of the point-dipole model, in which (1) the sample is treated as an isotropic infinite half space, (2) only the vertical component of the electric field is considered, (3) a quasi-electrostatic approximation is imposed, and (4) retardation effects are ignored. More complex versions of the point-dipole model have been developed over the years, yielding improved consistency with experimental results [65].

In its simplest form, the AFM tip can be approximated as a polarized sphere with dielectric constant ε_t and radius a located at position $(0, 0, z + a)$. In most cases, ε_t is assumed to be constant at the frequency of interest. The polarizability of the sphere is given by [65, 81]

$$\alpha = 4\pi a^3 \frac{\varepsilon_t - \varepsilon_i}{\varepsilon_t + 2\varepsilon_1},$$

where ε_i is the dielectric constant of the surrounding medium. In our case, the tip will always be in an ambient environment, so we can take $\varepsilon_i = 1$ and

$$\alpha = 4\pi a^3 \frac{\varepsilon_t - 1}{\varepsilon_t + 2}.$$

The dipole moment induced in the AFM tip is therefore $p_0 = \alpha E_0$, where E_0 represents the illuminating electric field. Now, we consider the following sample: the infinite half space $z < 0$. This dipole moment acts on the sample, creating a mirror dipole (according to the classical mirror image method) with a moment $p' = p\beta$, where $\beta = (\varepsilon_s - 1)/(\varepsilon_s + 1)$ is the dielectric surface response function (as in the r_p model discussed in the previous subsection) and ε_s is the dielectric constant of the sample.

This mirror acts back on the tip AFM tip's dipole, amplifying its dipole moment. This increase in dipole moment, in turn, affects the mirror's dipole, creating a reciprocal effect. Consequently, the overall dipole moment can be as an infinite geometric series:

$$p = p_0 \sum_{n=0}^{\infty} g^n = \frac{p_0}{1 - g}$$

where the factor g represents the enhancement in the relative probe-dipole moment after each reflection. It is therefore related to the dielectric surface response function $g \propto \beta$. After some algebra, it can be shown [76] that the total dipole moment takes the form $p = \alpha_{\text{eff}} E_0$, with:

$$\alpha_{\text{eff}} = \frac{\alpha(1 + \beta)}{1 - \frac{\alpha\beta}{[16\pi(a+z)]}}$$

where z is the distance between the sample surface and the tip at its lowest position in the oscillation. Thus, the scattered field is proportional to the polarizability. Finally, to compare this result with the experimental data, it is crucial to demodulate α_{eff} at higher harmonics of the probe oscillation frequency.

Although this simple model was successfully used to explain the non-linear near-field optical contrast obtained between different materials and the prediction of strong polaritonic resonances when the real part of the sample permittivity approaches -1 [107], it fails to reproduce the near-field obtained by performing a tip-surface approach curve (i.e., recording $|E_z|$ as a function of tip-sample distance z), and the spectral position of the resonance, among other effects [65, 108]. The reason behind these discrepancies is attributed to an underestimation of the lightning rod effect at the tip apex. Therefore, other theoretical approximations were developed to model the AFM tip more accurately. In particular, the finite-dipole model describes the AFM tip as an elongated spheroid instead of a point dipole, where only charges located near the lower vertex of the spheroid (with a total length less than the incident illumination wavelength) are taken into account for the near-field interaction. Within this approximation, it is possible to find an analytical approach that quantitatively describes the non-linear near-fields produced by the tip-sample interaction. However, this approach will not be discussed here, as it has not been used in this thesis. We suggest references [65, 81] to interested readers.

Numerical Methods: Finite-Element Method Simulations

In addition to analytical approaches, the modeling of both far-and near-field signals can be accomplished through full-wave simulations employing various algorithms, including finite element methods, finite-difference time-domain methods, and others. Commercially available solvers such as CST Microwave STUDIO, Lumerical FDTD, and COMSOL Multiphysics offer convenient modeling platforms and effective calculation algorithms.

The geometrical representation of the tip and sample can be achieved with minimal approximations. Previous studies have employed a signal demodulation procedure within simulations and obtained quantitatively consistent results [109, 110]. Alternatively, simpler methods that do not incorporate tip modulation have been implemented by considering the electric field intensity between the tip and the sample as proportional to the near-field signal. Particularly in investigations of plasmonic nanostructures, it is common practice to approximate the near-field signal using the electric field intensity near the sample surface [91] due to its computational efficiency.

Throughout this thesis, COMSOL Multiphysics will serve as a fundamental tool for numerical simulations. It will be consistently assumed that the polarizability of the dipole remains weakly affected by the polaritons excited in the sample. Crucially, the dipole is assumed to be not an induced dipole, but a dipolar source, so it creates fields without any need of external illumination. By calculating the amplitude of the near field, represented as $|E_z|$, as a function of the dipole's position (x, y), we can simulate near-field images. Throughout this thesis, we will mostly use three-dimensional models, although two-dimensional models will also be utilized where appropriate, for instance for the cross-sections of the fields shown in Chap. 1. As demonstrated, our simulated images generally exhibit a good agreement with both the analytical calculations and the experimental s-SNOM images, enabling us to conclude that the calculated field between the dipole and the samples, denoted as E_z, provides a valid numerical representation of the signals measured by s-SNOM. Furthermore, these calculated field profiles can be directly compared to validate theoretical predictions [111–113]. Notably, accurate dielectric function models are indispensable for the successful implementation of these numerical simulations, as they heavily rely on the dielectric properties of the modeled materials.

Further Analytical Methods

Despite the usefulness of the aforementioned analytic and numerical methods in obtaining quantitatively consistent results, it is imperative to construct an analytical model that describes the momentum-dependent dispersion of polaritons and other quantities of interest for each specific structure under investigation. This not only enables predictive capabilities beyond the mere agreement with experiments, but also provides deeper insights into various quantities of interest such as lifetime, propagation directions, group velocity, decay, and more. Such an analytical model is essential for a comprehensive understanding of the behavior of polaritons, allowing for their effective control and manipulation. In Chap. 3, we will develop the initial analytical model for the momentum-dependent dispersion of polaritons in thin biaxial slab structures, based on Maxwell's equations. This model neglects the tip-sample interaction, similarly to the r_p approximation that was outlined at the beginning of this section, yet it will be shown to provide excellent agreement with experimental data, as well as predictions of novel phenomena, later corroborated in experiments in all the subsequent chapters, particularly in Chaps. 5, 6, and 7. In Chaps. 8 and 9 we will develop analytical models for multilayer structures, also based on Maxwell's equations. In addition, Chaps. 7 and 8 contain analytical models of the polariton propagation in real space based on the dyadic Green's function.

Similar to numerical simulations, all these analytical models heavily rely on the dielectric properties of the materials being modeled. Hence, the availability of precise dielectric function models is crucial for successful execution of analytical calculations. In Chap. 4, we extract the accurate permittivity of α-MoO_3 through a combination of far-field FTIR spectroscopy and near-field s-SNOM nanoimaging, serving as the key ingredient in the evaluation of all the models developed throughout this thesis.

References

1. P. Blake et al., Graphene-based liquid crystal device. Nano Lett. **8**, 1704–1708 (2008). https://doi.org/10.1021/nl080649i
2. R.V. Gorbachev et al., Hunting for monolayer boron nitride: optical and Raman signatures. Small **7**, 465–468 (2011). https://onlinelibrary.wiley.com/doi/abs/10.1002/smll.201001628
3. S. Dai et al., Tunable phonon polaritons in atomically thin van der Waals crystals of boron nitride. Science **343**, 1125–1129 (2014). http://science.sciencemag.org/content/343/6175/1125
4. W. Ma et al., In-plane anisotropic and ultra-low-loss polaritons in a natural van der Waals crystal. Nature **562**, 557 (2018). https://www.nature.com/articles/s41586-018-0618-9
5. A.K. Geim, I.V. Grigorieva, van der Waals heterostructures. Nature **499**, 419–425 (2013). https://www.nature.com/articles/nature12385
6. K.S. Novoselov, A. Mishchenko, A. Carvalho, A.H.C. Neto, 2D materials and van der Waals heterostructures. Science **353**, aac9439 (2016). https://www.science.org/doi/abs/10.1126/science.aac9439
7. S. Das, J.A. Robinson, M. Dubey, H. Terrones, M. Terrones, Beyond graphene: progress in novel two-dimensional materials and van der Waals solids. Annu. Rev. Mater. Res. **45**, 1–27 (2015). https://doi.org/10.1146/annurev-matsci-070214-021034
8. K.S. Novoselov et al., Electric field effect in atomically thin carbon films. Science **306**, 666–669 (2004). https://www.science.org/doi/abs/10.1126/science.1102896
9. C. Lee et al., Anomalous lattice vibrations of single- and few-layer MoS_2. ACS Nano **4**, 2695–2700 (2010). https://doi.org/10.1021/nn1003937
10. Z. Cai, B. Liu, X. Zou, H.-M. Cheng, Chemical vapor deposition growth and applications of two-dimensional materials and their heterostructures. Chem. Rev. **118**, 6091–6133 (2018). https://doi.org/10.1021/acs.chemrev.7b00536

11. R. Frisenda et al., Recent progress in the assembly of nanodevices and van der Waals heterostructures by deterministic placement of 2D materials. Chem. Soc. Rev. **47**, 53–68 (2018). https://doi.org/10.1039/C7CS00556C
12. M. Yi, Z. Shen, A review on mechanical exfoliation for the scalable production of graphene. J. Mater. Chem. A **3**, 11700–11715 (2015). https://doi.org/10.1039/C5TA00252D
13. M.M. Benameur et al., Visibility of dichalcogenide nanolayers. Nanotechnology **22**, 125706 (2011). https://doi.org/10.1088/0957-4484/22/12/125706
14. D. Golla et al., Optical thickness determination of hexagonal boron nitride flakes. Appl. Phys. Lett. **102**, 161906 (2013). https://pubs.aip.org/aip/apl/article/102/16/161906/311730/Optical-thickness-determination-of-hexagonal-boron
15. A. Castellanos-Gomez et al., Deterministic transfer of two-dimensional materials by all-dry viscoelastic stamping. 2D Mater. **1**, 011002 (2014). https://doi.org/10.1088/2053-1583/1/1/011002
16. Wikipedia. *Fourier-transform infrared spectroscopy* (2023). https://en.wikipedia.org/wiki/Fourier-transform_infrared_spectroscopy. Accessed 27 June 2023
17. T.G. Folland, L. Nordin, D. Wasserman, J.D. Caldwell, Probing polaritons in the mid- to far-infrared. J. Appl. Phys. **125**, 191102 (2019). https://doi.org/10.1063/1.5090777
18. W. Herres, J. Gronholz, *Understanding FTIR Data Processing Part 1: Data Acquisition and Fourier Transformation 1 in Bruker Analytlsche Messtechnlk GmbH* (2007). https://mmrc.caltech.edu/FTIR/Literature/General/Understanding%20FTIR.pdf
19. H. Raether, *Surface Plasmons on Smooth and Rough Surfaces and on Gratings* (Springer, Heidelberg, 1988). https://scirp.org/reference/referencespapers.aspx?referenceid=1193551
20. T.G. Folland et al., Probing hyperbolic polaritons using infrared attenuated total reflectance micro-spectroscopy. MRS Commun. **8**, 1418–1425 (2018)
21. K. Pufahl et al., Controlling nanoscale air-gaps for critically coupled surface polaritons by means of non-invasive white-light interferometry. Appl. Phys. Lett. **113**. 161103 (2018). https://doi.org/10.1063/1.5049230
22. J.H. Burnett, S.G. Kaplan, E. Stover, A. Phenis, Refractive index measurements of Ge in *Infrared Sensors, Devices, and Applications VI*, ed. by P.D. LeVan, A.K. Sood, P. Wijewarnasuriya, A.I. D'Souza, vol. 9974 (SPIE, 2016). https://doi.org/10.1117/12.2237978
23. N.C. Passler et al., Second-harmonic generation from critically coupled surface phonon polaritons. ACS Photon. **4**, 1048–1053 (2017). https://doi.org/10.1021/acsphotonics.7b00118
24. N.C. Passler et al., Hyperbolic shear polaritons in low-symmetry crystals. Nature **602**, 595–600 (2022). https://doi.org/10.1038/s41586-021-04328-y
25. N.C. Passler et al., Strong coupling of epsilon-near-zero phonon polaritons in polar dielectric heterostructures. Nano Lett. **18**, 4285–4292 (2018). https://doi.org/10.1021/acs.nanolett.8b01273
26. N.C. Passler, A. Heßler, M. Wuttig, T. Taubner, A. Paarmann, Surface polariton-like s-polarized waveguide modes in switchable dielectric thin films on polar crystals. Adv. Opt. Mater. **8**, 1901056 (2020). https://onlinelibrary.wiley.com/doi/abs/10.1002/adom.201901056
27. F.J. Garcia-Vidal, L. Martń-Moreno, J.B. Pendry, Surfaces with holes in them: new plasmonic metamaterials. J. Opt. A Pure Appl. Opt. **7**, S97 (2005). https://dx.doi.org/10.1088/1464-4258/7/2/013
28. C.R. Williams et al., Highly confined guiding of terahertz surface plasmon polaritons on structured metal surfaces. Nat. Photon. **2**, 175–179 (2008). https://doi.org/10.1038/nphoton.2007.301
29. J.D. Caldwell et al., Sub-diffractional volume-confined polaritons in the natural hyperbolic material hexagonal boron nitride. Nat. Commun. **5**, 5221 (2014). https://doi.org/10.1038/ncomms6221
30. F.J. Alfaro-Mozaz et al., Deeply subwavelength phonon-polaritonic crystal made of a van der Waals material. Nat. Commun. **10**, 42 (2019). https://doi.org/10.1038/s41467-018-07795-6
31. O.G. Matveeva et al., Twist-tunable polaritonic nanoresonators in a van der Waals crystal. npj 2D Mater. Appl. **7**, 31 (2023). https://doi.org/10.1038/s41699-023-00387-z

32. G. Binnig, H. Rohrer, C. Gerber, E. Weibel, Tunneling through a controllable vacuum gap. Appl. Phys. Lett. **40**, 178–180 (1982)
33. G. Binnig, H. Rohrer, Scanning tunneling microscopy. Surf. Sci. **126**, 236–244 (1983). https://www.sciencedirect.com/science/article/pii/0039602883907161
34. G. Binnig, H. Rohrer, C. Gerber, E. Weibel, Surface studies by scanning tunneling microscopy. Phys. Rev. Lett. **49**, 57–61 (1982). https://link.aps.org/doi/10.1103/PhysRevLett.49.57
35. G. Binnig, C.F. Quate, C. Gerber, Atomic force microscope. Phys. Rev. Lett. **56**, 930–933 (1986). https://link.aps.org/doi/10.1103/PhysRevLett.56.930
36. F.J. Giessibl, Atomic resolution of the silicon (111)-(7×7) surface by atomic force microscopy. Science **267**, 68–71 (1995). https://www.science.org/doi/abs/10.1126/science.267.5194.68
37. E. Synge, XXXVIII. A suggested method for extending microscopic resolution into the ultramicroscopic region. Lond. Edinb. Dublin Philos. Mag. J. Sci. **6**, 356–362 (1928). https://doi.org/10.1080/14786440808564615
38. E. Betzig, J.K. Trautman, Near-field optics: microscopy, spectroscopy, and surface modification beyond the diffraction limit. Science **257**, 189–195 (1992). https://www.science.org/doi/abs/10.1126/science.257.5067.189
39. F. Keilmann, R. Hillenbrand, Near-field microscopy by elastic light scattering from a tip. Philos. Trans. R. Soc.Lond. Ser. A Math. Phys. Eng. Sci. **362**, 787–805 (2004). https://royalsocietypublishing.org/doi/abs/10.1098/rsta.2003.1347
40. S. Hunsche, M. Koch, I. Brener, M. Nuss, THz near-field imaging. Optics Commun. **150**, 22–26 (1998). https://www.sciencedirect.com/science/article/pii/S0030401898000443
41. J. Wessel, Surface-enhanced optical microscopy. J. Opt. Soc. Am. B **2**, 1538–1541 (1985). https://opg.optica.org/josab/abstract.cfm?URI=josab-2-9-1538
42. F. Zenhausern, M.P. O'Boyle, H.K. Wickramasinghe, Apertureless near-field optical microscope. Appl. Phys. Lett. **65**, 1623–1625 (1994). https://pubs.aip.org/aip/apl/article/65/13/1623/523284/Apertureless-near-field-optical-microscope
43. Y. Inouye, S. Kawata, Near-field scanning optical microscope with a metallic probe tip. Opt. Lett. **19**, 159–161 (1994). https://opg.optica.org/ol/abstract.cfm?URI=ol-19-3-159
44. L. Novotny, B. Hecht, *Principles of Nano-optics* (Cambridge University Press, 2006)
45. R. Hillenbrand, F. Keilmann, Material-specific mapping of metal/semiconductor/dielectric nanosystems at 10 nm resolution by backscattering near-field optical microscopy. Appl. Phys. Lett. **80**, 25–27 (2002). https://pubs.aip.org/aip/apl/article/80/1/25/515098/Material-specific-mapping-of-metal-semiconductor
46. T. Taubner, R. Hillenbrand, F. Keilmann, Performance of visible and mid-infrared scattering-type near-field optical microscopes. J. Microscopy **210**, 311–314 (2003). https://onlinelibrary.wiley.com/doi/abs/10.1046/j.1365-2818.2003.01164.x
47. M.B. Raschke et al., Apertureless near-field vibrational imaging of block-copolymer nanostructures with ultrahigh spatial resolution. ChemPhysChem **6**, 2197–2203 (2005). https://chemistryeurope.onlinelibrary.wiley.com/doi/abs/10.1002/cphc.200500218
48. F. Zenhausern, Y. Martin, H.K. Wickramasinghe, Scanning interferometric apertureless microscopy: optical imaging at 10 Angstrom resolution. Science **269**, 1083–1085 (1995). https://www.science.org/doi/abs/10.1126/science.269.5227.1083
49. R.M. Stöckle, Y.D. Suh, V. Deckert, R. Zenobi, Nanoscale chemical analysis by tip-enhanced Raman spectroscopy. Chem. Phys. Lett. **318**, 131–136 (2000). https://www.sciencedirect.com/science/article/pii/S0009261499014517
50. A.V. Zayats, V. Sandoghdar, Apertureless scanning near-field second-harmonic microscopy. Optics Commun. **178**, 245–249 (2000). https://www.sciencedirect.com/science/article/pii/S0030401800006556
51. I.I. Smolyaninov et al., Near-field second-harmonic microscopy of thin ferroelectric films. Optics Lett. **25**, 835–837 (2000). https://opg.optica.org/ol/abstract.cfm?URI=ol-25-11-835
52. E.J. Sánchez, L. Novotny, X.S. Xie, Near-field fluorescence microscopy based on two-photon excitation with metal tips. Phys. Rev. Lett. **82**, 4014–4017 (1999). https://link.aps.org/doi/10.1103/PhysRevLett.82.4014

53. T. Taubner, R. Hillenbrand, Nanoscale-resolved subsurface imaging by scatteringtype near-field optical microscopy. Optics Express **13**, 8893–8899 (2005). https://opg.optica.org/oe/abstract.cfm?URI=oe-13-22-8893
54. N. Anderson, P. Anger, A. Hartschuh, L. Novotny, Subsurface Raman imaging with nanoscale resolution. Nano Lett. **6**, 744–749 (2006). https://doi.org/10.1021/nl0600023
55. E. Wolf, M. Nieto-Vesperinas, Analyticity of the angular spectrum amplitude of scattered fields and some of its consequences. J. Opt. Soc. Am. A **2**, 886–890 (1985). https://opg.optica.org/josaa/abstract.cfm?URI=josaa-2-6-886
56. W.D. Newman et al., Observation of long-range dipole-dipole interactions in hyperbolic metamaterials. Sci. Adv. **4**, eaar5278 (2018). https://www.science.org/doi/abs/10.1126/sciadv.aar5278
57. B. Knoll, F. Keilmann, Enhanced dielectric contrast in scattering-type scanning near-field optical microscopy. Optics Commun. **182**, 321–328 (2000). https://www.sciencedirect.com/science/article/pii/S0030401800008269
58. F.L. Ruta, A.J. Sternbach, A.B. Dieng, A.S. McLeod, D.N. Basov, Quantitative nanoinfrared spectroscopy of anisotropic van der Waals materials. Nano Lett. **20**, 7933–7940 (2020). https://doi.org/10.1021/acs.nanolett.0c02671
59. A.J. Huber, F. Keilmann, J. Wittborn, J. Aizpurua, R. Hillenbrand, Terahertz near-field nanoscopy of mobile carriers in single semiconductor nanodevices. Nano Lett. **8**, 3766–3770 (2008). https://doi.org/10.1021/nl802086x
60. Z. Fei et al., Infrared nanoscopy of dirac plasmons at the graphene–SiO_2 interface. Nano Lett. **11**, 4701–4705 (2011). https://doi.org/10.1021/nl202362d
61. P.K. Gupta, S.C. Mehendale, Mid-infrared optically pumped molecular lasers. Hyperfine Interact. **37**, 243–273 (1987). https://doi.org/10.1007/BF02395713
62. J. Faist et al., Quantum cascade laser. Science **264**, 553–556 (1994). https://www.science.org/doi/abs/10.1126/science.264.5158.553
63. Y. Yao, A.J. Hoffman, C.F. Gmachl, Mid-infrared quantum cascade lasers. Nat. Photon. **6**, 432–439 (2012). https://doi.org/10.1038/nphoton.2012.143
64. R. Hillenbrand, B. Knoll, F. Keilmann, Pure optical contrast in scattering-type scanning near-field microscopy. J. Microsc. **202**, 77–83 (2001). https://onlinelibrary.wiley.com/doi/abs/10.1046/j.1365-2818.2001.00794.x
65. X. Chen et al., Modern scattering-type scanning near-field optical microscopy for advanced material research. Adv. Mater. **31**, 1804774 (2019). https://onlinelibrary.wiley.com/doi/abs/10.1002/adma.201804774
66. H.-G. Von Ribbeck, M.T. Wenzel, R. Jacob, L.M. Eng, Scattering near-field microscopy in the THz region using a free-electron laser in *35th International Conference on Infrared, Millimeter, and Terahertz Waves* (2010)
67. I.D. Barcelos et al., Probing polaritons in 2D materials with synchrotron infrared nanospectroscopy. Adv. Opt. Mater. **8**, 1901091 (2020). https://onlinelibrary.wiley.com/doi/abs/10.1002/adom.201901091
68. S. Mastel et al., Terahertz nanofocusing with cantilevered terahertz-resonant antenna tips. Nano Lett. **17**, 6526–6533 (2017). https://doi.org/10.1021/acs.nanolett.7b01924
69. L.D. Landáu, Y.M. Lífshits, *Electrodynamics of Continuous Media* (Pergamon Press/Addison-Wesley, Oxford/Reading, MA, 1960), p. 324. https://www.elsevier.com/books/electrodynamics-of-continuous-media/landau/978-0-08-057060-0
70. P. Ginzburg et al., Spontaneous emission in non-local materials. Light Sci. Appl. **6**, e16273–e16273 (2017). https://doi.org/10.1038/lsa.2016.273
71. M.B. Lundeberg et al., Tuning quantum nonlocal effects in graphene plasmonics. Science **357**, 187–191 (2017). https://www.science.org/doi/abs/10.1126/science.aan2735
72. R. Brown, N.R. Walet, F. Guinea, Edge modes and nonlocal conductance in graphene superlattices. Phys. Rev. Lett. **120**, 026802 (2018). https://link.aps.org/doi/10.1103/PhysRevLett.120.026802
73. S. Kawata, Y. Inouye, Scanning probe optical microscopy using a metallic probe tip. Ultramicroscopy **57**, 313–317 (1995). https://www.sciencedirect.com/science/article/pii/030439919400159K

74. M. Labardi, S. Patanè, & M. Allegrini, Artifact-free near-field optical imaging by apertureless microscopy. Appl. Phys. Lett. **77**, 621–623 (2000). https://pubs.aip.org/aip/apl/article/77/5/621/113442/Artifact-free-near-field-optical-imaging-by
75. R. Hillenbrand, F. Keilmann, Complex optical constants on a subwavelength scale. Phys. Rev. Lett. **85**, 3029–3032 (2000). https://link.aps.org/doi/10.1103/PhysRevLett.85.3029
76. N. Ocelic, A. Huber, R. Hillenbrand, Pseudoheterodyne detection for background-free nearfield spectroscopy. Appl. Phys. Lett. **89**, 101124 (2006). https://pubs.aip.org/aip/apl/article/89/10/101124/327139/Pseudoheterodyne-detection-for-background-free
77. A.J. Sternbach et al., Artifact free time resolved near-field spectroscopy. Optics Express **25**, 28589–28611 (2017). https://opg.optica.org/oe/abstract.cfm?URI=oe-25-23-28589
78. B. Deutsch, R. Hillenbrand, L. Novotny, Near-field amplitude and phase recovery using phase-shifting interferometry. Optics Express **16**, 494–501 (2008). https://opg.optica.org/oe/abstract.cfm?URI=oe-16-2-494
79. M. Schnell, P.S. Carney, R. Hillenbrand, Synthetic optical holography for rapid nanoimaging. Nat. Commun. **5**, 3499 (2014). https://doi.org/10.1038/ncomms4499
80. B. Deutsch, M. Schnell, R. Hillenbrand, P.S. Carney, Synthetic optical holography with nonlinear-phase reference. Optics Express **22**, 26621–26634 (2014). https://opg.optica.org/oe/abstract.cfm?URI=oe-22-22-26621
81. N. Ocelic, Quantitative near-field phonon-polariton spectroscopy (2007). https://mengkun.weebly.com/uploads/3/9/6/5/39652852/5._ocelic-phd_thesis-final.pdf
82. X.G. Xu, L. Gilburd, G.C. Walker, Phase stabilized homodyne of infrared scattering type scanning near-field optical microscopy. Appl. Phys. Lett. **105**, 263104 (2014). https://doi.org/10.1063/1.4905207
83. I. Stefanon et al., Heterodyne detection of guided waves using a scattering-type scanning near-field optical microscope. Optics Express **13**, 5553–5564 (2005). https://opg.optica.org/oe/abstract.cfm?URI=oe-13-14-5553
84. L. Gomez et al., Apertureless scanning near-field optical microscopy: a comparison between homodyne and heterodyne approaches. J. Opt. Soc. Am. B **23**, 823–833 (2006). https://opg.optica.org/josab/abstract.cfm?URI=josab-23-5-823
85. J.E. Hall et al., Heterodyne apertureless near-field scanning optical microscopy on periodic gold nanowells. Optics Express **15**, 4098–4105 (2007). https://opg.optica.org/oe/abstract.cfm?URI=oe-15-7-4098
86. D.E. Tranca, C. Stoichita, R. Hristu, S.G. Stanciu, G.A. Stanciu, A study on the image contrast of pseudo-heterodyned scattering scanning near-field optical microscopy. Optics Express **22**, 1687–1696 (2014). https://opg.optica.org/oe/abstract.cfm?URI=oe-22-2-1687
87. C. Moreno, J. Alda, E. Kinzel, G. Boreman, Phase imaging and detection in pseudoheterodyne scattering scanning near-field optical microscopy measurements. Appl. Optics **56**, 1037–1045 (2017). https://opg.optica.org/ao/abstract.cfm?URI=ao-56-4-1037
88. J. Chen et al., Optical nano-imaging of gate-tunable graphene plasmons. Nature **487**, 77 (2012). https://www.nature.com/articles/nature11254
89. Z. Fei et al., Gate-tuning of graphene plasmons revealed by infrared nano-imaging. Nature **487**, 82 (2012). https://www.nature.com/articles/nature11253
90. A.J. Giles et al., Ultralow-loss polaritons in isotopically pure boron nitride. Nat. Mater. **17**, 134–139 (2018). https://doi.org/10.1038/nmat5047
91. A.J. Huber, B. Deutsch, L. Novotny, R. Hillenbrand, Focusing of surface phonon polaritons. Appl. Phys. Lett. **92**, 203104 (2008). https://doi.org/10.1063/1.2930681
92. P. Alonso-González et al., Controlling graphene plasmons with resonant metal antennas and spatial conductivity patterns. Science **344**, 1369–1373 (2014). https://www.science.org/doi/abs/10.1126/science.1253202
93. S. Dai et al., Internal nanostructure diagnosis with hyperbolic phonon polaritons in hexagonal boron nitride. Nano Lett. **18**, 5205–5210 (2018). https://doi.org/10.1021/acs.nanolett.8b02162
94. B. Hauer et al., Exploiting phonon-resonant near-field interaction for the nanoscale investigation of extended defects. Adv. Funct. Mater. **30**, 1907357 (2020). https://onlinelibrary.wiley.com/doi/abs/10.1002/adfm.201907357

95. P. Alonso-González et al., Experimental verification of the spectral shift between near- and far-field peak intensities of plasmonic infrared nanoantennas. Phys. Rev. Lett. **110**, 203902 (2013). https://link.aps.org/doi/10.1103/PhysRevLett.110.203902
96. F. Huth et al., Nano-FTIR absorption spectroscopy of molecular fingerprints at 20 nm spatial resolution. Nano Lett. **12**, 3973–3978 (2012). https://doi.org/10.1021/nl301159v
97. T. Neuman et al., Mapping the near fields of plasmonic nanoantennas by scattering-type scanning near-field optical microscopy. Laser Photon. Rev. **9**, 637–649 (2015). https://onlinelibrary.wiley.com/doi/abs/10.1002/lpor.201500031
98. M. Schnell et al., Controlling the near-field oscillations of loaded plasmonic nanoantennas. Nat. Photon. **3**, 287–291 (2009). https://doi.org/10.1038/nphoton.2009.46
99. S. Dai et al., Subdiffractional focusing and guiding of polaritonic rays in a natural hyperbolic material. Nat. Commun. **6**, 6963 (2015). https://www.nature.com/articles/ncomms7963
100. P. Li et al., Hyperbolic phonon-polaritons in boron nitride for near-field optical imaging and focusing. Nat. Commun. **6**, 7507 (2015). https://doi.org/10.1038/ncomms8507
101. Z. Zheng et al., A mid-infrared biaxial hyperbolic van der Waals crystal. Sci. Adv. **5**, eaav8690 (2019). https://advances.sciencemag.org/content/5/5/eaav8690
102. S.A. Maier, *Plasmonics: Fundamentals and Applications* (Springer New York, New York, 2007). https://link.springer.com/book/10.1007/0-387-37825-1
103. G. Ford, W. Weber, Electromagnetic interactions of molecules with metal surfaces. Phys. Rep. **113**, 195–287 (1984). https://www.sciencedirect.com/science/article/pii/037015738490098X
104. N.C. Passler, A. Paarmann, Generalized 4 $\times$ 4 matrix formalism for light propagation in anisotropic stratified media: study of surface phonon polaritons in polar dielectric heterostructures. J. Opt. Soc. Am. B **34**, 2128–2139 (2017). https://opg.optica.org/josab/abstract.cfm?URI=josab-34-10-2128
105. R.Q. Yang, Infrared laser based on intersubband transitions in quantum wells. Superlattices Microstruct. **17**, 77–83 (1995). https://www.sciencedirect.com/science/article/pii/S0749603685710178
106. C.F. Bohren, D.R. Huffman, *Absorption and Scattering of Light by Small Particles* (Wiley and Sons, 2004). ISBN: 978-0-471-29340-8
107. R. Hillenbrand, T. Taubner, F. Keilmann, Phonon-enhanced light–matter interaction at the nanometre scale. Nature **418**, 159–162 (2002). https://www.nature.com/articles/nature00899#citeas
108. A. Cvitkovic, N. Ocelic, R. Hillenbrand, Analytical model for quantitative prediction of material contrasts in scattering-type near-field optical microscopy. Optics Express **15**, 8550–8565 (2007). https://opg.optica.org/oe/abstract.cfm?URI=oe-15-14-8550
109. R. Esteban, R. Vogelgesang, K. Kern, Full simulations of the apertureless scanning near field optical microscopy signal: achievable resolution and contrast. Optics Express **17**, 2518–2529 (2009). https://opg.optica.org/oe/abstract.cfm?URI=oe-17-4-2518
110. V.E. Babicheva, S. Gamage, M.I. Stockman, Y. Abate, Near-field edge fringes at sharp material boundaries. Optics Express **25**, 23935–23944 (2017). https://opg.optica.org/oe/abstract.cfm?URI=oe-25-20-23935
111. A.Y. Nikitin et al., Real-space mapping of tailored sheet and edge plasmons in graphene nanoresonators. Nat. Photon. **10**, 239–243 (2016). https://doi.org/10.1038/nphoton.2016.44
112. T. Wang et al., Phonon-polaritonic bowtie nanoantennas: controlling infrared thermal radiation at the nanoscale. ACS Photon. **4**, 1753–1760 (2017). https://doi.org/10.1021/acsphotonics.7b00321
113. A.J. Giles et al., Imaging of anomalous internal reflections of hyperbolic phonon-polaritons in hexagonal boron nitride. Nano Lett. **16**, 3858–3865 (2016). https://doi.org/10.1021/acs.nanolett.6b01341

Part II
Enabling Predictive Capabilities for Hyperbolic Polaritons in van der Waals Materials

As discussed at the end of the introductory Chaps. 1 and 2, the discovery of in-plane hyperbolic phonon polaritons in α-MoO_3 placed this vdW material at a privileged stage for studying and manipulating light at the nanoscale, beyond what had been achieved with other materials supporting PhPs. Yet, to accurately predict the IR response of α-MoO_3 and thus to enable predictive capabilities for advanced optical devices, it was imperative to develop an accurate dielectric function model, as well as to derive an analytical dispersion of polaritons in biaxial slabs. Both of them will be the subject of study of this part of the thesis. In Chap. 3 we analytically derive the general dispersion relation of electromagnetic modes in biaxial slabs, and we reduce it to a few limiting cases which are of great interest for the study of anisotropic polaritons in van der Waals slabs. In Chap. 4 we report on the accurate extraction of the permittivity of α-MoO_3 by combining far- and near-field studies. Building upon the predictive capabilities enabled in this part, advanced studies and devices exploiting the potential of strongly anisotropic vdW slabs will be developed in the subsequent chapters.

Chapter 3
Dispersion of Polaritons in Biaxial Slabs

Abstract In this chapter, organized in a tutorial style, we present a detailed derivation of the dispersion relation of electromagnetic modes in a biaxial slab. We show that our general dispersion relation successfully reduces to several known limiting cases of interest, such as uniaxial and isotropic slabs or semi-infinite crystals, among others. We manage to reduce the general dispersion relation to simple analytical expressions for short wavelength of the modes and small slab thicknesses, which are of great interest for the study of anisotropic polaritons in van der Waals (vdW) slabs. To demonstrate the validity of our analytical approximations, we compare them to full-wave simulations, finding an excellent agreement.

Most of the results reported in this chapter were developed in collaboration with Kirill Voronin (Moscow Institute of Physics and Technology, Russia), and published in "Analytical approximations for the dispersion of electromagnetic modes in slabs of biaxial crystals", by Gonzalo Álvarez-Pérez, Kirill V. Voronin et al., in Physical Review B 100, 235408 (2019).

When light enters an anisotropic medium, it splits into two waves that follow different paths and exhibit dissimilar wavelengths, polarizations and velocities. Among the simplest forms of anisotropic media are uniaxial materials, which are characterized by two different refractive indices and a single axis of asymmetry, known as the optic axis. In these materials, one of the waves has similar dispersion characteristics to those of light in free space, while the other wave deviates from this behavior. Consequently, these modes are commonly referred to as ordinary and extraordinary waves, respectively. This property of anisotropic media gives rise to exotic phenomena, like double refraction, making them highly valuable in a range of applications, including optics, mineralogy, and polarization-sensitive devices.

On the other hand, biaxial media have three distinct refractive indices and two distinct optic axes, thus earning their designation as biaxial. Unlike in uniaxial media, and as will be shown later in this chapter, in biaxial media none of the waves has the dispersion of light in free space, but in turn both of them display dissimilar velocities, directions, and polarization states while traveling through the crystal. It is due to this property that both waves are classified as extraordinary.

G. Álvarez Pérez, *Fundamentals of Nano-Optics in Hyperbolic van der Waals Materials*, Springer Theses, https://doi.org/10.1007/978-3-031-75767-9_3

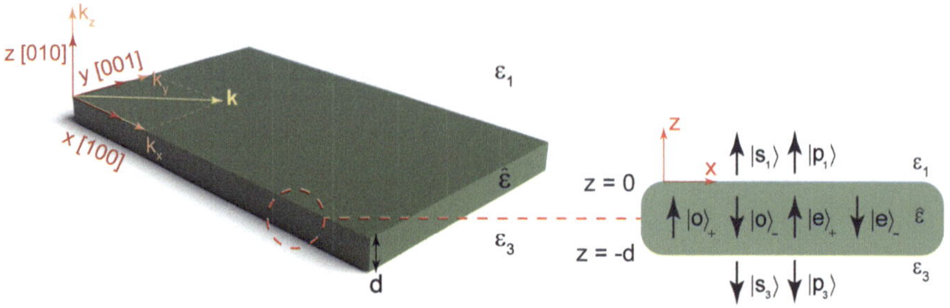

Fig. 3.1 Schematics of the biaxial slab. One of the main crystal axes [010] is perpendicular to the faces of the slab and coincides with the z axis, while the axes [100] and [001] belong to a plane parallel to the faces of the slab and are directed along the coordinate axes x and y, respectively. The mode propagates at an arbitrary angle φ with respect to the x axis

However, this distinctive characteristic of biaxial crystals comes at a cost: understanding and analyzing electromagnetic phenomena in biaxial media is considerably more intricate than in the case of uniaxial materials. In fact, biaxial media had received little attention in the nano-optics community until the discovery of the biaxial vdW material alpha-phase molybdenum trioxide as the first material supporting strongly in-plane anisotropic polaritons [1, 2]. A prior study had reported a rigorous analytical solution for the dispersion of surface waves on the boundaries of biaxial crystals [3]. Studies on the electromagnetic modes in biaxial slabs had been mainly the subject of a numerical analysis [4–6] or some particular configurations, such as grounded crystal slabs with modes' fixed propagation directions [7]. Yet, having an analytical dispersion for the electromagnetic eigenmodes in biaxial slabs is crucial for making accurate predictions of the polaritonic properties, such as the wavelength, propagation direction, decay length or lifetime. In this chapter [17], we provide the derivation of the dispersion relation governing the electromagnetic modes propagating in a biaxial slab.

To that end, we consider a biaxial slab of finite thickness and with arbitrary dielectric tensor $\hat{\varepsilon}$. The slab is surrounded by two semi-infinite isotropic media with arbitrary dielectric permittivities ε_1 and ε_3 (see Fig. 3.1a). We assume that one of the principal crystal axes is perpendicular to the faces of the slab, while the mode propagates along the crystal slab at an arbitrary angle φ with respect to the other principal axes, which lays in a plane parallel to the faces of the slab.

In order to understand how light propagates inside a biaxial slab, we first derive the electric fields inside an infinite biaxial crystal.

3.1 Electromagnetic Modes in Infinite Biaxial Crystals

Let us consider an infinite, nonmagnetic biaxial medium with dielectric permittivity tensor $\hat{\varepsilon}$. The coordinate system $\{x, y, z\}$ is chosen is such a way that $\hat{\varepsilon}$ is diagonal (see Fig. 3.1a), so that

$$\hat{\varepsilon} = \begin{pmatrix} \varepsilon_x & 0 & 0 \\ 0 & \varepsilon_y & 0 \\ 0 & 0 & \varepsilon_z \end{pmatrix}. \tag{3.1}$$

To accurately decompose the electromagnetic fields in the biaxial medium, we need to define appropriate basis vectors. To that end, we follow a standard procedure, (see for example Ref. [3]): we represent the electric and magnetic fields in the biaxial medium in the form of plane waves:

$$\mathbf{E} = E_0 \mathbf{e}\, e^{i\mathbf{kr}-i\omega t}, \qquad \mathbf{H} = H_0 \mathbf{h}\, e^{i\mathbf{kr}-i\omega t}, \tag{3.2}$$

where $\mathbf{e}$ and $\mathbf{h}$ are unknown dimensionless field basis vectors, E_0 and H_0 are arbitrary field amplitude coefficients, ω is the angular frequency, $\mathbf{k}$ is the wave vector, and $\mathbf{r}$ is the radius vector.

Now we consider Maxwell's equations:

$$\nabla \times \mathbf{E} = -\frac{1}{c}\frac{\partial \mathbf{H}}{\partial t}, \tag{3.3a}$$

$$\nabla \times \mathbf{H} = \frac{1}{c}\frac{\partial \mathbf{D}}{\partial t}. \tag{3.3b}$$

Taking the curl in Eq. (3.3a), then substituting Eq. (3.3b) into it, considering the time-dependence of the fields given in (3.2), and finally using the identity $\nabla^2 \mathbf{E} = \nabla(\nabla \cdot \mathbf{E}) - \nabla \times (\nabla \times \mathbf{E})$, we obtain a vectorial equation for the electric fields, $\mathbf{E}$:

$$\frac{\omega^2}{c^2}\hat{\varepsilon}\mathbf{E} = \nabla \cdot (\nabla \cdot \mathbf{E}) - \nabla^2 \mathbf{E}, \tag{3.4}$$

where ∇^2 is the Laplacian operator. Substituting $\mathbf{E}$ from Eq. (3.2) into Eq. (3.4), we obtain a linear homogeneous system of equations for the three components of the unknown basis vector, $\mathbf{e}$:

$$\mathcal{M}\,\mathbf{e} = \begin{pmatrix} \Delta_x & q_x q_y & \pm i q_x q_z \\ q_x q_y & \Delta_y & \pm i q_y q_z \\ \pm i q_x q_z & \pm i q_y q_z & \Delta_z \end{pmatrix} \begin{pmatrix} e_x \\ e_y \\ e_z \end{pmatrix} = 0, \tag{3.5}$$

where $k_0 = \omega/c$ is the free space wave vector, $q_{x,y} = k_{x,y}/k_0$ are the in-plane components of the normalized wave vector, and q_z is the out-of-plane component of the normalized wave vector, so that $k_z = \pm i q_z k_0$. The $+(-)$ sign must be taken for the wave propagating along (opposite to) the z-axis, while Δ_i are defined as

$$
\begin{aligned}
\Delta_x &= \varepsilon_x - q_y^2 + q_z^2, \\
\Delta_y &= \varepsilon_y - q_x^2 + q_z^2, \\
\Delta_z &= \varepsilon_z - q_x^2 - q_y^2.
\end{aligned}
\tag{3.6}
$$

Our choice of the dependence of the basis vectors upon the coordinates (the propagation along z is treated differently from the propagation along x and y) is dictated by the geometry of the problem (see Fig. 3.1), consistently with the standard waveguide theory.

The system (3.5) has nontrivial solutions only when det $(\mathcal{M}) = 0$, that gives the well-known Fresnel's equation for biaxial media [8, 9]:

$$
\begin{aligned}
&q_z^2 \left[q_z^2 \varepsilon_z + \varepsilon_z(\varepsilon_x + \varepsilon_y) - q_x^2(\varepsilon_x + \varepsilon_z) - q_y^2(\varepsilon_y + \varepsilon_z) \right] \\
&\quad + (\varepsilon_z - q_x^2 - q_y^2)(\varepsilon_x \varepsilon_y - q_x^2 \varepsilon_x - q_y^2 \varepsilon_y) = 0.
\end{aligned}
\tag{3.7}
$$

Equation (3.7) is a quadratic equation in terms of the squared z-component of the wave vector, q_z. Its solutions, q_{ez} and q_{oz}, read

$$
q_{o,ez}^2 = \frac{1}{2} \left\{ \frac{\varepsilon_x + \varepsilon_z}{\varepsilon_z} q_x^2 + \frac{\varepsilon_y + \varepsilon_z}{\varepsilon_z} q_y^2 - (\varepsilon_x + \varepsilon_y) \right\} \pm \frac{1}{2} \sqrt{D},
\tag{3.8}
$$

with D being the discriminant

$$
D = \left(\varepsilon_x - \varepsilon_y + \frac{\varepsilon_z - \varepsilon_x}{\varepsilon_z} q_x^2 - \frac{\varepsilon_z - \varepsilon_y}{\varepsilon_z} q_y^2 \right)^2 + 4 \frac{(\varepsilon_z - \varepsilon_x)(\varepsilon_z - \varepsilon_y)}{\varepsilon_z^2} q_x^2 q_y^2.
\tag{3.9}
$$

In Eq. (3.8), the sign "+" and "−" correspond to the labels "o", and "e", respectively.[1] Substituting Eq. (3.8) into the system (3.5), we find all three components of the two eigenvectors **e**. Since the system (3.5) is homogeneous, one of the components of the eigenvectors must be fixed. Without loss of generality, fixing the y-component to $e_y = q_x$ for the root "o" and to $e_y = q_y$ for the root "e", we find

$$
\mathbf{e}_o = \frac{1}{q} \begin{pmatrix} -q_y(1 - \Delta_1 \Delta_z) \\ q_x \\ \mp i q_x q_y q_{oz} \Delta_1 \end{pmatrix}, \qquad
\mathbf{e}_e = \frac{1}{q} \begin{pmatrix} q_x \frac{\Delta_2 - q_y^2}{\Delta_x^e} \\ q_y \\ \frac{\Delta_2}{\mp i q_{ez}} \end{pmatrix},
\tag{3.10}
$$

[1] The choice of "o" and "e" is not evident at this point, as both solutions are extraordinary in biaxial media. We will see in this chapter that "o" and "e" transform into the ordinary and extraordinary waves in uniaxial media, respectively.

where the factor $1/q$ (with q the normalized in-plane wave vector, $q^2 = \frac{k_x^2+k_y^2}{k_0^2}$) stands for the normalization, and

$$\Delta_1 = \frac{\Delta_x^o - q_x^2}{\Delta_z \Delta_x^o + q_x^2 q_{oz}^2}, \quad \Delta_2 = \frac{\Delta_x^e \Delta_y^e - q_x^2 q_y^2}{\Delta_x^e - q_x^2}. \tag{3.11}$$

From Eq. (3.7) we can easily find the asymptotes of the isofrequency curves for large $q_{x,y}$. Tending both q_x and q_y to infinity, and setting $q_z = 0$, we find

$$\frac{q_x}{q_y} = \sqrt{-\frac{\varepsilon_y}{\varepsilon_x}}. \tag{3.12}$$

In thin vdW slabs these asymptotes yield the direction of the propagation of the polaritonic "rays", excited by localized sources [1, 10, 11].

In the particular case of a uniaxial crystal (with the axis C pointing parallel to the z-axis, $C \| Oz$), $\varepsilon_x = \varepsilon_y = \varepsilon_\perp$ and $\varepsilon_z = \varepsilon_\|$, the derived basis vectors (3.10) can be straightforwardly transformed into the basis vectors for the ordinary and extraordinary waves. Taking into account that the z components of the wave vectors (3.8) are reduced to the well-known expressions for the ordinary and extraordinary waves

$$q_{oz}^2 = q^2 - \varepsilon_\perp, \quad q_{ez}^2 = \frac{\varepsilon_\perp}{\varepsilon_\|} q^2 - \varepsilon_\perp, \tag{3.13}$$

we find that $\Delta_1 = 0$, $\Delta_2 = \varepsilon_\perp + q_{ez}^2$ and obtain the basis vectors

$$\mathbf{e}_o = \frac{1}{q} \begin{pmatrix} -q_y \\ q_x \end{pmatrix}, \qquad \mathbf{e}_e = \frac{1}{q} \begin{pmatrix} q_x \\ q_y \\ \frac{\varepsilon_\perp + q_{ez}^2}{\mp i q_{ez}} \end{pmatrix}. \tag{3.14}$$

In case of an isotropic medium, $\varepsilon_\perp = \varepsilon_\| = \varepsilon$, the z components of the wave vectors degenerate $q_{oz}^2 = q_{ez}^2 = q_z^2 = q^2 - \varepsilon$ and the basis vectors (3.10) reduce to the ones for the s- and p-polarized waves ($\mathbf{e}_o \to \mathbf{e}_s$ and $\mathbf{e}_e \to \mathbf{e}_p$):

$$\mathbf{e}_s = \frac{1}{q} \begin{pmatrix} -q_y \\ q_x \\ 0 \end{pmatrix}, \qquad \mathbf{e}_p = \frac{1}{q} \begin{pmatrix} q_x \\ q_y \\ \frac{q^2}{\mp i q_z} \end{pmatrix}. \tag{3.15}$$

3.2 Electromagnetic Modes in a Biaxial Slab of Finite Thickness

Here, we derive the dispersion relation for polaritons in a biaxial slab of thickness d and permittivity $\hat{\varepsilon}$, occupying the region $0 > z > -d$ between two dielectric half-spaces with permittivities ε_1 (region "1", $z > 0$) and ε_3 (region "3", $z < -d$).

General Form of the Dispersion Relation of Electromagnetic Modes in a Biaxial Slab of Finite Thickness

Let us first represent the electric fields above ($z > 0$) and below ($z < -d$) the slab, in the isotropic media "1" and "3", respectively. In these regions we can describe the fields in the form of the s- and p-polarized plane waves. For compactness, from now on, we will use Dirac notation, in which the s- and p-polarization basis vectors read

$$|s_{1,3}\rangle_{\pm} = \frac{1}{q}\begin{pmatrix} -q_y \\ q_x \\ 0 \end{pmatrix} e^{ik_x x + ik_y y}, \qquad |p_{1,3}\rangle_{\pm} = \frac{1}{q}\begin{pmatrix} q_x \\ q_y \\ \frac{q^2}{\mp i q_{1,3z}} \end{pmatrix} e^{ik_x x + ik_y y}, \tag{3.16}$$

where $q_{1,3z} = \sqrt{q_x^2 + q_y^2 - \varepsilon_{1,3}} > 0$ is the out-of-plane component of the normalized wave vector. Here and in the definition of the p-polarization basis vector, $|p_{1,3}\rangle_{\pm}$, the $+(-)$ sign should be taken for the wave propagating along (opposite to) the z-axis, while in case of the s-polarization $|s_{1,3}\rangle_{+}$ and $|s_{1,3}\rangle_{-}$ are degenerated. For convenience, we also introduce the in-plane subvectors of the vectors given by Eq. (3.16):

$$|s\rangle = \frac{1}{q}\begin{pmatrix} -q_y \\ q_x \end{pmatrix} e^{ik_x x + ik_y y}, \qquad |p\rangle = \frac{1}{q}\begin{pmatrix} q_x \\ q_y \end{pmatrix} e^{ik_x x + ik_y y}. \tag{3.17}$$

The fields of the mode propagating along the slab in the upper and lower media can be compactly written as the sum of the s- and p-polarized plane waves:

$$\begin{aligned} \mathbf{E}_1 = \mathbf{E}_1(x, y, z) &= \sum_{\beta=s,p} a_\beta^1 |\beta_1\rangle_+ e^{ik_z z}, \\ \mathbf{E}_3 = \mathbf{E}_3(x, y, z) &= \sum_{\beta=s,p} a_\beta^3 |\beta_3\rangle_- e^{-ik_z z}, \end{aligned} \tag{3.18}$$

with unknown amplitudes $a_\beta^{1,3}$.

In contrast, the electric fields inside the biaxial slab ($0 > z > -d$) should be represented with the help of the basis vectors found in Sect. 3.1 as

$$\mathbf{E}_2 = \mathbf{E}_2(x, y, z) = \sum_{\gamma=o,e} a_\gamma^{2\downarrow} |\gamma\rangle_+ e^{ik_{\gamma z}z} + a_\gamma^{2\uparrow} |\gamma\rangle_- e^{-ik_{\gamma z}z}, \tag{3.19}$$

where $|\gamma\rangle_\pm$ denotes $|o\rangle_\pm$ and $|e\rangle_\pm$, being the polarization basis vector in the biaxial slab

$$|o\rangle_\pm = \frac{1}{q}\begin{pmatrix} -q_y(1-\Delta_1\Delta_z) \\ q_x \\ \mp i q_x q_y q_{oz}\Delta_1 \end{pmatrix} e^{ik_x x + ik_y y},$$
$$|e\rangle_\pm = \frac{1}{q}\begin{pmatrix} q_x \frac{\Delta_2 - q_y^2}{\Delta_x^e} \\ q_y \\ \frac{\Delta_2}{\mp i q_{ez}} \end{pmatrix} e^{ik_x x + ik_y y}. \tag{3.20}$$

The factors $a_\gamma^{2\uparrow}$ and $a_\gamma^{2\downarrow}$ represent the unknown amplitudes of the plane waves travelling along and opposite to the z-axis, respectively. Analogously to isotropic regions, we can introduce the in-plane subvectors $|o\rangle$ and $|e\rangle$ of the vectors $|o\rangle_\pm$, $|e\rangle_\pm$, respectively. These subvectors can be compactly written as

$$|o\rangle = |s\rangle + \frac{q_y c_1}{q}|u_x\rangle, \qquad |e\rangle = |p\rangle + \frac{q_x c_2}{q}|u_x\rangle, \tag{3.21}$$

where $|u_x\rangle = (1, 0)^T$, and

$$c_1 = \Delta_1\Delta_z, \qquad c_2 = \left(\frac{\Delta_2 - q_y^2}{\Delta_x^e} - 1\right). \tag{3.22}$$

To find the magnetic fields we use Maxwell's equation, $\nabla \times \mathbf{E} = -\frac{1}{c}\frac{\partial \mathbf{H}}{\partial t}$. In case of a plane wave it simplifies to

$$\mathbf{H} = \mathbf{q}_i \times \mathbf{E}, \tag{3.23}$$

where $\mathbf{q}_i$ is the normalized wave vector of a corresponding plane wave. Then, we can apply the boundary conditions, which imply the continuity of the in-plane components of both electric and magnetic fields on the faces of the film (at $z = 0$ and at $-d$):

$$\begin{aligned} &\mathbf{E}_{1t}(z=0) = \mathbf{E}_{2t}(z=0), && \mathbf{H}_{1t}(z=0) = \mathbf{H}_{2t}(z=0), \\ &\mathbf{E}_{2t}(z=-d) = \mathbf{E}_{3t}(z=-d), && \mathbf{H}_{2t}(z=-d) = \mathbf{H}_{3t}(z=-d), \end{aligned} \tag{3.24}$$

where the subscript "t" in Eqs. (3.24) stands for the in-plane subvectors. According to Eq. (3.23), the in-plane subvectors of the magnetic field can be written as $\mathbf{H}_t = \mathbf{e}_z \times \mathbf{q}_i \times \mathbf{E}$.

Using the field representation (3.18), (3.20), we can rewrite the boundary condition at $z = 0$ in Eqs. (3.24) in a more explicit way:

$$\sum_{\beta=s,p} a_\beta^1 |\beta\rangle = \sum_{\gamma=e,o} a_\gamma^{2\downarrow} |\gamma\rangle + a_\gamma^{2\uparrow} |\gamma\rangle, \tag{3.25}$$

$$\mathbf{e}_z \times \mathbf{q}_{1+} \times \sum_{\beta=s,p} a_\beta^1 |\beta_1\rangle_+ = \mathbf{e}_z \times \sum_{\gamma=e,o} \left(a_\gamma^{2\downarrow} \mathbf{q}_{\gamma+} \times |\gamma\rangle_+ + a_\gamma^{2\uparrow} \mathbf{q}_{\gamma-} \times |\gamma\rangle_- \right), \tag{3.26}$$

where $\mathbf{q}_{1,3\pm} = \left(q_x, q_y, \pm i q_{1,3z}\right)^T$ and $\mathbf{q}_{\gamma\pm} = \left(q_x, q_y, \pm i q_{\gamma z}\right)^T$. To simplify Eq. (3.26), let us introduce auxiliary three-dimensional vectors, $|\beta_{1,3}\rangle'_\pm$ and $|\gamma\rangle'_\pm$:

$$\begin{aligned} |\beta_{1,3}\rangle'_\pm &= -\mathbf{e}_z \times \mathbf{q}_{1,3\pm} \times |\beta_{1,3}\rangle_\pm, \\ |\gamma\rangle'_\pm &= -\mathbf{e}_z \times \mathbf{q}_{\gamma\pm} \times |\gamma\rangle_\pm, \end{aligned} \tag{3.27}$$

where $\beta = s, p$ and $\gamma = o, e$. Calculating the vector products in Eq. (3.27) we obtain the following explicit relations for the in-plane two-dimensional (2D) subvectors $|s_{1,3}\rangle'$ and $|p_{1,3}\rangle'$:

$$|s_{1,3}\rangle' = -Y_s^{1,3} |s\rangle, \qquad |p_{1,3}\rangle' = -Y_p^{1,3} |p\rangle, \tag{3.28}$$

with $Y_\beta^{1,3}$ the admittances for the s- and p- polarized waves:

$$Y_s^1 = i q_{1z}, \qquad Y_p^1 = \frac{\varepsilon_1}{i q_{1z}}, \qquad Y_s^3 = -i q_{3z}, \qquad Y_p^3 = -\frac{\varepsilon_3}{i q_{3z}}. \tag{3.29}$$

Since according to Eq. (3.27), the z components of the three-dimensional vectors $|\gamma\rangle'_\pm$ are 0, we keep the same notation for their 2D in-plane subvectors: $|o\rangle'_\pm$ and $|e\rangle'_\pm$, which read explicitly as

$$|o\rangle'_\pm = \pm i q_{oz} \left[|s\rangle + q_y \Delta_1 |a\rangle \right], \qquad |e\rangle'_\pm = \frac{\Delta_2}{\pm i q_{ez}} |p\rangle \pm i q_{ez} |b\rangle, \tag{3.30}$$

with auxiliary vectors

$$|a\rangle = \frac{1}{q} \begin{pmatrix} \Delta_z + q_x^2 \\ q_x q_y \end{pmatrix}, \qquad |b\rangle = \frac{1}{q} \begin{pmatrix} q_x (c_2 + 1) \\ q_y \end{pmatrix}. \tag{3.31}$$

As a result, using definition (3.27) and Eqs. (3.28), we obtain a simple form of the Eq. (3.26):

$$\sum_{\beta=s,p} a_\beta^1 Y_\beta^1 |\beta\rangle = \sum_{\gamma=o,e} \left(a_\gamma^{2\downarrow} |\gamma\rangle'_+ + a_\gamma^{2\uparrow} |\gamma\rangle'_- \right). \tag{3.32}$$

If we multiply (3.25) and (3.32) by $\langle\beta|$ [here, only the exponential of the bra-vector should be complex conjugated, for example, $\langle s| = (-q_y \quad q_x)e^{-ik_x x - ik_y y}$] and taking into account that $\langle\beta|\beta'\rangle = \delta_{\beta,\beta'}$, we get the following system of equations corresponding to the boundary condition at the interface $z = 0$:

$$\begin{aligned} &\sum_{\gamma=o,e} \left(a_\gamma^{2\downarrow} \langle\beta|\gamma\rangle + a_\gamma^{2\uparrow} \langle\beta|\gamma\rangle \right) - a_\beta^1 = 0, \\ &\sum_{\gamma=o,e} \left(a_\gamma^{2\downarrow} \langle\beta|\gamma\rangle'_+ + a_\gamma^{2\uparrow} \langle\beta|\gamma\rangle'_- \right) - a_\beta^1 Y_\beta^1 = 0. \end{aligned} \tag{3.33}$$

Analogously, for the interface $z = -d$, with the help of the auxiliary vectors $|\gamma\rangle'_\pm$ we find

$$\begin{aligned} &\sum_{\gamma=o,e} \left(a_\gamma^{2\downarrow} \langle\beta|\gamma\rangle e^{q_{\gamma z} k_0 d} + a_\gamma^{2\uparrow} \langle\beta|\gamma\rangle e^{-q_{\gamma z} k_0 d} \right) - a_\beta^3 = 0, \\ &\sum_{\gamma=o,e} \left(a_\gamma^{2\downarrow} \langle\beta|\gamma\rangle'_+ e^{q_{\gamma z} k_0 d} + a_\gamma^{2\uparrow} \langle\beta|\gamma\rangle'_- e^{-q_{\gamma z} k_0 d} \right) - a_\beta^3 Y_\beta^3 = 0. \end{aligned} \tag{3.34}$$

Equations (3.33) and (3.34) form a system of eight linear equations with eight unknowns. By defining $\xi^{\gamma\downarrow} = e^{q_{\gamma z} k_0 d}$ and $\xi^{\gamma\uparrow} = e^{-q_{\gamma z} k_0 d}$ with $\gamma = o, e$ (for the waves propagating along and opposite to the z axis, respectively), we have:

$$\begin{pmatrix} -1 & 0 & \langle s|o\rangle & \langle s|o\rangle & \langle s|e\rangle & \langle s|e\rangle & 0 & 0 \\ 0 & -1 & \langle p|o\rangle & \langle p|o\rangle & \langle p|e\rangle & \langle p|e\rangle & 0 & 0 \\ -Y_s^1 & 0 & \langle s|o\rangle'_+ & \langle s|o\rangle'_- & \langle s|e\rangle'_+ & \langle s|e\rangle'_- & 0 & 0 \\ 0 & -Y_p^1 & \langle p|o\rangle'_+ & \langle p|o\rangle'_- & \langle p|e\rangle'_+ & \langle p|e\rangle'_- & 0 & 0 \\ 0 & 0 & \langle s|o\rangle\xi^{o\downarrow} & \langle s|o\rangle\xi^{o\uparrow} & \langle s|e\rangle\xi^{e\downarrow} & \langle s|e\rangle\xi^{e\uparrow} & -1 & 0 \\ 0 & 0 & \langle p|o\rangle\xi^{o\downarrow} & \langle p|o\rangle\xi^{o\uparrow} & \langle p|e\rangle\xi^{e\downarrow} & \langle p|e\rangle\xi^{e\uparrow} & 0 & -1 \\ 0 & 0 & \langle s|o\rangle'_+\xi^{o\downarrow} & \langle s|o\rangle'_-\xi^{o\uparrow} & \langle s|e\rangle'_+\xi^{e\downarrow} & \langle s|e\rangle'_-\xi^{e\uparrow} & -Y_s^3 & 0 \\ 0 & 0 & \langle p|o\rangle'_+\xi^{o\downarrow} & \langle p|o\rangle'_-\xi^{o\uparrow} & \langle p|e\rangle'_+\xi^{e\downarrow} & \langle p|e\rangle'_-\xi^{e\uparrow} & 0 & -Y_p^3 \end{pmatrix} \begin{pmatrix} a_s^1 \\ a_p^1 \\ a_o^{2\downarrow} \\ a_o^{2\uparrow} \\ a_e^{2\downarrow} \\ a_e^{2\uparrow} \\ a_s^3 \\ a_p^3 \end{pmatrix} = 0. \tag{3.35}$$

Using the explicit expressions for the vectors $|s\rangle$ and $|p\rangle$, $|o\rangle$ and $|e\rangle$, and $|o\rangle'_\pm$, $|e\rangle'_\pm$, given by Eqs. (3.17), (3.21), and (3.30), respectively, the scalar products in Eq. (3.35) can be explicitly calculated as

$$\begin{aligned}
\langle s|o\rangle &= \eta_1,\\
\langle p|o\rangle &= \eta_2,\\
\langle s|e\rangle &= \eta_3,\\
\langle p|e\rangle &= \eta_4,\\
\langle s|o\rangle'_\pm &= \pm i q_{oz}\eta_1,\\
\langle p|o\rangle'_\pm &= \pm i q_{oz}\eta_2\frac{\varepsilon_z}{\Delta_z},\\
\langle s|e\rangle'_\pm &= \pm i q_{ez}\eta_3,\\
\langle p|e\rangle'_\pm &= \frac{\Delta_2}{\pm i q_{ez}} \pm i q_{ez}\eta_4,
\end{aligned} \tag{3.36}$$

where $\eta_1 = 1 - \frac{c_1 q_y^2}{q^2}$, $\eta_2 = \frac{q_x q_y c_1}{q^2}$, $\eta_3 = \frac{-q_x q_y c_2}{q^2}$, and $\eta_4 = 1 + \frac{q_x^2 c_2}{q^2}$.

The homogeneous system (3.35) has non-trivial solutions only when its determinant is equal to zero. The zeros of the determinant yield the dispersion relation for the modes in the biaxial slab. In general, the dispersion relation can be analyzed numerically, but in the limit of a small slab thickness, $k_0 d \ll 1$, as well as in the short-wavelength limit (large values of q), it can be written in a compact analytical form, as will be shown below, in Sects. 3.4 and 3.5, respectively. Before considering these interesting limits, we will ensure that our dispersion relation analytically reproduces some known examples.

Electromagnetic Modes in a Uniaxial Slab of Finite Thickness

Consider a uniaxial crystal with the axis C pointing along the z-axis, $C\|Oz$, so that $\varepsilon_x = \varepsilon_y = \varepsilon_\perp$ and $\varepsilon_z = \varepsilon_\|$. In this case $\Delta_1 = 0$ and $\Delta_2 = q_{ez}^2 + \varepsilon_\perp$, yielding $c_1 = c_2 = 0$, $\eta_2 = \eta_3 = 0$ and $\eta_1 = \eta_4 = 1$. Then the scalar products given by Eq. (3.36) greatly simplify:

$$\begin{aligned}
\langle s|o\rangle &= \langle p|e\rangle = 1,\\
\langle p|o\rangle &= \langle s|e\rangle = \langle p|o\rangle'_\pm = \langle s|e\rangle'_\pm = 0,\\
\langle s|o\rangle'_\pm &= \pm i q_{oz},\\
\langle p|e\rangle'_\pm &= \frac{\Delta_2}{\pm i q_{ez}} \pm i q_{ez} = \frac{\varepsilon_\perp}{\pm i q_{ez}}.
\end{aligned} \tag{3.37}$$

Consequently, the system of equations (3.35) reduces to

$$\begin{pmatrix} -1 & 0 & 1 & 1 & 0 & 0 & 0 & 0 \\ -q_{1z} & 0 & q_{oz} & -q_{oz} & 0 & 0 & 0 & 0 \\ 0 & -1 & \xi^{o\downarrow} & \xi^{o\uparrow} & 0 & 0 & 0 & 0 \\ 0 & q_{3z} & q_{oz}\xi^{o\downarrow} & -q_{oz}\xi^{o\uparrow} & 0 & 0 & 0 & 0 \\ 0 & 0 & 0 & 0 & -1 & 0 & 1 & 1 \\ 0 & 0 & 0 & 0 & -\frac{\varepsilon_1}{q_{z1}} & 0 & \frac{\Delta_2}{q_{ez}} & -\frac{\Delta_2}{q_{ez}} \\ 0 & 0 & 0 & 0 & 0 & -1 & \xi^{e\downarrow} & \xi^{e\uparrow} \\ 0 & 0 & 0 & 0 & 0 & \frac{\varepsilon_3}{q_{3z}} & \frac{\Delta_2}{q_{ez}}\xi^{e\downarrow} & -\frac{\Delta_2}{q_{ez}}\xi^{e\uparrow} \end{pmatrix} \begin{pmatrix} a_s^1 \\ a_s^3 \\ a_o^{2\downarrow} \\ a_o^{2\uparrow} \\ a_p^1 \\ a_p^3 \\ a_e^{2\downarrow} \\ a_e^{2\uparrow} \end{pmatrix} = 0. \quad (3.38)$$

Vanishing of the determinant of the matrix in Eq. (3.38) results in two separate equations, yielding (after some straightforward algebra) the dispersion of the ordinary and extraordinary modes:

$$\text{ordinary:} \quad \tanh(q_{oz}k_0 d) = -\frac{q_{oz}(q_{1z} + q_{3z})}{q_{1z}q_{3z} + q_{oz}^2}, \quad (3.39)$$

$$\text{extraordinary:} \quad \tanh(q_{ez}k_0 d) = -\frac{q_{ez}\varepsilon_\perp(q_{1z}\varepsilon_3 + q_{3z}\varepsilon_1)}{q_{1z}q_{3z}\varepsilon_\perp^2 + q_{ez}^2\varepsilon_1\varepsilon_3}. \quad (3.40)$$

Electromagnetic Modes in an Isotropic Slab of Finite Thickness

The dispersion of the modes in an isotropic slab can be easily derived from the dispersion of the modes in the uniaxial slab, by setting $\varepsilon_\perp = \varepsilon_\parallel = \varepsilon_2$. By doing so, the z components of the wave vectors q_{oz}, q_{ez} degenerate to $q_{2z}^2 = q^2 - \varepsilon_2$ and Eqs. (3.39), (3.40) transform into the well-known expressions for electromagnetic TE and TM modes in a conventional slab waveguide, respectively:

$$\text{TE:} \quad \tanh(q_{2z}k_0 d) = -\frac{q_{2z}(q_{1z} + q_{3z})}{q_{1z}q_{3z} + q_{2z}^2}, \quad (3.41)$$

$$\text{TM:} \quad \tanh(q_{2z}k_0 d) = -\frac{q_{2z}\varepsilon_2(q_{1z}\varepsilon_3 + q_{3z}\varepsilon_1)}{q_{1z}q_{3z}\varepsilon_2^2 + q_{2z}^2\varepsilon_1\varepsilon_3}. \quad (3.42)$$

Note that Eq. (3.42) is the generalization of Eq. (1.1), introduced at the beginning of Chap. 1 to calculate the dispersion of light in a dielectric waveguide made of SiO_2.

3.3 Very Thick Slabs: Electromagnetic Modes at the Surface of a Biaxial Crystal

Let us consider now another extreme case, assuming that the thickness of the slab tends to infinity, $d \to \infty$. Then our dispersion relation should split into two independent dispersion relations describing surface modes at the interfaces between the biaxial crystal and two isotropic media with dielectric permittivities ε_1 and ε_3. To obtain these dispersion relations in a simple analytical form, we multiply the third and fifth columns of the determinant of the system (3.35) by $\xi^{o\uparrow}$ and $\xi^{e\uparrow}$, respectively. Then tending $d \to \infty$ in the determinant, and assuming that both q_{oz} and q_{ez} have a non-vanishing real part, we see that all the matrix elements proportional to $\xi^{o,e\uparrow}$ (third and fifth elements in the four first rows and fourth and sixth ones in the four last rows) vanish. As a result, the 8×8 determinant becomes a product of the two determinants 4×4, each of them describing the surface modes at the 1–2 ($z = 0$) and 2–3 ($z = -d$) interfaces. Without loss of generality, let us consider only one of these determinants 4×4, corresponding to the interface 1–2. Zeroing the determinant, we have

$$\begin{vmatrix} -1 & 0 & \langle s|o\rangle & \langle s|e\rangle \\ 0 & -1 & \langle p|o\rangle & \langle p|e\rangle \\ -Y_s^1 & 0 & \langle s|o'\rangle_- & \langle s|e'\rangle_- \\ 0 & -Y_p^1 & \langle p|o'\rangle_- & \langle p|e'\rangle_- \end{vmatrix} = 0. \tag{3.43}$$

Then, using the Gauss method, we can reduce the dimension of the matrix to 2×2, as

$$\begin{vmatrix} Y_s^1\langle s|o\rangle - \langle s|o'\rangle_- & Y_s^1\langle s|e\rangle - \langle s|e'\rangle_- \\ Y_p^1\langle p|o\rangle - \langle p|o'\rangle_- & Y_p^1\langle p|e\rangle - \langle p|e'\rangle_- \end{vmatrix} = 0. \tag{3.44}$$

To write the dispersion relation in a compact form, we express ε_z from Frensel's equation for biaxial slabs (3.7) as a function of q_{oz}:

$$\varepsilon_z = \frac{\varepsilon_x\varepsilon_y q^2 + \left(q_{oz}^2 - q^2\right)\left(\varepsilon_x q_x^2 + \varepsilon_y q_y^2\right)}{q_{oz}^2\left(\varepsilon_x + \varepsilon_y + q_{oz}^2 - q^2\right) + \varepsilon_x\varepsilon_y - \varepsilon_x q_x^2 - \varepsilon_y q_y^2}. \tag{3.45}$$

Then, using the identities $q^2 = \varepsilon_1 + q_{1z}^2$ and $q^2 = q_x^2 + q_y^2$, we substitute ε_z from Eq. (3.45) into the expressions for $Y_{s,p}^1$ in Eq. (3.29) and scalar products given by Eq. (3.36), appearing in the elements of the matrix in Eq. (3.44). After some algebraic operations, Eq. (3.44) reproduces the dispersion relation for surface waves on boundaries of biaxial crystals, derived in Ref. [3]:

$$
\begin{aligned}
(q_{1z}+q_{oz})(q_{1z}+q_{ez})\left(\varepsilon_x\varepsilon_y-\varepsilon_x q_x^2-\varepsilon_y q_y^2-\varepsilon_1 q_{oz}q_{ez}\right)\\
-q_{oz}q_{ez}(\varepsilon_1-\varepsilon_x)\left(\varepsilon_1-\varepsilon_y\right)=0.
\end{aligned}
\tag{3.46}
$$

Electromagnetic Modes at the Surface of a Uniaxial Crystal with the Axis Perpendicular to the Interface

Consider a uniaxial crystal with the axis C along the z-axis, thus directed perpendicularly to the interface of the crystal. Defining, as before, $\varepsilon_x=\varepsilon_y=\varepsilon_\perp$ and $\varepsilon_z=\varepsilon_\|$, and taking into account that according to Eq. (3.13), $\varepsilon_\perp = q^2-q_{oz}^2$, Eq. (3.46) simplifies as:

$$
q_{oz}(q_{1z}+q_{oz})(q_{1z}+q_{ez})(\varepsilon_\perp q_{oz}+\varepsilon_1 q_{ez})+q_{oz}q_{ez}(q_{1z}+q_{oz})^2(q_{1z}-q_{oz})^2=0.
\tag{3.47}
$$

Since $\mathrm{Re}(q_{oz})>0$ and $\mathrm{Re}(q_{1z})>0$, and therefore $q_{oz}(q_z+q_{oz})\neq 0$, we can divide Eq. (3.47) by $q_{oz}(q_z+q_{oz})$. Then it transforms into:

$$
q_{ez}\varepsilon_1(q_{oz}+q_{ez})+q_{1z}\varepsilon_\perp(q_{oz}+q_{ez})=0.
\tag{3.48}
$$

Assuming that $q_{ez}+q_{oz}\neq 0$, we obtain the dispersion relation for the surface wave at the boundary of a uniaxial crystal:

$$
\frac{q_{ez}}{\varepsilon_\perp}+\frac{q_{1z}}{\varepsilon_1}=0,
\tag{3.49}
$$

where $q_{1z}=\sqrt{q^2-\varepsilon_1}$ and $q_{ez}=\sqrt{\frac{\varepsilon_\perp}{\varepsilon_\|}q^2-\varepsilon_\perp}$. Deriving q from this equation, the dispersion relation takes the well-known form (see e.g. Ref. [12]):

$$
q=\sqrt{\varepsilon_1\varepsilon_\|\frac{\varepsilon_1-\varepsilon_\perp}{\varepsilon_1^2-\varepsilon_\|\varepsilon_\perp}}.
\tag{3.50}
$$

In the isotropic case, $\varepsilon_\|=\varepsilon_\perp=\varepsilon$, Eq. (3.49) simplifies to the dispersion relation for surface waves at the interface between two isotropic media:

$$
q=\sqrt{\frac{\varepsilon_1\varepsilon}{\varepsilon_1+\varepsilon}}.
\tag{3.51}
$$

Electromagnetic Modes at the Surface of a Uniaxial Crystal with In-Plane Crystal Axis

Consider a uniaxial crystal with the C axis along the y-axis, thus lying in the plane of the interface of the crystal. Redefining here $\varepsilon_x = \varepsilon_z = \varepsilon_\perp$ and $\varepsilon_y = \varepsilon_\parallel$, Eq. (3.8) for $q_{o,ez}$ transforms to (similar to Eq. (3.13)):

$$q_{oz}^2 = q^2 - \varepsilon_\perp, \qquad \varepsilon_\perp q_{ez}^2 = \varepsilon_\perp q_x^2 - \varepsilon_\parallel q_y^2 - \varepsilon_\perp \varepsilon_\parallel. \tag{3.52}$$

Substituting the expression for q_{ez} from Eq. (3.52) into Eq. (3.46) and dividing it by q_{ez}, we get the well-known dispersion relation for Dyakonov surface waves [13]:

$$(q_{1z} + q_{oz})(q_{1z} + q_{ez})(\varepsilon_1 q_{oz} + \varepsilon_\perp q_{ez}) + q_{oz}(\varepsilon_1 - \varepsilon_\perp)\left(\varepsilon_1 - \varepsilon_\parallel\right) = 0. \tag{3.53}$$

3.4 Electromagnetic Modes in the Limit of an Ultrathin Biaxial Slab

As discussed in Chap. 1, polaritons in ultra-thin slabs and monolayers (for instance, plasmon polaritons in graphene [14] or hyperbolic phonon polaritons in thin slabs of polar dielectrics, such as h-BN [15]) have recently attracted considerable attention due to their outstanding properties. Therefore, the limit of a vanishing slab thickness $d \to 0$ is of a great practical interest. Let us illustrate how our general dispersion relation, given by the determinant of the system (3.35), can be simplified for ultra-thin slabs. Analogously to the methodology used for isotropic slabs [16], we can approximate the slab of a finite thickness by a 2D conducting sheet with effective conductivity $\hat{\sigma} = \frac{\omega d \hat{\varepsilon}}{4\pi i}$. To that end, let us assume that all the components of the tensor $\hat{\varepsilon}$ are large, i.e. $|\varepsilon_i| \gg 1 (i = x, y, z)$.

Then, retaining in Eq. (3.8) the first non-vanishing terms depending upon q_x and q_y in the expressions for the normalized z components of the wave vectors, $q_{o,ez}$, Eq. (3.8) can be greatly simplified:

$$\begin{aligned} q_{o,ez}^2 = &\frac{1}{2}\left\{-(\varepsilon_x + \varepsilon_y) + \frac{\varepsilon_x + \varepsilon_z}{\varepsilon_z} q_x^2 + \frac{\varepsilon_y + \varepsilon_z}{\varepsilon_z} q_y^2\right\} \\ &\pm \frac{1}{2}\left\{\varepsilon_x - \varepsilon_y + \frac{\varepsilon_z - \varepsilon_x}{\varepsilon_z} q_x^2 - \frac{\varepsilon_z - \varepsilon_y}{\varepsilon_z} q_y^2\right\}, \end{aligned} \tag{3.54}$$

where "+" and "−" should be taken for the labels "o", and "e", respectively. Equations (3.54) then further simplify to

$$\begin{aligned} q_{oz}^2 &= -\varepsilon_y + q_x^2 + \frac{\varepsilon_y}{\varepsilon_z} q_y^2, \\ q_{ez}^2 &= -\varepsilon_x + q_y^2 + \frac{\varepsilon_x}{\varepsilon_z} q_x^2. \end{aligned} \tag{3.55}$$

Using Eq. (3.55), we find

$$c_1 = -\frac{\varepsilon_z}{q_y^2}, \qquad c_2 = \frac{\varepsilon_z}{q_x^2} \frac{(\varepsilon_x - \varepsilon_y)}{(\varepsilon_x - \varepsilon_z)}, \tag{3.56}$$

so that the scalar products (3.36) can be written as

$$\begin{aligned}
\langle s|o\rangle &= \frac{q_x^2}{q^2}, \\
\langle p|o\rangle &= \frac{q_x q_y}{q^2}, \\
\langle s|e\rangle &= \varepsilon_z \frac{\varepsilon_x - \varepsilon_y}{\varepsilon_x - \varepsilon_z} \frac{q_y}{q_x q^2}, \\
\langle p|e\rangle &= -\varepsilon_z \frac{\varepsilon_x - \varepsilon_y}{\varepsilon_x - \varepsilon_z} \frac{1}{q^2}, \\
\langle s|o\rangle'_\pm &= \pm i q_{oz} \langle s|o\rangle, \\
\langle p|o\rangle'_\pm &= \pm i q_{oz} \langle p|o\rangle, \\
\langle s|e\rangle'_\pm &= \pm i q_{ez} \langle s|e\rangle, \\
\langle p|e\rangle'_\pm &= \pm i q_{ez} \langle p|e\rangle.
\end{aligned} \tag{3.57}$$

Additionally, assuming the small thickness of the slab, we can simplify the elements of the matrix in Eq. (3.35) by expanding the exponentials ξ^γ into a Taylor series in $k_0 d$ and retaining the first non-vanishing terms. We have $\xi^{\gamma\downarrow} = e^{q_{\gamma z} k_0 d} = 1 + q_{\gamma z} k_0 d$ and $\xi^{\gamma\uparrow} = e^{-q_{\gamma z} k_0 d} = 1 - q_{\gamma z} k_0 d$. To simplify the determinant of the system (3.35), we sum up its third and fifth columns with the fourth and sixth columns, respectively, and then subtract the fourth and sixth columns (both multiplied by a factor 1/2), from the third and fifth columns, respectively. Then using row operations in the Gauss method, we eliminate the two first and two last columns in the obtained determinant: we multiply the first (second) row by Y_s^1 (Y_p^1) and subtract it from the third (fourth) row and analogously for fifth (sixth) and seventh (eighth) rows. As a result, we get the following equation:

$$\begin{vmatrix}
i\langle s|o\rangle & -Y_s^1\langle s|o\rangle & i\langle s|e\rangle & -Y_s^1\langle s|e\rangle \\
i\langle p|o\rangle & -Y_p^1\langle p|o\rangle & i\langle p|e\rangle & -Y_p^1\langle p|e\rangle \\
-Y_s^3 k_0 d\langle s|o\rangle & \left(2\alpha_y + Y_s^1 - Y_s^3\right)\langle s|o\rangle & -Y_s^3 k_0 d\langle s|e\rangle & \left(2\alpha_x + Y_s^1 - Y_s^3\right)\langle s|e\rangle \\
-Y_p^3 k_0 d\langle p|o\rangle & \left(2\alpha_y + Y_p^1 - Y_p^3\right)\langle p|o\rangle & -Y_p^3 k_0 d\langle p|e\rangle & \left(2\alpha_x + Y_p^1 - Y_p^3\right)\langle p|e\rangle
\end{vmatrix} = 0, \tag{3.58}$$

where $\alpha_{x,y} = \frac{2\pi\sigma_{x,y}}{c} = \frac{k_0 d \varepsilon_{x,y}}{2i}$ are the normalized 2D effective conductivity components. Using the smallness of $k_0 d$, on the one hand, and the assumed large values of the components of the tensor $\hat{\varepsilon}$, on the other hand, the determinant (3.58) can be further simplified. Namely, a more detailed analysis (which we omit here) shows that the elements proportional to $k_0 d$ (the first and third elements of the third and fourth rows) yield the contribution of a higher order of smallness and thus can be neglected. As a result, the determinant (3.58) factorizes into a product of the two determinants of the sub-matrices 2×2:

$$\begin{vmatrix} \langle s|o\rangle & \langle s|e\rangle \\ \langle p|o\rangle & \langle p|e\rangle \end{vmatrix} \cdot \begin{vmatrix} \left(2\alpha_y + Y_s^1 - Y_s^3\right) q_x & \left(2\alpha_x + Y_s^1 - Y_s^3\right) q_y \\ \left(2\alpha_y + Y_p^1 - Y_p^3\right) q_y & -\left(2\alpha_x + Y_p^1 - Y_p^3\right) q_x \end{vmatrix} = 0. \tag{3.59}$$

Taking into account that the first determinant in Eq. (3.59) gives $-\frac{\varepsilon_z}{q^2}\frac{\varepsilon_x - \varepsilon_y}{\varepsilon_x - \varepsilon_z} \neq 0$, the dispersion relation is found by zeroing the second determinant:

$$\left\{\alpha_x q_y^2 + \alpha_y q_x^2 + \frac{q_x^2 + q_y^2}{2}\left(iq_{1z} + iq_{3z}\right)\right\} \left\{\alpha_x q_x^2 + \alpha_y q_y^2 + \frac{q_x^2 + q_y^2}{2} \cdot \left(\frac{\varepsilon_1}{iq_{1z}} + \frac{\varepsilon_3}{iq_{3z}}\right)\right\} = q_x^2 q_y^2 \left(\alpha_x - \alpha_y\right)^2. \tag{3.60}$$

This dispersion relation, written for biaxial slabs of a small but nonzero thickness (the effective conductivities $\alpha_{x,y}$ are thickness-dependent), has been used for the analysis of hyperbolic phonon polaritons in thin slabs of α-MoO_3 [1]. Nevertheless, to our knowledge, it had not been consistently derived for a nonvanishing slab thickness before the work described in this chapter was carried out and published in Ref. [17]. For 2D anisotropic sheets (of zero thickness) Eq. (3.60) is exact and had been reported in Refs. [18, 19]. As expected, the asymptotes of the dispersion relation (3.60) ($q_{x,y} \rightarrow \infty$) coincide with the ones, following from Fresnel's equation [compare with Eq. (3.12)]:

$$\frac{q_x}{q_y} = \sqrt{-\frac{\alpha_y}{\alpha_x}} = \sqrt{-\frac{\varepsilon_y}{\varepsilon_x}}. \tag{3.61}$$

In case of an isotropic 2D sheet $\alpha_x = \alpha_y = \alpha$ and Eq. (3.60) splits into two independent equations describing the dispersion of the TE and TM modes [20] propagating along the sheet:

$$\begin{aligned} \text{TE}: &\quad q_{1z} + q_{3z} - 2i\alpha = 0, \\ \text{TM}: &\quad \frac{\varepsilon_1}{q_{1z}} + \frac{\varepsilon_3}{q_{3z}} + 2i\alpha = 0. \end{aligned} \tag{3.62}$$

To demonstrate the validity of the simplified dispersion relation (3.60), we compare in Fig. 3.3 the refractive indices of a mode found from (3.60) (solid curves) to those found from full-wave simulations (points). Figure 3.3a and b represents the result for two different illustrative sets of parameters: in Fig. 3.3a only one of the in-plane dielectric permittivities is negative, $\varepsilon_x < 0$ (while $\varepsilon_y, \varepsilon_z > 0$), and in Fig. 3.3b both in-plane permittivities are negative, $\varepsilon_x, \varepsilon_y < 0$ (while $\varepsilon_z > 0$). Both in Fig. 3.3a and b the propagation of the mode at different angles φ (see Fig. 3.2) is considered. As expected, the agreement between the analytical approximation and rigorous numeric simulations improves for smaller values of $k_0 d$, although in Fig. 3.3b the agreement is good in the whole range of $k_0 d$ shown. Impressively, the agreement between the analytical and numerical results is in general excellent for all the shown propagation directions, even though neither $k_0 d$ is very small nor the values of ε_x, ε_y, and ε_z are very large, as it was initially assumed for the derivation of Eq. (3.60).

In this context, however, it is worth noting that non-local effects can sometimes play a significant role in ultrathin slabs, as discussed in [21]. These effects are not accounted for in the current analytic theory and numerical calculations, presenting an interesting avenue for further research.

Fig. 3.2 Schematics of the ultra-thin crystal slab

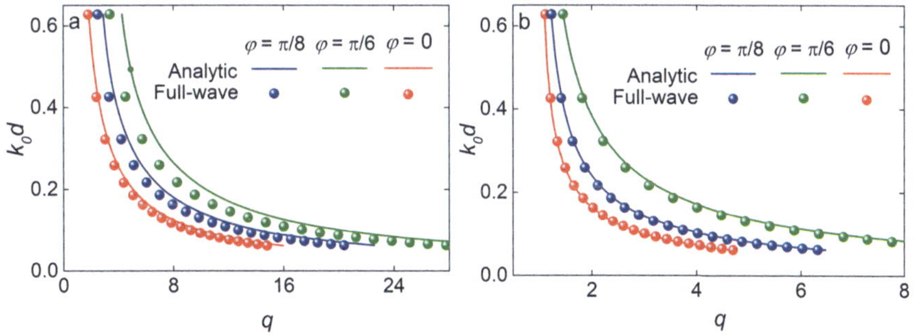

Fig. 3.3 Comparison of the dispersion relation given by Eq. (3.60) and by the full-wave numerical simulations. In both panels $\varepsilon_1 = \varepsilon_3 = 1$. Permittivity tensor components: (**a**) $\varepsilon_x = -2$, $\varepsilon_y = 2$, $\varepsilon_z = 2$. (**b**) $\varepsilon_x = -7$, $\varepsilon_y = -3$, $\varepsilon_z = 2$

3.5 Quasistatic Limit: Highly Confined Electromagnetic Modes in a Biaxial Slab

The general dispersion relation given by the zeroing of the determinant in Eq. (3.35) can be also greatly simplified under the quasistatic limit, valid for highly confined modes, $|q| \gg 1$. Such simplification is similar to the one made for the dispersion of the modes in uniaxial crystal slabs [22], as described in the introductory Chap. 1—see Eq. (1.16). For large q the expressions for the z components of the wave vectors inside the slab can be approximated as:

$$q_{oz}^2 = q^2 - \frac{\varepsilon_x \varepsilon_z q_y^2 + \varepsilon_y \varepsilon_z q_x^2 - \varepsilon_x \varepsilon_y q^2}{\varepsilon_z q^2 - \varepsilon_x q_x^2 - \varepsilon_y q_y^2}, \qquad q_{ez}^2 = \frac{\varepsilon_x}{\varepsilon_z} q_x^2 + \frac{\varepsilon_y}{\varepsilon_z} q_y^2, \tag{3.63}$$

where we have retained the second-order (the highest-order) term in q both in q_{ez}^2 and q_{oz}^2, as well as the zeroth-order term in the expression for q_{oz}^2, to avoid uncertainty in Δ_1 (since $\Delta_x^o = q_x^2$ and $\Delta_z = q_{oz}^2 = q^2$, we have $\Delta_1 = \frac{0}{0}$). Substituting the Eqs. (3.63) into Eqs. (3.11), (3.22), we have

$$\begin{aligned} \Delta_1 &= \frac{1}{q_{oz}^2} \frac{\varepsilon_x - \varepsilon_y}{\varepsilon_z - \varepsilon_x}, \qquad \Delta_2 = q_{ez}^2, \\ c_1 &= \frac{\varepsilon_x - \varepsilon_y}{\varepsilon_x - \varepsilon_z}, \qquad c_2 = \frac{\varepsilon_y - \varepsilon_x}{q_{ez}^2 - q_{oz}^2}, \end{aligned} \tag{3.64}$$

where we have neglected all small amendments in the expressions for each constant. Then we obtain the simplified expressions for the scalar products (3.36):

$$\begin{aligned} \langle s|o\rangle &= \frac{\varepsilon_z}{\varepsilon_x - \varepsilon_z} \frac{q_{ez}^2 - q_{oz}^2}{q_{oz}^2}, \\ \langle p|o\rangle &= \frac{q_x q_y}{q_{oz}^2} \frac{\varepsilon_x - \varepsilon_y}{\varepsilon_x - \varepsilon_z}, \\ \langle s|e\rangle &= \frac{q_x q_y}{q_{oz}^2} \frac{\varepsilon_x - \varepsilon_y}{q_{ez}^2 - q_{oz}^2}, \\ \langle p|e\rangle &= 1, \\ \langle s|o'\rangle_\pm &= \pm i q_{oz} \langle s|o\rangle, \\ \langle p|o'\rangle_\pm &= \pm i \varepsilon_z \frac{q_x q_y}{q_{oz}^3} \frac{\varepsilon_x - \varepsilon_y}{\varepsilon_z - \varepsilon_x}, \\ \langle s|e'\rangle_\pm &= \pm i q_{ez} \langle s|e\rangle, \\ \langle p|e'\rangle_\pm &= \mp i q_{ez} \frac{\varepsilon_z}{q_{oz}^2}. \end{aligned} \tag{3.65}$$

To simplify the matrix (3.35) and to eliminate the two first and two last columns, we use the same column and row operations as in Sect. 3.4 and obtain the following equation (containing the 4×4 matrix):

$$\begin{vmatrix} \langle s|o'\rangle_+ & -Y_s^1\langle s|o\rangle & \langle s|e'\rangle_+ & -Y_s^1\langle s|e\rangle \\ \langle p|o'\rangle_+ & -Y_p^1\langle p|o\rangle & \langle p|e'\rangle_+ & -Y_p^1\langle p|e\rangle \\ \langle s|o'\rangle_+ C_o - Y_s^3\langle s|o\rangle S_o & \langle s|o'\rangle_+ S_o - Y_s^3\langle s|o\rangle C_o & \langle s|e'\rangle_+ C_e - Y_s^3\langle s|e\rangle S_e & \langle s|e'\rangle_+ S_e - Y_s^3\langle s|e\rangle C_e \\ \langle p|o'\rangle_+ C_o - Y_p^3\langle p|o\rangle S_o & \langle p|o'\rangle_+ S_o - Y_p^3\langle p|o\rangle C_o & \langle p|e'\rangle_+ C_e - Y_p^3\langle p|e\rangle S_e & \langle p|e'\rangle_+ S_e - Y_p^3\langle p|e\rangle C_e \end{vmatrix} = 0, \tag{3.66}$$

where, for compactness, we have introduced abbreviated notations for the hyperbolic functions:

$$C_{o,e} = \cosh(q_{o,ez} k_0 d), \quad S_{o,e} = \sinh(q_{o,ez} k_0 d). \tag{3.67}$$

Let us notice that the matrix elements containing either $\langle s|o'\rangle_+$ or $Y_s^{1,3}\langle s|o\rangle$ are of order $\sim q$, while the other matrix elements are of order $\sim \frac{1}{q}$, thus being much smaller in magnitude. Consequently, we can neglect the third and fourth elements of the first and third rows (the contribution of these elements to the determinant is of the second and fourth order of smallness in $\frac{1}{q}$). As a result, the determinant in Eq. (3.66) factorizes into a product of two determinants of the sub-matrices 2×2, so that Eq. (3.66) splits into the two following equations:

$$\begin{vmatrix} \langle s|o'\rangle_+ & -Y_s^1\langle s|o\rangle \\ \langle s|o'\rangle_+ C_o - Y_s^3\langle s|o\rangle S_o & \langle s|o'\rangle_+ S_o - Y_s^3\langle s|o\rangle C_o \end{vmatrix} = 0,$$

$$\begin{vmatrix} \langle p|e'\rangle_+ & -Y_p^1\langle p|e\rangle \\ \langle p|e'\rangle_+ C_e - Y_p^3\langle p|e\rangle S_e & \langle p|e'\rangle_+ S_e - Y_p^3\langle p|e\rangle C_e \end{vmatrix} = 0. \tag{3.68}$$

Simplifying all the admittances as $Y_p^1 = \frac{\varepsilon_1}{q_{1z}} \approx \frac{\varepsilon_1}{q} \approx \frac{\varepsilon_1}{q_{oz}}$, $Y_p^3 = -\frac{\varepsilon_3}{q_{3z}} \approx -\frac{\varepsilon_3}{q} \approx -\frac{\varepsilon_3}{q_{oz}}$ and $Y_s^1 \approx -Y_s^3 \approx iq \approx iq_{oz}$, we can easily calculate both determinants (3.68). Zeroing the first determinant does not give any physically reasonable solutions since

$$-\frac{2\varepsilon_z^2\left(q_{ez}^2 - q_{oz}^2\right)^2 e^{q_{oz}k_0 d}}{\left(\varepsilon_x - \varepsilon_z\right)^2 q_{oz}^2} \neq 0. \tag{3.69}$$

Therefore, the dispersion relation follows from the vanishing of the second determinant in Eq. (3.68):

$$\tanh(q_{ez} k_0 d) = -\frac{(\varepsilon_1 + \varepsilon_3)\,\varepsilon_z q_{oz} q_{ez}}{\varepsilon_1 \varepsilon_3 q_{oz}^2 + \varepsilon_z^2 q_{ez}^2}. \tag{3.70}$$

To write Eq. (3.70) in a convenient form, let us define

$$\rho = i\sqrt{\frac{\varepsilon_z q^2}{\varepsilon_x q_x^2 + \varepsilon_y q_y^2}} = i\sqrt{\frac{\varepsilon_z}{\varepsilon_x \cos^2\varphi + \varepsilon_y \sin^2\varphi}}, \tag{3.71}$$

where φ is the angle between the x axis and the in-plane component of wave vector. Then using Eq. (3.63) (neglecting here the second term in q_{oz}^2), Eq. (3.70) can be written as

$$\tan\left(\frac{qk_0 d}{\rho}\right) = \frac{\rho\frac{\varepsilon_1+\varepsilon_3}{\varepsilon_z}}{1 - \frac{\rho^2\varepsilon_1\varepsilon_3}{\varepsilon_z^2}}. \tag{3.72}$$

Taking into account that $\arctan\left(\frac{x+y}{1-xy}\right) = \arctan(x) + \arctan(y)$, we get a simple expression for the normalized in-plane wave vector q of electromagnetic modes[2] in a biaxial slab in the short-wavelength limit, $q \gg 1$:

$$q = \frac{\rho}{k_0 d}\left[\arctan\left(\frac{\varepsilon_1\rho}{\varepsilon_z}\right) + \arctan\left(\frac{\varepsilon_3\rho}{\varepsilon_z}\right) + \pi l\right], \quad l = 0, 1, 2... \tag{3.73}$$

We can verify that Eq. (3.73) transforms into the dispersion of modes in a uniaxial slab (with the axis C along the z-axis), setting $\varepsilon_x = \varepsilon_y = \varepsilon_\perp$ and $\varepsilon_z = \varepsilon_\parallel$. Then defining $\psi = -\rho = -i\sqrt{\frac{\varepsilon_\parallel}{\varepsilon_\perp}}$, and taking into account that $\frac{\varepsilon_{1,3}\rho}{\varepsilon_\parallel} = \frac{\varepsilon_{1,3}}{\psi\varepsilon_\perp}$, we reproduce the dispersion relation, used for the analysis of hyperbolic phonon polaritons in h-BN uniaxial crystal slabs [22]:

$$q = -\frac{\psi}{k_0 d}\left[\arctan\left(\frac{\varepsilon_1}{\psi\varepsilon_\perp}\right) + \arctan\left(\frac{\varepsilon_3}{\psi\varepsilon_\perp}\right) + \pi l\right], \quad l = 0, 1, 2... \tag{3.74}$$

The same results can be straightforwardly derived from the exact Eq. (3.40), in the limit of large q. Eq. (3.73) also reduces to Eq. (3.74) when the propagation of the mode coincides either with the x axis (in this case we should set $\varepsilon_x = \varepsilon_\perp$ and $\varepsilon_z = \varepsilon_\parallel$) or with the y-axis (in this case we should set $\varepsilon_y = \varepsilon_\perp$ and $\varepsilon_z = \varepsilon_\parallel$). In the two latter particular cases, anisotropic polaritons in α-MoO_3 slabs were studied via Eq. (3.74) in Ref. [24].

[2] Note that Eq. (3.70) is the dispersion relation of purely TM modes with large momenta. To analytically calculate the dispersion relation of TE modes in biaxial layers, a different approximation must be used since, in general, the dispersion relation of polaritons in arbitrarily thick biaxial layers, given in Eq. (3.35), does not split into two independent equations separately describing the dispersion relations for the TE and TM modes. We show the existence of TE modes in thin biaxial layers and calculate their dispersion in Ref. [23].

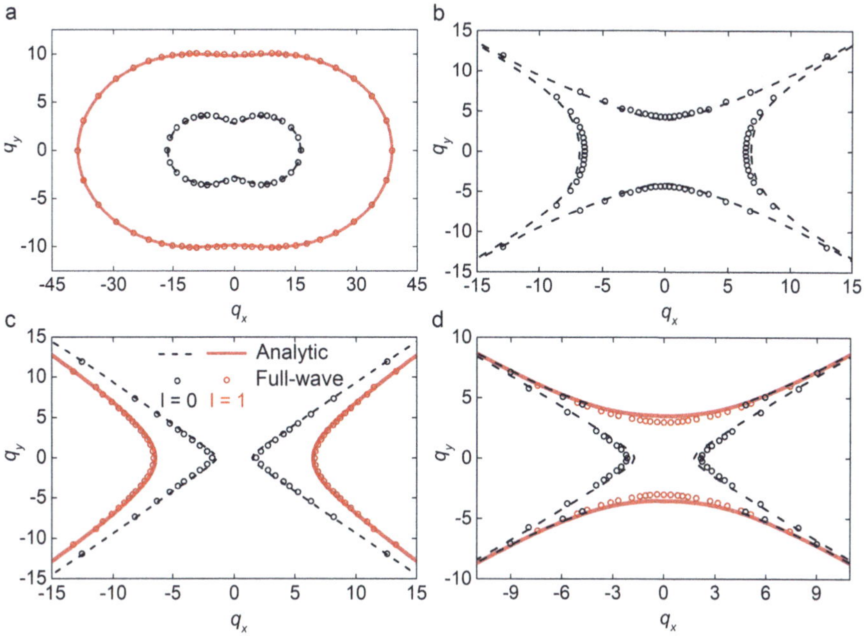

Fig. 3.4 Comparison between the isofrequency curves of the modes in a biaxial crystal slab. The isofrequency curves found according to Eq. (3.73) are plotted by the dashed black curves ($l = 0$) and solid red curves ($l = 1$), while those found from the full-wave simulations are plotted by the black and red dots, respectively. In all panels $\lambda = 1\,\mu$m, $d = 100$ nm and $\varepsilon_1 = \varepsilon_3 = 1$. Permittivity tensor components: (**a**) $\varepsilon_x = -0.1$, $\varepsilon_y = -1$, $\varepsilon_z = 2$. (**b**) $\varepsilon_x = -0.1$, $\varepsilon_y = -1$, $\varepsilon_z = -2$. (**c**) $\varepsilon_x = -2$, $\varepsilon_y = 2$, $\varepsilon_z = 2$. (**d**) $\varepsilon_x = -2$, $\varepsilon_y = 2$, $\varepsilon_z = -2$

To verify the validity of our analytical approximation, we compare the isofrequency curves obtained from Eq. (3.73) with those obtained from full-wave electromagnetic simulations. As an example, we take the slab thickness $d = 100$ nm and the free-space wavelength $\lambda = 1\,\mu$m. We consider the following four different combinations of the (purely real) permittivity tensor components:

$$\begin{aligned} &(\text{a})\ \varepsilon_x < 0,\ \ \varepsilon_y < 0,\ \ \varepsilon_z > 0,\\ &(\text{b})\ \varepsilon_x < 0,\ \ \varepsilon_y < 0,\ \ \varepsilon_z < 0,\\ &(\text{c})\ \varepsilon_x < 0,\ \ \varepsilon_y > 0,\ \ \varepsilon_z > 0,\\ &(\text{d})\ \varepsilon_x < 0,\ \ \varepsilon_y > 0,\ \ \varepsilon_z < 0. \end{aligned} \tag{3.75}$$

Figure 3.4 shows the isofrequency curves extracted from both Eq. (3.73) (red and black curves) and full-wave numeric simulations (dots). For the parametric sets (a), (c), (d) corresponding to the volume modes, the isofrequency curves for the two lowest modes are shown: $l = 0$ (black dashed curve and black dots) and $l = 1$ (red solid curve and red dots). In contrast, for the set of parameters (b), the mode exponentially decays inside and outside the slab (so that it has a surface wave character[3]) and therefore only the solution with $l = 0$ makes sense. In all panels of Fig. 3.4 we see an excellent agreement between the numeric simulations and analytical approximations for large q ($q \gtrsim 10$) and even very reasonable agreement for q comparable to 1. This agreement, particularly for the case of small and moderate values of q, unambiguously evidences that Eq. (3.73) can be used in a wide space of parameters for the characterization of diverse modes in both natural and artificial biaxial crystal slabs.

3.6 Conclusions

To summarize, we have presented an analytical derivation of the electromagnetic modes that can be guided along biaxial crystal slabs. We have provided simple expressions for the dispersion of the modes in the limit of an ultra-thin slab and for the case of large k-vectors of the modes. Both limits are currently of great importance for studying highly-confined anisotropic polaritons in vdW biaxial crystal slabs and, particularly, for the interpretation of the state-of-the-art near-field experiments. These will be shown all throughout this thesis, where the results developed here, specially the dispersion under the limit of large k-vectors, will be consistently employed.

References

1. W. Ma et al., In-plane anisotropic and ultra-low-loss polaritons in a natural van der Waals crystal. Nature **562**, 557 (2018). https://www.nature.com/articles/s41586-018-0618-9
2. Z. Zheng et al., A mid-infrared biaxial hyperbolic van der Waals crystal. Sci. Adv. **5**, eaav8690 (2019). https://advances.sciencemag.org/content/5/5/eaav8690
3. E.E. Narimanov, Dyakonov waves in biaxial anisotropic crystals. Phys. Rev. A **98**, 013818 (2018). https://link.aps.org/doi/10.1103/PhysRevA.98.013818
4. T. Maldonado, T. Gaylord, Hybrid guided modes in biaxial planar waveguides. J. Light. Technol. **14**, 486–499 (1996). http://ieeexplore.ieee.org/stamp/stamp.jsp?tp=&arnumber=485612&isnumber=10352
5. M.S. Kharusi, Uniaxial and biaxial anisotropy in thin-film optical waveguides. J. Opt. Soc. Am. **64**, 27–35 (1974). https://www.osapublishing.org/josa/abstract.cfm?uri=josa-64-1-27
6. N. Jovanovic, W. Papousek, A 2×2 matrix algebra for electromagneticwave propagation in stratified biaxial media. Int. J. Electron. Commun. **55**, 123–126 (2001). https://www.sciencedirect.com/science/article/pii/S1434841104700072
7. Z. Chen, Z. Shen, Surface waves propagating on grounded anisotropic dielectric slab. Appl. Sci. **8**, 102 (2018). https://www.mdpi.com/2076-3417/8/1/102

[3] In Ref. [23], we derive a rigorous criterion based on the out-of-plane momentum to classify electromagnetic modes in biaxial slabs into surface and volume-like.

8. L.D. Landáu, Y.M. Lífshits, *Electrodynamics of Continuous Media* (Pergamon Press/Addison-Wesley, Oxford/Reading, MA, 1960), p. 324. https://www.elsevier.com/books/electrodynamics-of-continuous-media/landau/978-0-08-057060-0
9. M. Born et al., *Principles of Optics: Electromagnetic Theory of Propagation, Interference and Diffraction of Light*, 7th edn. (Cambridge University Press, 1999)
10. S. Dai et al., Subdiffractional focusing and guiding of polaritonic rays in a natural hyperbolic material. Nat. Commun. **6**, 6963 (2015). https://www.nature.com/articles/ncomms7963
11. P. Li et al., Infrared hyperbolic metasurface based on nanostructured van der Waals materials. Science **359**, 892–896 (2018). https://science.sciencemag.org/content/359/6378/892.abstract
12. V. Agranovich, D. Mills, *Surface Polaritons – Electromagnetic Waves at Surfaces and Interfaces* (Elsevier Science, 1982). https://www.elsevier.com/books/surface-polaritons/agranovich/978-0-444-86165-8
13. M.I. D'yakonov, New type of electromagnetic wave propagating at an interface. J. Exp. Theor. Phys. **94**, 119–123 (1988). http://www.jetp.ac.ru/cgi-bin/dn/e_067_04_0714.pdf
14. P. Gonçalves, N. Peres, *An Introduction to Graphene Plasmonics* (World Scientific Publishing, 2016). https://www.worldscientific.com/worldscibooks/10.1142/9948
15. J.D. Caldwell et al., Photonics with hexagonal boron nitride. Nat. Rev. Mater. **4**, 552–567 (2019). https://www.nature.com/articles/s41578-019-0124-1
16. A.Y. Nikitin, *Graphene Plasmonics. World Scientific Handbook of Metamaterials and Plasmonics. World Scientific Series in Nanoscience and Nanotechnology* (Cambridge University Press, 2017), pp. 307–338. https://www.worldscientific.com/worldscibooks/10.1142/10642-vol4#t=toc
17. G. Álvarez-Pérez, K.V. Voronin, V.S. Volkov, P. Alonso-GonzáLlez, A.Y. Nikitin, Analytical approximations for the dispersion of electromagnetic modes in slabs of biaxial crystals. Phys. Rev. B **100**, 235408 (2019). https://link.aps.org/doi/10.1103/PhysRevB.100.235408
18. J.S. Gómez-Díaz, M. Tymchenko, A. Alù, Hyperbolic plasmons and topological transitions over uniaxial metasurfaces. Phys. Rev. Lett. **114**, 233901 (2015). https://link.aps.org/doi/10.1103/PhysRevLett.114.233901
19. O. Yermakov et al., Hybrid waves localized at hyperbolic metasurfaces. Phys. Rev. B **91**, 235423 (2015). https://journals.aps.org/prb/abstract/10.1103/PhysRevB.91.235423
20. G.W. Hanson, Dyadic Green's functions and guided surface waves for a surface conductivity model of graphene. J. Appl. Phys. **103**, 064302 (2008). https://aip.scitation.org/doi/abs/10.1063/1.2891452?journalCode=jap
21. C.R. Gubbin, De Liberato, S. Quantum theory of longitudinal-transverse polaritons in nonlocal thin films. Phys. Rev. Appl. **17**, 014037 (2022). https://link.aps.org/doi/10.1103/PhysRevApplied.17.014037
22. S. Dai et al., Tunable phonon polaritons in atomically thin van der Waals crystals of boron nitride. Science **343**, 1125–1129 (2014). http://science.sciencemag.org/content/343/6175/1125
23. K.V. Voronin, G. Álvarez-Pérez, C. Lanza, P. Alonso-González, A.Y. Nikitin, Fundamentals of polaritons in strongly anisotropic thin crystal layers. ACS Photonics **11**(2), 550–560 (2024). https://doi.org/10.1021/acsphotonics.3c01428
24. Z. Zheng et al., Phonon polaritons: highly confined and tunable hyperbolic phonon polaritons in van der Waals semiconducting transition metal oxides. Adv. Mater. **30**, 1870091 (2018). https://onlinelibrary.wiley.com/doi/abs/10.1002/adma.201870091

Chapter 4
Infrared Permittivity of α-MoO_3 from Near- and Far-Field Correlative Studies

Abstract In this chapter, the accurate mid-infrared (mid-IR) dielectric function of alpha-phase molybdenum trioxide (α-MoO_3) is reported by modeling far-field polarized mid-IR reflectance spectra acquired on a single thick flake of this material. The far-field model is refined by contrasting the experimental dispersion and damping of phonon polaritons (PhPs), revealed by polariton interferometry using scattering-type scanning near-field optical microscopy (s-SNOM) on thin flakes of α-MoO_3, with analytical and transfer-matrix calculations, as well as full-wave simulations. Through these correlative efforts, exceptional quantitative agreement is attained to both far- and near-field properties for multiple flakes, thus providing strong verification of the accuracy of this model, while offering a novel approach to extracting dielectric functions of nanomaterials. In addition, by employing density functional theory (DFT), insights into the various vibrational states dictating the dielectric function model and the intriguing optical properties of α-MoO_3 are provided.

The far-field experiments reported in this chapter were carried out in collaboration with Dr. Thomas G. Folland and Prof. Joshua D. Caldwell during a short stay of the candidate at Vanderbilt University, USA. The near-field experiments were performed in collaboration with Ana Isabel Fernández-Tresguerres Mata, while sample preparation was done by Javier Taboada Gutiérrez (University of Oviedo, Spain). DFT calculations were carried out by Prof. Ion Errea (University of the Basque Country, Spain). The work was published in "Infrared Permittivity of the Biaxial van der Waals Semiconductor α-MoO_3 from Near- and Far-Field Correlative Studies" by Gonzalo Álvarez-Pérez et al., in Advanced Materials 32 (29), 1908176 (2020). The final note on the far-IR dielectric function of α-MoO_3 was published in "Nanoscale-Confined Terahertz Polaritons in a van der Waals Crystal" by Thales V. A. G. de Oliveira et al., in Advanced Materials 33 (2), 2005777 (2021).

As mentioned in the introductory Chaps. 1 and 2, the discovery of in-plane HPhPs in α-MoO_3 in the mid-IR range in 2018 [1, 2] established this material as an excellent platform for both directly accessing exotic optical effects and controlling light at the nanoscale. However, to successfully interpret optical studies and design nano-

G. Álvarez Pérez, *Fundamentals of Nano-Optics in Hyperbolic van der Waals Materials*, Springer Theses, https://doi.org/10.1007/978-3-031-75767-9_4

optical devices based on α-MoO_3, it is crucial to being able to accurately predict its mid-IR optical response, which in turn heavily relies on the dielectric permittivity parameters. Hence, the accurate measurement of the mid-IR dielectric function model held paramount importance for exploiting the full potential of α-MoO_3 for mid-IR nanophotonics.

Prior work using α-MoO_3 had relied on a dielectric function that was estimated from previously reported optical phonon frequencies [2–4]. This approach (1) assumed the accuracy of those prior reports of the phonon frequencies, and (2) could not accurately extract the high- and low-frequency dielectric constants nor the phonon damping constants. These parameters are critical not only for building an accurate model, but also for providing insights into how the crystal lattice dictates the optical response of the material. In this chapter, we will show how model fitting of polarized far-field IR spectroscopy on single thick flakes of α-MoO_3 can be used to overcome these limitations, identifying the phonon frequencies, high- and low-frequency dielectric constants, and damping rates. Moreover, in an effort to refine this far-field approach and to compensate for any errors induced by the limited lateral scale of the flake ($150 \times 650\,\mu$m), which prevents conventional ellipsometry measurements, we will compare near-field measurements of the damping and the dispersion of PhPs propagating on a thin flake of α-MoO_3 to full-wave numerical simulations, analytical dispersions and transfer-matrix calculations that use the initially extracted dielectric function as input. This correlative approach yields exceptional quantitative agreement between the model and far- and near-field results through an iterative process, combining both correlated efforts and dramatically improving the fit. This perfect agreement strongly verifies the accuracy and robustness of the resulting dielectric function. Additionally, we will show how we can gain insights into the vibrational modes dictating the oscillators on which this dielectric function model is based and into the intriguing optical properties of the material by employing DFT to calculate the permittivity. This chapter [50] therefore reports an accurate and predictive model for the mid-IR dielectric function of α-MoO_3, an emerging vdW material for nanophotonics, while also offering an innovative approach to extracting dielectric functions of nanomaterials, where the use of traditional methods is challenging [5–7] or even not possible.

4.1 Far-Field Fourier Transform Infrared Micro-Spectroscopy of a Thick α-MoO_3 Flake

Traditionally, the dielectric function of a material is extracted using far-field polarized reflection and transmission or ellipsometry measurements. To extract the three principal values $\varepsilon_x(\omega)$, $\varepsilon_y(\omega)$, and $\varepsilon_z(\omega)$ of α-MoO_3 in the mid- and far-IR—between $\approx$9 and 23 μm (1111–435 cm^{-1})—we initially carry out Fourier transform infrared (FTIR) micro-spectroscopy (see Sect. 2.2 for more details on the technique). This technique can capture the optical response of relatively small single

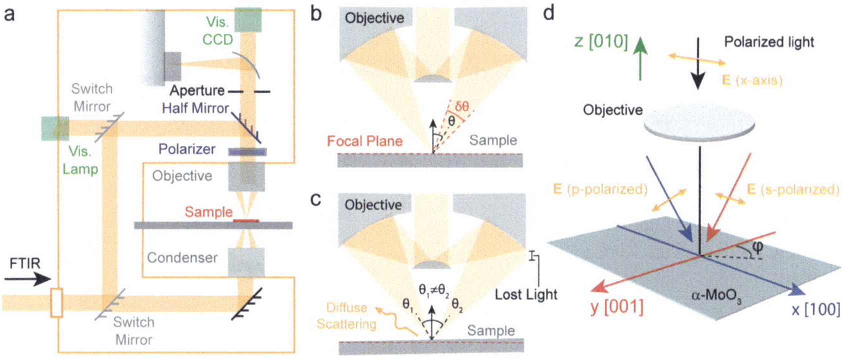

Fig. 4.1 Far-field FTIR reflection measurements of the optical properties of an α-MoO_3 crystal slab. (**a**) Schematics illustrating the setup for an FTIR microscope and beam path including (**b**) a Cassegrain objective with incidence angle θ and angular spread $\delta\theta$, accounted for in our fitting process. In addition to angular spread, if the sample is not flat or oriented parallel with the focal plane, as in (**c**), then some of the light will not be collected through specular reflection outside the numerical aperture of the lens, or diffuse scattering. This is accounted for by fitting a spectrally flat scaling factor for the spectrum, which is close to 1 in our model. In addition to angular spread and imperfections in the specular reflection signal, the Cassegrain objective also produces a mixed polarization state of incoming light, as shown in (**d**). In this schematic, light is prepared propagating in the z axis toward the objective, and is polarized along the x axis. However, when it enters the objective it is reflected and re-distributed along different azimuthal incidence angles φ. The effect this has on the polarization state can be understood by considering the light propagating in the x–z and y–z planes, whilst the electric field is maintained in the x-direction. The light propagating in the x–z plane being incident with p-polarization, and light propagating in the y–z plane with s polarization (to ensure both have the electric field still oriented along the x axis). This mixing of s and p polarized light is accounted for in our model, and we fit the total combination of s- and p-polarized light with a value close to 1:1 mixing of s and p

crystals across a broad spectral bandwidth, as required for accurate and unique determination of the dielectric function of α-MoO_3. The process for using FTIR micro-spectroscopy to determine the dielectric function involves taking accurate reflection and transmission spectra from a flake, and then using a Kramers-Krönig consistent model to fit to the experimental results. This model is typically chosen as it provides useful insights into the vibrational and/or electronic states of the material.

Our far-field measurements were undertaken using a Bruker Hyperion 2000 microscope coupled to a Bruker Vertex70v FTIR spectrometer, equipped with a traditional broadband MCT detector (400–8000 cm^{-1}), and a wide range FIR beam-splitter (30–6000 cm^{-1}) (see Fig. 4.1). Off-normal (×36 Cassegrain, 25° average incidence angle) polarized reflection and transmission incidence spectra were obtained. Both KRS5 and polyethylene wire grid polarizers were used in order to optimize the spectral throughput of the system at the relevant phonon frequencies. The spectra were collected with a 2 cm^{-1} spectral resolution and spatial resolution defined by the internal adjustable aperture of the microscope that was set to the size of the specific crystal of interest. All measurements were performed in reference to a gold film.

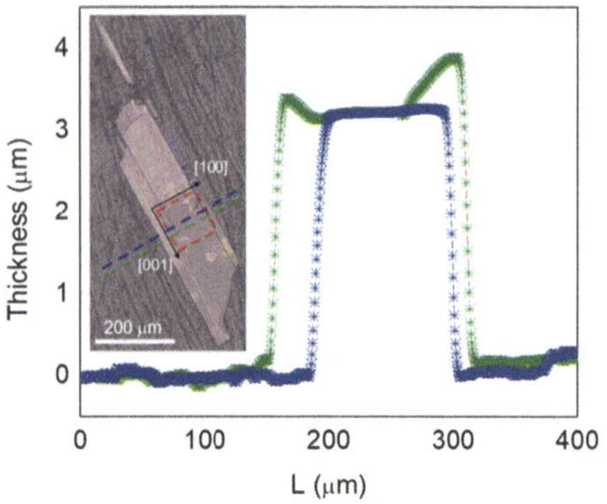

Fig. 4.2 Determination of the thickness of the α-MoO_3 flake used for FTIR spectroscopy. The measurement was done using a contact profilometer, along the lines depicted in blue and green in the inset. The red square shows the area in which the FTIR spectra were acquired

While in principle FTIR micro-spectroscopy can determine the optical properties of extremely small flakes—down to $10 \times 10\,\mu m$—, in practice this limits the spectral range, accuracy, and signal-to-noise ratio of the spectra, in comparison to measurements on large films, due to the diffraction limit of light. As α-MoO_3 is a biaxial crystal with optic phonons that occur in the mid-IR ($\lambda > 9.9$ μm), we acquired polarized FTIR reflection and transmission spectra from the same relatively large area of an α-MoO_3 single flake (red square in the inset of Fig. 4.2), to ensure that good quality data were obtained across the full spectral range of interest, with the polarizer aligned along both in-plane principal axes.

The α-MoO_3 flake in which the FTIR spectra were acquired had lateral dimensions of about 150 μm by 650 μm. It was exfoliated from bulk α-MoO_3 crystals and then transferred onto an AgCl substrate (see inset in Fig. 4.2), which is highly transparent across the entire spectral range of interest (see Sect. A.1 in Appendix A for the experimental extraction of the AgCl dielectric function). The thickness was determined using a contact profilometer, yielding a value of 3 ± 0.2 μm (Fig. 4.2). Even though profilometers can be very accurate, we have considered an error of 200 nm bearing in mind (1) the pressure exerted by the stylus onto the AgCl substrate, which is very soft, and (2) that the thickness is not perfectly constant in the area in which the FTIR spectra were acquired.

Reflectance Spectra of a Thick α-MoO_3 Flake As a result of the strong in-plane anisotropy in the optical response of α-MoO_3, the reflectance spectra along the two in-plane crystal axes, [100] and [001], are clearly different (Fig. 4.3). There are several high-reflectivity bands covering different frequency ranges, indicating the existence of RBs that, as commented previously, are originated by the polar nature of the bonds in the so-called polar crystals, which include α-MoO_3. In particular, as described in Chap. 1, these bands result from the breaking of the degeneracy between the longitudinal optical (LO) and transverse optical (TO) phonons, and the latter becoming IR active. Consequently, a strong absorption occurs at the TO phonon, and the real part of the frequency-dependent dielectric permittivity becomes negative within the RB, leading to the well-reported high reflectivity of polar materials within this spectral region [8]. Thus, we can use the reflectance data

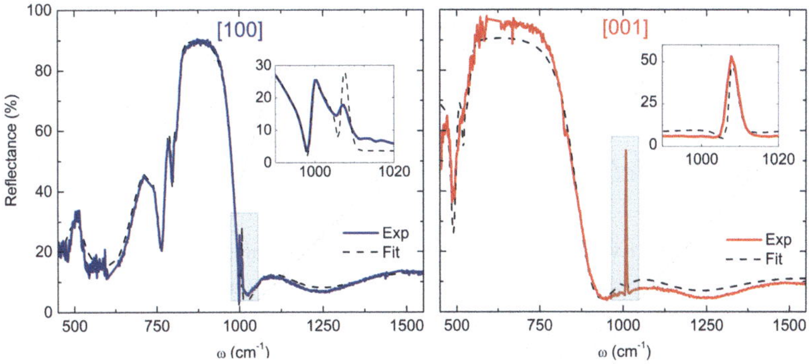

Fig. 4.3 Far-field FTIR reflectance spectra of an α-MoO$_3$ thick crystal slab. The polarization is aligned along the [100] direction, or x axis **(left)**, and along the [001] direction, or y axis **(right)**, shown in Fig. 4.2. Dashed lines represent the initial fit used to extract the α-MoO$_3$ dielectric function, then refined by near-field spectroscopy. The insets show the spectral region close to $1000\,\mathrm{cm}^{-1}$ for clarity

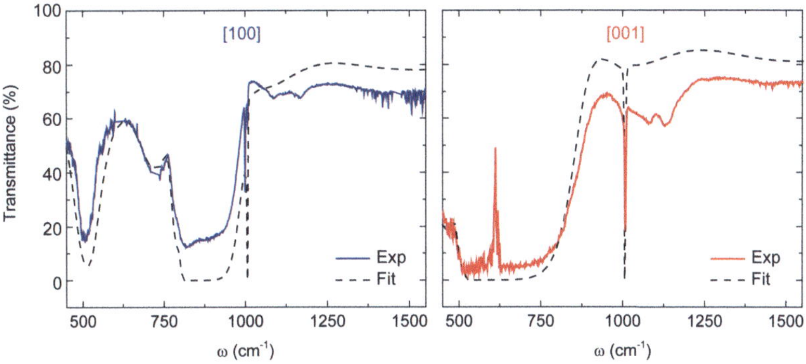

Fig. 4.4 Far-field FTIR transmission spectra of an α-MoO$_3$ thick crystal slab. The polarization is aligned along the [100] direction, or x axis **(left)**, and along the [001] direction, or y axis **(right)**, shown in Fig. 4.2. Dashed lines represent the initial fit used to extract the α-MoO$_3$ dielectric function, then refined by near-field spectroscopy

to approximate the position of the TO and LO phonons by the onset of the RB on the low- and high-frequency sides, respectively, providing the starting parameters for extraction of the IR dielectric function.

Transmission Spectra of a Thick α-MoO$_3$ Flake Transmission spectra were also taken through our sample of α-MoO$_3$, as shown in Fig. 4.4. Whilst the spectral features of the transmission spectra are generally well reproduced, we note that the amplitudes of the transmission spectra are significantly offset. This is largely due to the collection of IR light which is not transmitted through the crystal and prevents the use of these spectra for fitting. We also observe broad absorption bands between

1050 and $1150\,cm^{-1}$ which are not accounted for in our model. These are likely associated with weak multi-phonon processes in the α-MoO_3 flake.

Permittivity Model and Fitting

Although the permittivity of such an anisotropic crystal is a complicated function of frequency ω, in practice it can be approximated by model functions over the frequency range of interest. As shown in the introductory Chap. 1, a widespread choice for polar dielectrics is the well-known Lorentz (or TO–LO) model for coupled oscillators [9, 10]:

$$\varepsilon_i(\omega) = \varepsilon_i^{\infty} \prod_{j=1}^{N} \frac{\omega_{\mathrm{LO},ij}^2 - \omega^2 - i\gamma_{\mathrm{LO},ij}\omega}{\omega_{\mathrm{TO},ij}^2 - \omega^2 - i\gamma_{\mathrm{TO},ij}\omega}, \tag{4.1}$$

where $\varepsilon_i(\omega)$ denotes the i-th principal component of the permittivity tensor, and $i = x, y, z$ stand for the three principal axes of the crystal, which correspond to the crystalline directions of α-MoO_3, [100], [001], and [010], respectively. ε_i^{∞} represents the high-frequency dielectric constant, while $\omega_{\mathrm{LO},ij}$ and $\omega_{\mathrm{TO},ij}$ refer to the j-th LO and TO phonon frequencies, along the i-th direction. $\gamma_{\mathrm{TO},ij}$ represents the damping factor of the Lorentzian line shape that is derived from the phonon scattering rate and is inversely proportional to the phonon scattering lifetime. Finally, j is the subscript denoting the different phonon pairs along the same axis. According to the number of high-reflectivity bands in the reflectance spectra in Fig. 4.1, we consider three oscillators for $\varepsilon_x(\omega)$ and one oscillator for $\varepsilon_y(\omega)$ and $\varepsilon_z(\omega)$, respectively.

Fitting of Far-Field Experimental Data All dielectric function extraction was performed using the WVASE software (J.A. Woolam). The free parameters, ε_i^{∞}, $\omega_{\mathrm{LO},ij}$, $\omega_{\mathrm{TO},ij}$, and γ_{ij} in Eq. (4.1), are initially adjusted so that the simulated reflectance more accurately reproduces the experimental far-field FTIR spectra (Fig. 4.3). To do so, the system is modeled as a three-layer structure (air/α-MoO_3/AgCl).

Unlike in IR ellipsometry, there are a series of non-idealities which need to be accounted for when extracting the dielectric function from FTIR microscopy data [10]. First, we account for the incident angle and spread of the Cassegrain objective (25° weighted average, 10° spread). Second, light from a Cassegrain objective is distributed over the full 180° azimuthal incidence spread, which means that s- and p-polarized light must be mixed during the fitting process. Finally, due to non-uniformity of the crystal surface, not all incident light was collected, so a spectrally uniform linear scaling factor of 0.92 was used to change the percentage of collected light. Each of these three considerations was required to get an accurate fit of the model. Whilst ideally both reflection and transmission data would be considered, chromatic aberration induced by the dispersion of the substrate prevented the use of transmission data in our fitting procedure.

Due to the large dimensionality of the parameter space and the fit being multi-valued (see the following subsection), we performed ab initio calculations of the dielectric tensor using DFT. See Appendix B for the derivation of the DFT dielectric function of α-MoO_3 and middle panel in Fig. 4.12 for the result. Starting values were based on those from the ab initio model, and the fit was performed accounting for results along both crystal axes at the same time. We proceeded by varying each parameter systematically, and once sufficiently close a global fit was performed to obtain the fully optimized parameters.

Using this model and our experimental data, we can in principle extract all the parameters that define the IR dielectric function in Eq. (4.1), as previously employed for measuring the dielectric function of other anisotropic vdW materials, as hBN [11, 12] or quartz [13, 14].

However, this approach relies heavily upon reflection spectra collected at low angles, which is not very sensitive to absorption and thus damping, making the fitting procedure weakly sensitive to the phonon damping γ_{ij}. Furthermore, experimental non-idealities influence the fitted $\varepsilon_i(\omega)$ $(i = x, y, z)$ and LO phonon energy values, which are discussed in the next subsection.

Fit Sensitivity of α-MoO_3 Far-Field Data In this section we will assess fit *uniqueness* to demonstrate the limitations of the far-field model and, particularly, the limitations when it comes to phonon damping. A completely *unique* parameter can be fit without coupling to any other parameters in the model, i.e. its value will be determined by some unique feature in the spectrum (for example a peak frequency). Fit uniqueness studies were performed by fixing the parameter under study at several values close to the global minimum, and then fitting all other parameters. We can then examine how much the mean squared error (MSE) increases relative to the global minimum value. Rapid increases in the MSE to small variations in the parameter value indicate a *unique* and well fitted parameter, and vice versa. In this case we choose to analyze the parameters associated with the primary RB in the x direction: ε_i^{∞}, $\omega_{LO,ij}$, $\omega_{TO,ij}$, and γ_{ij} (results shown in Fig. 4.5).

The TO phonon clearly possesses the most unique fit value, with extremely strong increase in MSE away from the fitted frequency. The LO phonon frequency ω_{LO}, ε^{∞} and damping values γ show less dramatic variations in the MSE when compared to their global minimum value. In order to quantitatively compare all 4 parameters that fit our studies, we can use a figure of merit. We define the FOM as the change in the fit parameter when the MSE is increased by 1% of the global minimum, divided by the value at the global minimum. For ω_{TO} we get 0.4%, for the LO phonon we get 0.6%, for ε^{∞} we get 4%, and for γ we get 96%. This demonstrates how our far-field fit is extremely weakly sensitive to the phonon damping value. We also note that the ε^{∞} value is not as unique as the TO and LO phonon energies. This is because non-idealities introduced into the model (nominally the amount of reflection collected, as well as the precise thickness) have coupling for this parameter.

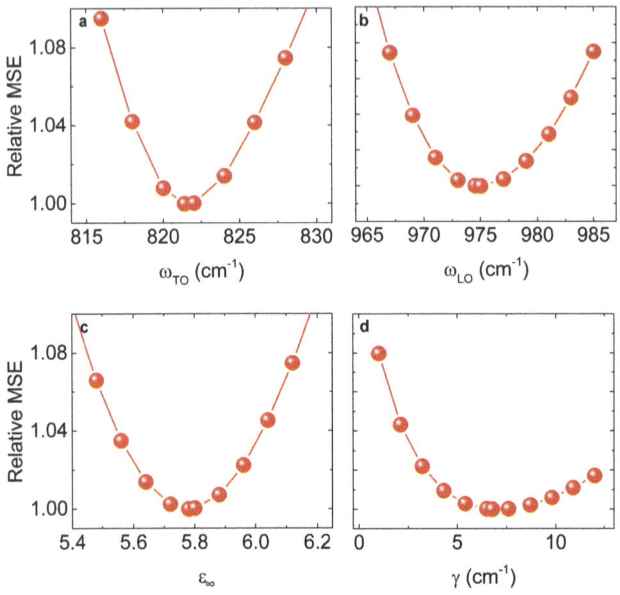

Fig. 4.5 Fit uniqueness for the (**a**) TO phonon energy (**b**) LO phonon energy (**c**) ε_∞ and (**d**) damping for the strong phonon band along the [100], or x, direction

While the fit uniqueness value suggests that TO and LO energy values should be accurately determined from far-field data, coupling between parameters can complicate matters. Here we consider the coupling between the ε^∞ value (which has dependence on experimental non-idealities) and the TO and LO phonon energies (Fig. 4.6). We see that the ε^∞ value couples to the LO phonon energy. This means that any error in the determination in ε^∞ introduced by the experimental non-idealities and any associated local minima will have a knock-on effect on the LO phonon energy. Most notably, any inaccuracy in the thickness of the flake, as well as the percentage of the first reflection collected, can influence the value of ε^∞.

4.2 Near-Field s-SNOM Infrared Nano-spectroscopy and Nano-imaging of a Thin α-MoO_3 Flake

Unlike FTIR, near-field polariton imaging by s-SNOM is extremely sensitive to both damping and LO phonon energies. As such, carrying out s-SNOM nanoimaging studies correlative to the FTIR spectroscopy allows us to jump out of local minima described in the previous subsection, and find the best and most realistic fit parameters. More specifically, near the LO phonon frequency, the PhP wavelength is highly sensitive to small variations in the permittivity, while the measured propagation length is dictated by damping. Therefore, by extracting the PhP wavelength and

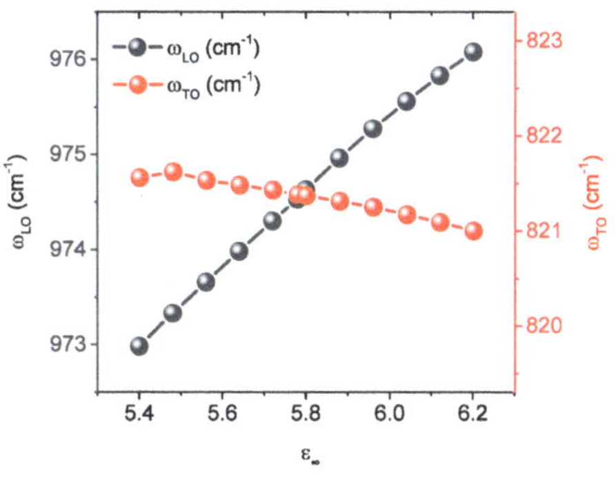

Fig. 4.6 Dependence between ε_∞ and the TO and LO phonon frequencies, $\omega_{\mathrm{LO},ij}$ and $\omega_{\mathrm{TO},ij}$ for $i = x$ and $j = 2$. The plot shows that the LO phonon frequency ω_{LO} is very sensitive to slight variations in the ε^∞ value, increasing for increasing values of ε^∞. The TO phonon frequency ω_{TO} is also dependent, although less, and it follows the opposite tendency: decreases for increasing values of ε^∞

propagation length from s-SNOM monochromatic measurements on thin flakes at multiple incident laser frequencies, these parameters can be determined with dramatically improved accuracy. Even so, s-SNOM is not highly sensitive to other parameters, such as the high-frequency permittivity ε^∞, and it is difficult to obtain measurements close to the TO phonon due to strong absorptive losses. Hence, our procedure for extracting the permittivity of α-MoO_3 is based on an iterative effort utilizing the two correlated measurements in the far- and near-field, respectively.

Near-Field Infrared Nano-imaging Using s-SNOM

IR nano-imaging was performed using a commercially available s-SNOM (from Neaspec) where a metallized cantilevered atomic force microscope (AFM) tip was used as a scattering near-field probe. The tip oscillates vertically at the mechanical resonant frequency (around 270 kHz) of the cantilever, with an amplitude of about 50 nm. The tip is side-illuminated with p-polarized IR light of frequency ω_0 (from tunable CO_2 and quantum cascade lasers) and electric field E_{inc}. As described in detail in Chap. 2, the Pt-coated tip acts as an IR antenna by concentrating the incident field into a nanoscale spot at its apex, which interacts with the sample surface and thus modifies the tip-scattered field E_{sca}. E_{sca} is recorded with a pseudo-heterodyne Michelson interferometer. Demodulation of the interferometric detector signal at the nth harmonics of the tip oscillation frequency yields the complex-valued near-field signals $\sigma_n = s_n e^{i\phi_n}$, with s_n being the near-field amplitude and ϕ_n being the near-field phase (in this work, $n = 2$). By recording the near-field signals as a function of the lateral tip position, we obtain near-field images or line trace. In the case of probing a material supporting polaritons, the

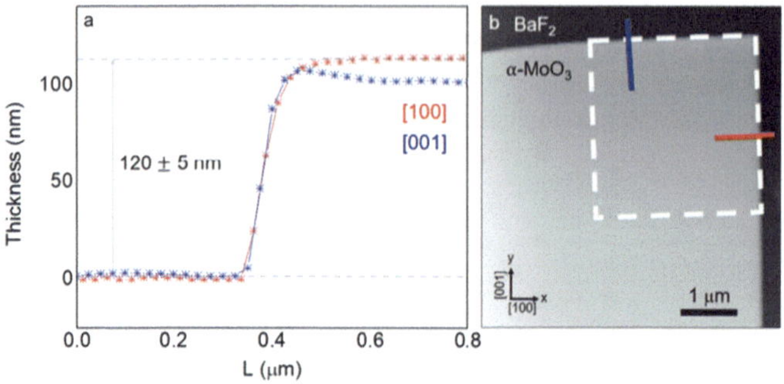

Fig. 4.7 **α-MoO_3 flake used for s-SNOM imaging.** (**a**) AFM profile along the two orthogonal in-plane directions: the flake thickness is 120 ± 5 nm. (**b**) AFM image of the 90- degree-like corner used for near-field imaging. The white square shows the area in which s- SNOM imaging was performed

Table 4.1 **Parameters for the IR dielectric function of α-MoO_3, extracted from far-field experiments.** The high-frequency permittivities are $\varepsilon_x^\infty = 5.78$, $\varepsilon_y^\infty = 6.07$ and $\varepsilon_z^\infty = 4.47$ (fixed to the ab-initio value) for the experiment. The flake thickness used in the fit was 2.9 μm

i	j	$\omega_{TO,ij}$ (cm^{-1})	$\omega_{LO,ij}$ (cm^{-1})	γ_{ij} (cm^{-1})
x	1	506.7	534.3	49.1
x	2	821.4	974.5	6.8
x	3	998.7	999.2	0.35
y	1	544.6	850.1	9.5
z	1	956.7	1006.7	0.65

nanoscale *hotspot* at the tip apex acts as a local source of polaritons. The tip-launched polaritons reflect at the flake edges and produce polariton interference, yielding fringes in the near-field images. The distance between the interference fringes corresponds to half the polariton wavelength, $\lambda_p/2$. Similar setups will be used in the following chapters, where this detailed discussion will be therefore omitted.

Figure 4.7 shows an AFM image of the α-MoO_3 flake used for near-field imaging. Since the specific geometry of the edges of the flake can affect the near field distributions of PhPs, the corner in which we tested the predictive capabilities of our model was 90-degree-like. Averaging line profiles along both in plane directions, we obtain a flake thickness of 120 ± 5 nm, where the error stands for the thickness deviation along the flake edge and from data processing. Such profiles also show that the corners are steep and regular along the [010] or z direction, as needed for homogeneous launching of PhPs.

The flake was transferred on top of BaF_2, which, as AgCl, is highly transparent across the entire spectral range of interest. The BaF_2 dielectric function was also extracted (see Sect. A.2 in Appendix A for details).

After the first far-field adjustment of the dielectric function parameters, shown in Table 4.1, they are provided as input to model near-field experimental data. For

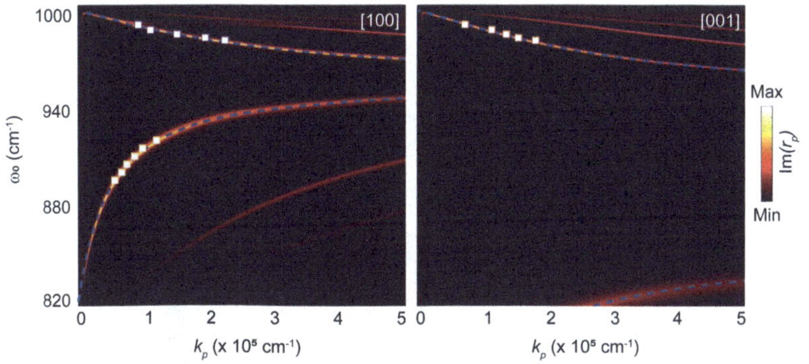

Fig. 4.8 Dispersion of PhPs on a thin α-**MoO**$_3$ slab. Transfer-matrix calculations (false color plot) and analytical dispersions (dashed blue lines) of PhPs propagating on a 120-nm-exfoliated α-MoO_3 flake on top of BaF_2 along the [100] (left panel) and [001] (right panel) crystal directions. These calculations use the extracted permittivity from FTIR reflectance measurements as input. The white squares indicate experimental data from monochromatic s-SNOM imaging

this purpose, we use the analytical dispersion of polaritons propagating in a biaxial slab embedded between two semi-infinite media, derived in Chap. 3 and reproduced here for completeness:

$$q = \frac{\rho}{k_0 d}\left[\arctan\left(\frac{\varepsilon_1 \rho}{\varepsilon_z}\right) + \arctan\left(\frac{\varepsilon_3 \rho}{\varepsilon_z}\right) + \pi l\right], \quad l = 0, 1, 2... \tag{4.2}$$

where $q = k_p/k_0$ is the normalized in-plane momentum ($k_p^2 = k_x^2 + k_y^2$), ε_1 and ε_3 are the permittivities of the superstrate and substrate, respectively, d is the thickness of the α-MoO_3 flake, $k_0 = \omega/c$ is the wavevector in free space and

$$\rho = i\sqrt{\frac{\varepsilon_z q^2}{\varepsilon_x q_x^2 + \varepsilon_y q_y^2}} = i\sqrt{\frac{\varepsilon_z}{\varepsilon_x \cos^2\alpha + \varepsilon_y \sin^2\alpha}}, \tag{4.3}$$

with α being the angle between the x axis and the in-plane component vector. On the other hand, an instructive way to visualize both the dispersion and the damping is via a false-color plot of the imaginary part of Fresnel's reflection coefficient, $\mathrm{Im}(r_p)$, at real q and ω, obtained by means of transfer-matrix calculations [15]. Briefly, PhPs correspond to the divergences of the calculated reflectivity $r_p(k_p, \omega)$ of the anisotropic structure at complex momenta q. Both the analytical dispersion from Eq. (4.2) and that inferred from transfer-matrix calculations are compared to the experimental dispersion (Fig. 4.8), measured via polariton interferometry on an α-MoO_3 flake of thickness $d = 120$ nm.

The comparison of the analytical dispersion from Eq. (4.2) and that inferred from transfer-matrix calculations with the experimental dispersion (Fig. 4.8) allows for a finer tuning of the free parameters in Eq. (4.1). Following this process, an excellent

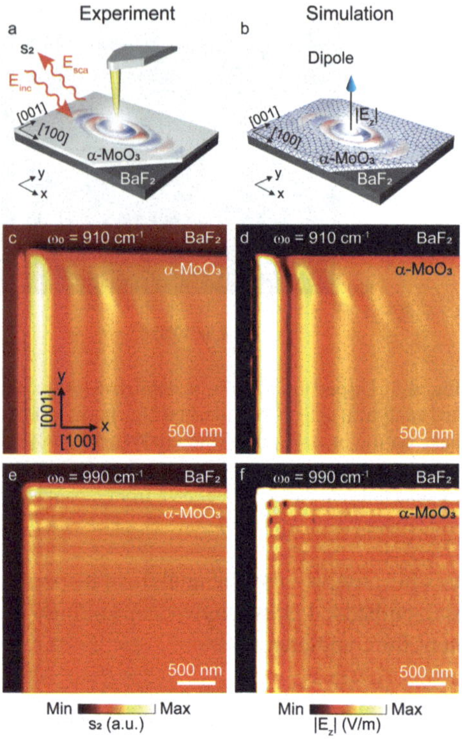

Fig. 4.9 Near-field simulated and experimental images of in-plane anisotropic PhPs on an α-MoO_3 flake. (**a**) Schematics of the s-SNOM experimental configuration and (**b**) the full-wave numerical simulation model mimicking the experiment. (**c**) Experimental near-field amplitude s_2 images and (**d**) simulated z component of the electric field $|E_z|$ of a 120-nm-thick α-MoO_3 flake on top of BaF_2 at incident frequency $\omega_0 = 910\,\text{cm}^{-1}$. (**e,f**) Corresponding images for incident frequency $\omega_0 = 990\,\text{cm}^{-1}$

agreement between the experiment and our model is obtained. Remarkably, the ability to successfully fit the polaritonic response of a thin α-MoO_3 flake placed on a different substrate from the one used for the FTIR spectra reinforces our claims of robust quantitative modeling of the far- and near-field IR response of this material, especially considering the strong thickness dependence of hyperbolic polaritons [16].

To accurately adjust the damping and unambiguously verify that our dielectric function successfully accounts for the polaritonic response of α-MoO_3, we analyze the PhP propagation length from both simulated and experimental near-field images. To do so, we use our dielectric function model as input for full-wave numerical simulations that mimic s-SNOM measurements (Fig. 4.9a,b).

Full-Wave Numerical Simulations

To perform full-wave numerical simulations, we approximate the s-SNOM tip by a dipole source with a constant dipole moment, as introduced in Chap. 2. We assume that the polarizability of the dipole is weakly affected by the PhPs excited in the α-MoO_3 flake, and their back-action onto the tip can be thus neglected. Therefore, we calculate the amplitude of the near field, $|E_z|$, as a function of the dipole position (x, y), using COMSOL Multiphysics.

The s-SNOM nanoimaging experiments that we aim to fit were carried out on a 120-nm-thick α-MoO_3 flake on top of BaF_2 at two representative frequencies: (1) $\omega_0 = 910\,\mathrm{cm}^{-1}$—within the lowest-frequency high-reflectivity band along the [100] crystal direction in Fig. 4.3 (hyperbolic band [1, 2] RB2, 821–963 cm^{-1}); see Fig. 4.9c,d—and (2) $\omega_0 = 990\,\mathrm{cm}^{-1}$—within the high-reflectivity band along both in-plane directions in Fig. 4.3e (elliptic band [1, 2] RB3, 957–1007 cm^{-1}); see Fig. 4.9e,f. Within RB2, we experimentally observe (Fig. 4.9c) fringes along the [100] crystal direction, which indicate the excitation of in-plane hyperbolic PhPs in α-MoO_3, consistently with what had been previously reported [1, 2].

Our approach consists on performing full-wave numerical simulations, varying the inputted permittivity parameters until they match the experimental PhP wavelength and propagation length. The optimal value of the permittivity parameters, in particular the damping, will be the one that gives the best agreement with the experiment.

The experimental PhP propagation length and wavelength are obtained by fitting line profiles of the polaritonic near field along the [100] and [001] crystal directions to oscillating signals with a factor accounting for dissipation described by an exponential decay of the PhPs field with distance x. The decay of polaritons away from an edge is due to a combination of damping $\mathrm{Im}(k_p) > 0$ and circular-wave geometrical spreading [1, 17]. Thus, the oscillating signal of the polaritonic near field can be fitted with:

$$\xi_{\mathrm{opt}}(x) = \xi_0 A \frac{e^{i2k_p x}}{\sqrt{x}} + B \frac{e^{ik_p x}}{x^a}, \qquad A, b, a > 0, \tag{4.4}$$

where k_p is the complex PhP wavevector. The first term is the returning (to the s-SNOM tip) field for a damped circular wave reflected from a straight edge, with the PhP travelling a distance $2x$. The second term interferes with the first, producing alternating fringe amplitudes. It arises because PhPs are not only generated/detected beneath the tip apex, but also at the edge of the flake travelling a distance x. Equation (4.4) can be easily transformed into

$$\xi_{\mathrm{opt}}(x) = \xi_0 A \frac{e^{\frac{-2x}{L_p}\sin\left(4\pi \frac{x-x_A}{\lambda_p}\right)}}{\sqrt{x}} + B \frac{e^{\frac{-x}{L_p}\sin\left(2\pi \frac{x-x_B}{\lambda_p}\right)}}{x^a}, \qquad A, b, L_p, \lambda_p, a > 0. \tag{4.5}$$

From the fitting procedure in Eq. (4.5) we directly extract the simulated and experimental PhP propagation length L_p and wavelength λ_p. To fit the simulated profiles, we fix $A = 0$ and $a = 1/2$, as PhPs in our full-wave numerical simulations are launched by a point dipole, since there are no edge-launched waves and just λ_p-periodic waves with circular geometrical decay are present. On the other hand, in the experimental profiles, both contributions are present: in this case, since the geometrical decay of the PhP travelling the tip—edge distance only once is not

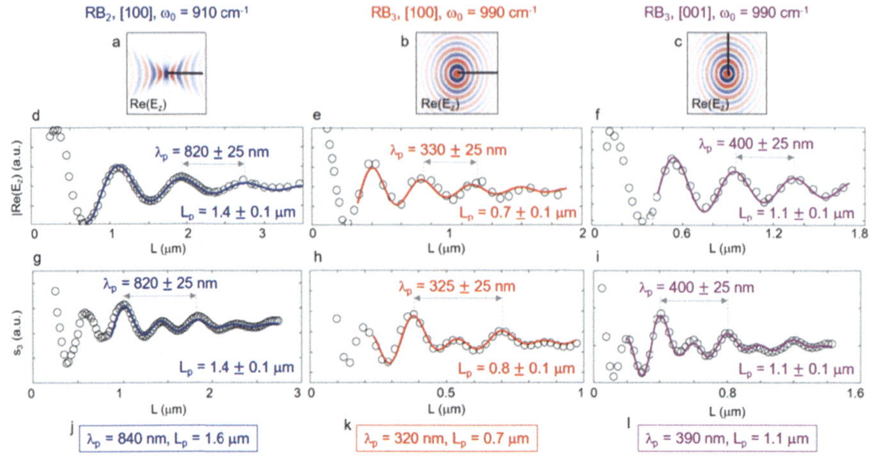

Fig. 4.10 (**a**,**b**,**c**) Field distribution of propagating PhPs launched by a point dipole oriented perpendicular to a 120-nm-thick α-MoO$_3$ layer for incident frequencies $\omega_0 = 910\,\text{cm}^{-1}$ (in RB2, left column) and $\omega_0 = 990\,\text{cm}^{-1}$ (in RB3, middle and right columns). (**d**,**e**,**f**) Profiles extracted along the lines depicted in (**a**,**b**,**c**) (dots). By performing a fitting to Eq. (4.5) with $A = 0$ and $a = 1/2$ (color lines), we find the PhP propagation lengths to be $L_p = 1.4 \pm 0.1$ μm (along [100] in RB2), $L_p = 0.7 \pm 0.1$ μm (along [100] in RB3) and $L_p = 1.1 \pm 0.1$ μm (along [001] in RB3). All simulations were performed using our dielectric model as input. (**g**,**h**,**i**) s-SNOM (amplitude) line traces (dots) along the near-field images shown in Fig. 4.9c,e. Fits to Eq. (4.5) setting $a \approx 1$ are shown (color lines). The obtained propagation lengths $L_p = t_0$ are $L_p = 1.4 \pm 0.1$ μm (along [100] in RB2), $L_p = 0.8 \pm 0.1$ μm (along [100] in RB3) and $L_p = 1.1 \pm 0.1$ μm (along [001] in RB3). (**j**,**k**,**l**) Theoretical values found from the analytical dispersion of PhPS, Eq. (4.2), yielding values in exceptional agreement with simulation and experiment: $\lambda_p = 1.6$ μm (along [100] in RB2), $\lambda_p = 0.7$ μm (along [100] in RB3) and $\lambda_p = 1.1$ μm (along [001] in RB3)

known a priori, we allow for a variable decay $a \approx 1$ [17]. Figure 4.10 shows the profiles and the fittings performed to extract the polaritonic wavelength and propagation length, as well as theoretical values calculated from Eq. (4.2), found from $L_p = 1/\text{Im}(k_p)$ and $\lambda_p = 2\pi/k_p$, respectively. All experimental, simulated and analytical values are in excellent agreement.

In the experiments, we obtain a PhP wavelength of $\lambda_p^x = 820 \pm 25$ nm and a propagation length of $L_p^x = 1400 \pm 100$ nm. To account for these experimental results with our dielectric function, we adjust the parameter γ_{x2} in the model—which determines the damping of PhPs in RB2—and run a full-wave numerical simulation using the adjusted dielectric function. From the image resulting from this simulation (Fig. 4.9d) we find the PhP wavelength and propagation length to be $L_p = 1400 \pm 100$ nm, thus in excellent agreement with the experimental results. Within RB3, we observe in the experimental image fringes along both in-plane directions (Fig. 4.9e), being the PhP wavelengths and propagation lengths $\lambda_p^x = 330 \pm 25$ nm and $L_p^x = 700 \pm 100$ mm, respectively along the [100] crystal direction, and $\lambda_p^y = 400 \pm 25$ nm and $L_p^y = 1100 \pm 100$ nm along the [001] direction. As before, to account for these experimental results with our dielectric function, we tune the parameter γ_{z1} in our model—determining the damping of PhPs in RB3—and run a full-wave numerical simulation using the adjusted dielectric function. From the resulting image (Fig. 4.9f), we find the PhP wavelengths and propagation lengths to be $L_p = 1100 \pm 100$ nm along [001] direction. Again, we obtain excellent agreement with the experiment, unambiguously demonstrating the validity of our model.

4.3 Further Experimental Verification: Attenuated Total Reflection Micro-Spectroscopy of a Thick α-MoO_3 Flake

As a further verification, we also carried out attenuated total reflectance (ATR) micro-spectroscopy of an α-MoO_3 flake on top of BaF_2 (Fig. 4.11). FTIR-ATR has been proven to be an effective method for studying the properties of anisotropic 2D materials [18], as it is sensitive to both dielectric and polaritonic resonant modes supported by these crystals.

This method is based on the Kretschmann-Raether configuration [19] for launching hyperbolic polaritons [20, 21]. As illustrated in Sect. 2.3 for a Ge prism, a high

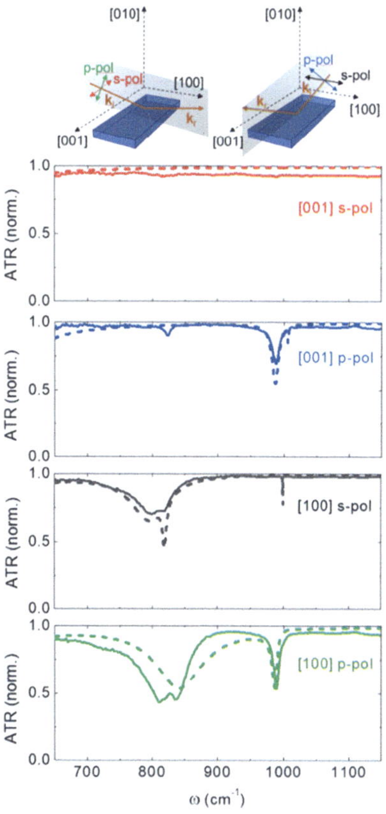

Fig. 4.11 ATR-FTIR spectra (solid lines) on an α-MoO_3 flake along both in-plane crystal axes. The flake is the one shown in Fig. 4.2. Dashed lines indicate the simulated spectra, using our permittivity model as input

index prism allows for coupling to modes with an effective index greater than one. One of the important parameters for this experiment is field orientation—as it is sensitive to both dielectric cavity resonances under s polarized light and polariton resonances under p-polarized light. The optical anisotropy of α-MoO_3 implies we also need to consider the azimuthal incident angle due to the in-plane anisotropy. This means we collect four spectra: s- and p-polarized along the two different crystal axes (as shown in Fig. 4.11). Within this experiment we are able to probe the polaritons launched in x and z axes, as we are limited by the optical transmittance of the Ge prism. The polariton/dielectric resonances manifest as absorption dips in the spectrum—with those located around 810 cm^{-1} related to the [100] phonon, and those around 1000 cm^{-1} associated with the [010] phonon. The simulations generally match our measured spectra, with the main differences being the presence of additional peaks in the green curves, and improved sharpness in the measured ATR spectra. This is most likely due to misalignment of the incident IR beam with the crystal axes in the ATR measurement, and the overestimation of damping for our bulk dielectric function, as discussed previously in the last part of Sect. 4.1. However, the measurement speaks to the usefulness of our dielectric function for predicting the behavior of α-MoO_3 samples.

The good agreement between calculated and simulated spectra (using our dielectric function) again demonstrates the validity of our model and its broad applicability for predictive simulations.

4.4 Experimental and Ab Initio Mid-IR Dielectric Function of α-MoO_3

We show in Table 4.2 the resulting frequencies and damping rates of the phonons of our model along all three principal axes and plot the extracted dielectric function in Fig. 4.12 (middle panel). The parameters provided in Table 4.2 are in agreement with previously far-field extracted phonon frequencies [3, 4] and near-field experimental results [1, 2, 22]. The latter, however, do not focus on the permittivity extraction but take the frequencies from the aforementioned far-field reports. In addition, unlike earlier results, our correlative FTIR-s-SNOM experiments allowed us to identify two additional weak phonon bands aligned along the x-axis ([100] crystal direction), and accurately determine the high-frequency dielectric permittivity along both the [100] and [001] directions. We note that the weak phonon spectrally located at 998 cm^{-1} results in a discontinuity in the hyperbolic polariton dispersion relation (Fig 4.8), which had not been predicted using earlier experimental or theoretical models. We also note that the apparent strong damping of the optic phonon located near 506 cm^{-1} is likely an artifact resulting from being spectrally near the cut-off of the FTIR mercury cadmium telluride (MCT) detector. As expected for a biaxial crystal, α-MoO_3 exhibits different permittivities along the three principal axes throughout the entire spectral range studied ($\varepsilon_x \neq \varepsilon_y \neq \varepsilon_z$). Furthermore, over the spectral

Table 4.2 Parameters for the IR dielectric function of $\alpha-MoO_3$, extracted from correlative far- and near-field experiments (left) and ab initio calculations (right). The high-frequency permittivities are ε_x^{∞} = 5.78, ε_y^{∞} = 6.07 and ε_z^{∞} = 4.47 (fixed to the ab-initio value) for the experiment and ε_x^{∞} = 5.86, ε_y^{∞} = 6.59 and ε_z^{∞} = 4.47 for the ab initio calculation. The flake thickness used in the fit was 2.9 μm. "–": not considered

		Far and near-field correlative studies			Ab initio calculations		
i	j	$\omega_{TO,ij}$ (cm^{-1})	$\omega_{LO,ij}$ (cm^{-1})	γ_{ij} (cm^{-1})	$\omega_{TO,ij}$ (cm^{-1})	$\omega_{LO,ij}$ (cm^{-1})	γ_{ij} (cm^{-1})
x	1	506.7	534.3	49.1	449	467	8.3
x	2	821.4	963.0	6.0	769	947	3.7
x	3	998.7	999.2	0.35	1016	1018	0.4
y	1	544.6	850.1	9.5	505	820	12
z	–	–	–	–	765	772	3.7
z	1	956.7	1006.9	1.5	976	1027	0.4

range extending from 545 to 1006 cm^{-1}, there is always at least one component i exhibiting a negative Re(ε_i) value, thus the crystal can support PhPs over this entire frequency range, having multiple hyperbolic regimes, each of them when at least one component i fulfills Re(ε_i) < 0.

To identify the vibrational character of the multiple optic phonons supported by α-MoO_3, and to further test the validity of our dielectric function model, we also calculated the permittivity tensor and vibrational eigenmodes using DFT. The calculations, the details of which can be found in Appendix B, were carried out by Prof. Ion Errea (University of the Basque Country, Spain). The resulting ab initio permittivity function qualitatively agrees with the experimentally extracted dielectric function (Fig. 4.12, lower panel). This underlines the potential of employing first-principles methods to characterize the IR permittivity of emerging materials for nanophotonics. However, slight spectral shifts in the phonon frequencies are observed in the calculations with respect to the experiment, which is typical for DFT with semi-local exchange-correlation functionals for semiconducting oxides [23]. A further discrepancy is that, in the ab initio-calculated dielectric function, RB2 and RB3 (Fig. 4.12) do not spectrally overlap due to small underestimation and overestimation of the phonon frequencies along the [100] and [010] directions, respectively, in contrast with the experimental results [1, 2]. We highlight that the ab initio model also predicts several phonon modes beyond the spectral range of our experimental measurements, extending into the THz (see section "THz/Far-IR Transmission Spectra Collected from α-MoO_3 Powder Using a Nb-Superconducting Bolometer"). The validity of these predicted modes was verified through experimentally measured far-IR transmission spectra collected from α-MoO_3 powder using a Nb-superconducting bolometer, as will be shown in the next section. While we were not able to determine the exact frequencies for the phonons in this spectral band, the combination of the ab initio model and THz transmission spectra suggested that this might be an interesting regime for future work. As a matter of fact, PhPs in

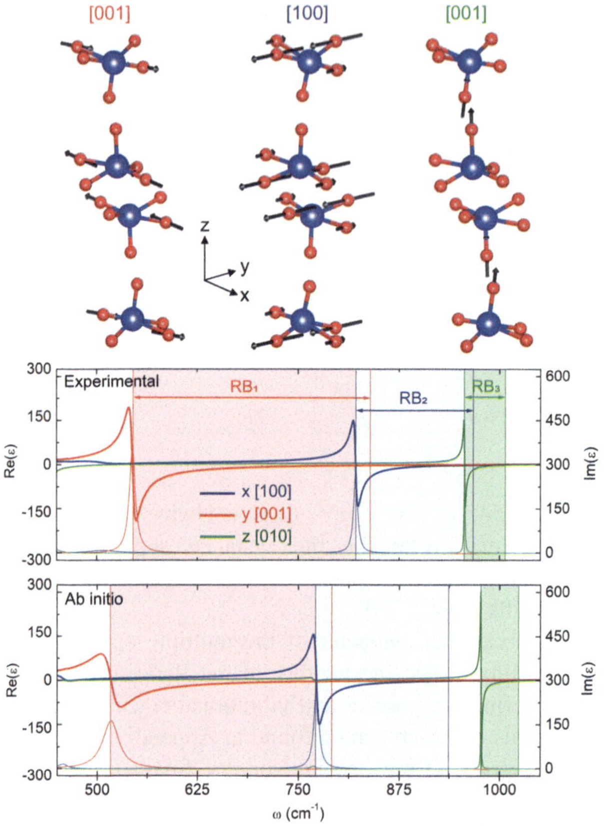

Fig. 4.12 Mid-IR dielectric function of α-MoO_3. Experimental **(middle panel)** and ab initio **(lower panel)** permittivity of α-MoO_3 along the three crystallographic axes. The shaded regions indicate the three Reststrahlen bands (RB1, RB2, and RB3), corresponding to phonons on each of the crystallographic axes: [001], [100], and [010]. The lattice vibrations of these phonons are illustrated in the **upper panel** (molybdenum/oxygen atoms in blue/red, respectively)

α-MoO_3 were explored in the THz regime following the initial mid-IR permittivity extraction. These results will be discussed in detail the next section.

4.5 THz/far-IR Permittivity of α-MoO_3

The far-IR or terahertz (THz) frequency range ($\nu = 0.1$–10 THz), also known as the THz gap, plays a crucial role in various emerging applications. These applications include the latest generation of mobile networks [24–26] or non-destructive sensing of organic molecules for medical purposes [27, 28]. To confine the electromagnetic field in the THz spectral range, specialized metallic structures like antennas or slits are typically employed. However, due to the relatively long wavelengths of THz radiation, these structures have relatively large physical dimensions. This limitation is particularly significant for the manipulation of THz light at the nanometer length scale and for the development of nanophotonic technologies, as it hinders the confinement of light to scales smaller than the wavelength.

In this context, PhP provide a superb solution for squeezing THz light into nanoscale dimensions. Specifically, vdW crystals like α-MoO_3 exhibit extreme confinement of electromagnetic fields and minimal optical losses. These very advantageous properties stem from their ultra-thin composition, as has been extensively discussed in this thesis within the context of the mid-IR spectrum. However, when it comes to THz radiation, the small size of α-MoO_3 and other vdW crystals compared to the long THz wavelengths makes direct far-field measurements in this spectral range substantially more impractical than in the mid-IR. Probing these excitations thus requires sophisticated techniques to confine and manipulate THz light at the nanometer length scale. Additionally, the scarcity of powerful, reliable, and high-quality THz light sources adds to the challenges of working within this spectral range. Nevertheless, in recent years, with advancements in THz detectors and sources [29, 30], this regime has become more accessible for scientific and technological applications. Consequently, the investigation of material properties at THz frequencies with nanometer resolution [31] has emerged as an important field in optics and materials sciences.

THz/Far-IR Transmission Spectra Collected from α-MoO₃ Powder Using a Nb-Superconducting Bolometer

As pointed out in the above discussion, the small size of α-MoO_3 crystals precludes direct far-field measurements in the far-IR. However, we can instead disperse a bulk powder across the surface of a THz/far-IR transparent material. Whilst this provides an ensemble averaged response of the powder, losing sensitivity to crystal axis, it does provide us with a rough estimate of the phonon energies, given by where the transmission/reflection is reduced/enhanced respectively. To perform these measurements, we dispersed α-MoO_3 crystals in IPA, via sonication for 30 min, and then drop cast on a piece of high density poly-ethylene (HDPE).

We acquired unpolarized reflection and transmission measurements at 40° angle of incidence, shown in Fig. 4.13. In general, the HDPE is weakly reflecting and moderately transmissive in the mid to far-IR. The powder sample generally shows reduced transmission and reflection—which is a consequence of scattering introduced by the crystals which have a size on the scale of the wavelength. However, at specific frequencies we observe a correlated dip in the transmission, and a peak in reflection. This suggests the presence of phonon modes at or close to these frequencies—notably 300, 349, 361 and 371 cm^{-1}. Whilst we cannot conclusively say that these correlated peaks and dips correspond to the exact phonon frequencies, the measurements initially suggested that this frequency range could be of interest for subsequent work.

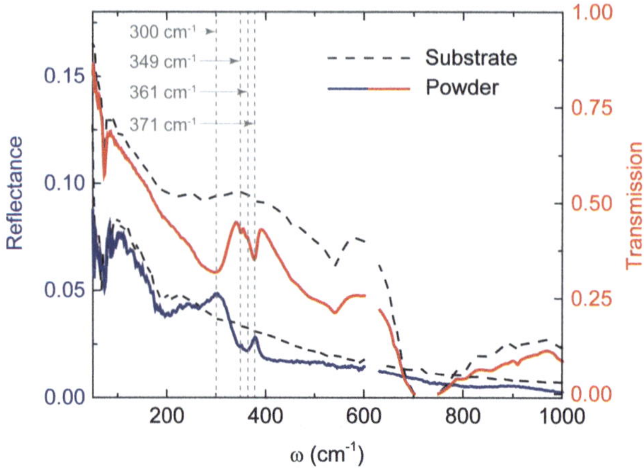

Fig. 4.13 α-**MoO**$_3$ **response in FIR/THz using reflection and transmission spectroscopy.** Lines indicate correlated peaks/troughs (300, 349, 361 and 371 cm^{-1}), which will be associated with phonon modes

s-SNOM Nanoimaging of HPhPs in α-MoO$_3$ Using a Free Electron Laser

The experimental demonstration of nanoscale-confined PhPs at THz frequencies, and in particular HPhPs, had remained elusive until recently. In collaboration with Thales V. A. G. de Oliveira et al. [32], and by combining a s-SNOM with a free-electron laser (FEL, Fig. 4.14a), we revealed a range of low-loss polaritonic excitations at frequencies from 8 to 12 THz in α-MoO$_3$. The research reported in [32] constituted the first visualization of PhPs in the THz frequency range. It also allowed extracting the far-IR permittivity of α-MoO$_3$ by combining near-field nanoimaging with transmission FTIR spectra collected from α-MoO$_3$ powder, described above, and ab initio calculations, described in Appendix B, thus following an analogous procedure to the one discussed along this chapter for extracting the mid-IR dielectric permittivity.

In particular, the PhP dispersion was quantitatively extracted by nanoimaging flakes with different thicknesses using a wide set of FEL frequencies. In Fig. 4.14b, we plot the experimental and analytical dispersion relations $\nu(k)$ for PhPs propagating along the [001] crystal direction (PhPs$_{[001]}$, symbols and analytical curve shaded in red) and along the [100] crystal direction (PhP$_{[100]}$, symbols and analytical curve shaded in blue). These measurements were performed on a set of four flakes with thicknesses d = 53, 131, 197, and 295 nm. Since we observe increasing k values (smaller wavelengths) by sweeping the illumination frequency from $\nu_{TO} \rightarrow \nu_{LO}$ (increasing frequency), we can conclude that the nanoimaged THz polaritons in this frequency range propagate with a positive phase velocity, in analogy with the mid-IR polaritons reported previously.

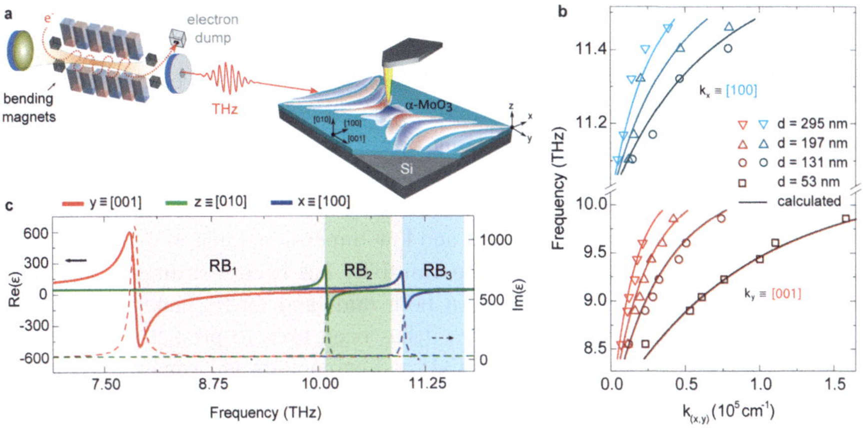

Fig. 4.14 Polariton nanoimaging with a free-electron laser coupled to a s-SNOM microscope and dispersion of THz polaritons in α-MoO$_3$ with hyperbolic propagation. (**a**) Illustration of the experiment. The s-SNOM tip is polarized by intense picosecond THz pulses provided by a widely tunable free-electron laser, which in turn launches polaritons in an α-MoO$_3$ slab that propagate away from the tip. (**b**) THz polariton dispersion of α-MoO$_3$ flakes of different thicknesses d. Data points and analytical dispersion are color-coded in shades of blue (for $PhP_{[100]}$), and red (for $PhP_{[001]}$). (**c**) Dielectric permittivity tensor of α-MoO$_3$ in the THz spectral range. The real (solid lines) and imaginary (dashed lines) parts of the permittivity tensor reveal three distinct RBs with negative permittivities along different crystal axes, shaded in red ([001]), green ([010]), and blue ([100])

The analytical dispersions derived from Eq. (4.2) show excellent agreement with the experimental data. This fitting, in combination with transmission FTIR spectra from α-MoO$_3$ powder, shown above, and ab initio calculations, allowed extracting the dielectric permittivity tensor of α-MoO$_3$ in the THz spectrum, previously unknown. As shown in Fig. 4.14c, we observe spectral bands (marked as RB1-3) wherein at least one of the principal components of the permittivity is negative. In particular, within the bands RB1 and RB3 this occurs along the [001] and [100] crystal directions, respectively, indicating their potential to support PhPs with in-plane hyperbolic propagation. Therefore, not only does this study enable extending all the outstanding properties of HPhPs in α-MoO$_3$ into the THz regime, but it also provides the necessary predictive capabilities for subsequent studies based on them.

4.6 Conclusions

In summary, we have extracted the full IR complex-valued dielectric function of the biaxial vdW semiconductor α-MoO$_3$ along all the three crystallographic directions. Its robustness is demonstrated by successfully numerically reproducing three different experimental measurements of flakes with significantly different thicknesses on different substrates. This, given the substantial impact of the substrate's refractive index on hyperbolic polariton propagation [33], further demonstrates the general nature of our reported dielectric function model. We achieved this by combining the virtues of far-field (broadband characterization and accurate determination of TO phonon energies) with near-field imaging (sensitivity to phonon damping and LO phonon energies) and advanced theoretical and numerical approaches, providing an alternative, more accurate approach to predicting the IR dielectric function—and thus the optical response—of 2D, nano and low-dimensional materials.

Following our work, the extracted permittivity has been instrumental in many theoretical and experimental studies on both mid- and far-IR nanophotonic and optoelectronic devices. To name a few, it has been used to predict and interpret the first reports on negative reflection [34] and anomalous and negative refraction [35, 36] of nanoscale-confined polaritons, the advent of twistoptics [37, 38] and recent results on twisted trilayers [39]. Works on hyperfocusing [40–42] and unidirectional excitation [43] of in-plane HPhPs, hybridization of HPhPs in with polar substrates [44, 45], twist-tunable polaritonic nanoresonators [46], polarization converters [47], bottom-up synthesis of α-MoO_3 nanoribbons [48], launching and manipulation of higher-order in-plane HPhPs [49] and many others. Based on the breadth of exotic phenomena this material exhibits, we envision potential for even more unprecedented planar nanophotonic studies and technologies using the extracted permittivity.

References

1. W. Ma et al., In-plane anisotropic and ultra-low-loss polaritons in a natural van der Waals crystal. Nature **562**, 557 (2018). https://www.nature.com/articles/s41586-018-0618-9
2. Z. Zheng et al., A mid-infrared biaxial hyperbolic van der Waals crystal. Sci. Adv. **5**, eaav8690 (2019). https://advances.sciencemag.org/content/5/5/eaav8690
3. M.A. Py, P.E. Schmid, J.T. Vallin, Raman scattering and structural properties of MoO_3. Il Nuovo Cimento B (1971-1996) **38**, 271–279 (1977). https://doi.org/10.1007/BF02723496
4. K. Eda, Longitudinal-transverse splitting effects in IR absorption spectra of MoO_3. J. Solid State Chem. **95**, 64–73 (1991). https://www.sciencedirect.com/science/article/pii/002245969190376S
5. A.A. Govyadinov et al., Recovery of permittivity and depth from near-field data as a step toward infrared nanotomography. ACS Nano **8**, 6911–6921 (2014). https://doi.org/10.1021/nn5016314
6. D. Hu et al., Probing optical anisotropy of nanometer-thin van der Waals microcrystals by near-field imaging. Nat. Commun. **8**, 1471 (2017). https://doi.org/10.1038/s41467-017-01580-7
7. D. Wintz et al., Guided modes of anisotropic van der Waals materials investigated by near-field scanning optical microscopy. ACS Photon. **5**, 1196–1201 (2018). https://doi.org/10.1021/acsphotonics.7b01518
8. S. Adachi, *Optical Constants of Crystalline and Amorphous Semiconductors* (Springer New York, New York, 1999). https://link.springer.com/book/10.1007/978-1-4615-5247-5
9. M. Schubert, T.E. Tiwald, C.M. Herzinger, Infrared dielectric anisotropy and phonon modes of sapphire. Phys. Rev. B **61**, 8187–8201 (2000). https://link.aps.org/doi/10.1103/PhysRevB.61.8187

10. T.G. Folland, L. Nordin, D. Wasserman, J.D. Caldwell, Probing polaritons in the mid- to far-infrared. J. Appl. Phys. **125**, 191102 (2019). https://doi.org/10.1063/1.5090777
11. J.D. Caldwell et al., Sub-diffractional volume-confined polaritons in the natural hyperbolic material hexagonal boron nitride. Nat. Commun. **5**, 5221 (2014). https://doi.org/10.1038/ncomms6221
12. A.J. Giles et al., Ultralow-loss polaritons in isotopically pure boron nitride. Nat. Mater. **17**, 134–139 (2018). https://doi.org/10.1038/nmat5047
13. R. Estevâm da Silva et al., Far-infrared slab lensing and subwavelength imaging in crystal quartz. Phys. Rev. B **86**, 155152 (2012). https://link.aps.org/doi/10.1103/PhysRevB.86.155152
14. C.J. Winta, M. Wolf, A. Paarmann, Low-temperature infrared dielectric function of hyperbolic α-quartz. Phys. Rev. B **99**, 144308 (2019). https://link.aps.org/doi/10.1103/PhysRevB.99.144308
15. N.C. Passler, A. Paarmann, Generalized 4×4 matrix formalism for light propagation in anisotropic stratified media: study of surface phonon polaritons in polar dielectric heterostructures. J. Opt. Soc. Am. B **34**, 2128–2139 (2017). https://opg.optica.org/josab/abstract.cfm?URI=josab-34-10-2128
16. J.D. Caldwell et al., Photonics with hexagonal boron nitride. Nat. Rev. Mater. **4**, 552–567 (2019). https://www.nature.com/articles/s41578-019-0124-1
17. A. Woessner et al., Highly confined low-loss plasmons in graphene.boron nitride heterostructures. Nat. Mater. **14**, 421–425 (2015). https://doi.org/10.1038/nmat4169
18. T.G. Folland et al., Probing hyperbolic polaritons using infrared attenuated total reflectance micro-spectroscopy. MRS Commun. **8**, 1418–1425 (2018)
19. H. Raether, *Surface Plasmons on Smooth and Rough Surfaces and on Gratings* (Springer, Heidelberg, 1988). https://scirp.org/reference/referencespapers.aspx?referenceid=1193551
20. Y. Mazor, B.Z. Steinberg, Longitudinal chirality, enhanced nonreciprocity, and nanoscale planar one-way plasmonic guiding. Phys. Rev. B **86**, 045120 (2012). https://link.aps.org/doi/10.1103/PhysRevB.86.045120
21. T.G. Folland et al., Reconfigurable infrared hyperbolic metasurfaces using phase change materials. Nat. Commun. **9**, 4371 (2018). https://doi.org/10.1038/s41467-018-06858-y
22. Z. Zheng et al., Phonon polaritons: highly confined and tunable hyperbolic phonon polaritons in van der Waals semiconducting transition metal oxides. Adv. Mater. **30**, 1870091 (2018). https://onlinelibrary.wiley.com/doi/abs/10.1002/adma.201870091
23. H. Ding, K.G. Ray, V. Ozolins, M. Asta, Structural and vibrational properties of α-MoO_3 from van der Waals corrected density functional theory calculations. Phys. Rev. B **85**, 012104 (2012). https://link.aps.org/doi/10.1103/PhysRevB.85.012104
24. S. Dang, O. Amin, B. Shihada, M.-S. Alouini, What should 6G be? Nat. Electron. **3**, 20–29 (2020). https://doi.org/10.1038/s41928-019-0355-6
25. K. Sengupta, T. Nagatsuma, D.M. Mittleman, Terahertz integrated electronic and hybrid electronic.photonic systems. Nat. Electron. **1**, 622–635 (2018). https://doi.org/10.1038/s41928-018-0173-2
26. A.Y. Nikitin et al., Real-space mapping of tailored sheet and edge plasmons in graphene nanoresonators. Nat. Photon. **10**, 239–243 (2016). https://doi.org/10.1038/nphoton.2016.44
27. B.B. Lahiri, S. Bagavathiappan, T. Jayakumar, J. Philip, Medical applications of infrared thermography: A review. Infrared Phys. Technol. **55**, 221–235 (2012). https://www.sciencedirect.com/science/article/pii/S1350449512000308?via%3Dihub
28. S.S. Dhillon et al., The 2017 terahertz science and technology roadmap. J. Phys. D Appl. Phys. **50**, 043001 (2017). https://doi.org/10.1088/1361-6463/50/4/043001
29. P.H. Siegel, THz technology: an overview. Int. J. High Speed Electron. Syst. **13**, 351–394 (2003). https://doi.org/10.1142/S0129156403001776
30. G.P. Gallerano, S.G. Biedron, Overview of terahertz radiation sources, tech. rep. (United States, 2004). http://inis.iaea.org/search/search.aspx?orig_q=RN:41080790
31. N.B. Lawler, D. Ho, C.W. Evans, V.P. Wallace, K.S. Iyer, Convergence of terahertz radiation and nanotechnology. J. Mater. Chem. C **8**, 10942–10955 (2020). https://doi.org/10.1039/D0TC01716G

32. T.V.A.G. De Oliveira et al., Nanoscale-confined terahertz polaritons in a van der Waals crystal. Adv. Mater. **33**, 2005777 (2021). https://onlinelibrary.wiley.com/doi/abs/10.1002/adma.202005777
33. A. Fali et al., Refractive index-based control of hyperbolic phonon-polariton propagation. Nano Lett. **19**, 7725–7734 (2019). https://doi.org/10.1021/acs.nanolett.9b02651
34. G. Álvarez-Pérez et al., Negative reflection of nanoscale-confined polaritons in a low-loss natural medium. Sci. Adv. **8**, eabp8486 (2022). https://www.science.org/doi/abs/10.1126/sciadv.abp8486
35. J. Duan et al., Planar refraction and lensing of highly confined polaritons in anisotropic media. Nat. Commun. **12**, 4325 (2021). https://doi.org/10.1038/s41467-021-24599-3
36. A.J. Sternbach et al., Negative refraction in hyperbolic hetero-bicrystals. Science **379**, 555–557 (2023). https://www.science.org/doi/abs/10.1126/science.adf1065
37. J. Duan et al., Twisted nano-optics: manipulating light at the nanoscale with twisted phonon polaritonic slabs. Nano Lett. **20**, 5323–5329 (2020). https://doi.org/10.1021/acs.nanolett.0c01673
38. M. Chen et al., Configurable phonon polaritons in twisted α-MoO_3. Nat. Mater. **19**, 1307–1311 (2020). https://doi.org/10.1038/s41563-020-0732-6
39. J. Duan et al., Multiple and spectrally robust photonic magic angles in reconfigurable α-MoO_3 trilayers. Nat. Mater. **22**, 867–872 (2023). https://doi.org/10.1038/s41563-023-01582-5
40. J. Martín-Sánchez et al., Focusing of in-plane hyperbolic polaritons in van der Waals crystals with tailored infrared nanoantennas. Sci. Adv. **7**, eabj0127 (2021). https://www.science.org/doi/abs/10.1126/sciadv.abj0127
41. Z. Zheng et al., Controlling and focusing in-plane hyperbolic phonon polaritons in α-MoO_3 with a curved plasmonic antenna. Adv. Mater. **34**, 2104164 (2022). https://onlinelibrary.wiley.com/doi/abs/10.1002/adma.202104164
42. L. Ma, A. Ge, L. Sun, F. Liu, W. Lu, Focusing of hyperbolic phonon polaritons by bent metal nanowires and their polarization dependence. ACS Photon. **10**, 1841–1849 (2023). https://doi.org/10.1021/acsphotonics.3c00230
43. Q. Zhang et al., Unidirectionally excited phonon polaritons in high-symmetry orthorhombic crystals. Sci. Adv. **8**, eabn9774 (2022). https://www.science.org/doi/10.1126/sciadv.abn9774
44. J. Duan et al., Enabling propagation of anisotropic polaritons along forbidden directions via a topological transition. Sci. Adv. **7**, eabf2690 (2021). https://www.science.org/doi/abs/10.1126/sciadv.abf2690
45. Q. Zhang et al., Hybridized hyperbolic surface phonon polaritons at α-MoO_3 and polar dielectric interfaces. Nano Lett. **21**, 3112–3119 (2021). https://doi.org/10.1021/acs.nanolett.1c00281
46. O.G. Matveeva et al., Twist-tunable polaritonic nanoresonators in a van der Waals crystal. npj 2D Mater. Appl. **7**, 31 (2023). https://doi.org/10.1038/s41699-023-00387-z
47. S. Abedini Dereshgi et al., Lithography-free IR polarization converters via orthogonal in-plane phonons in α-MoO_3 flakes. Nat. Commun. **11**, 5771 (2020). https://doi.org/10.1038/s41467-020-19499-x
48. S.-J. Yu et al., Ultrahigh-quality infrared polaritonic resonators based on bottom-up-synthesized van der Waals nanoribbons. ACS Nano **16**, 3027–3035 (2022). https://doi.org/10.1021/acsnano.1c10489
49. G. Lu et al., Launching and manipulation of higher-order in-plane hyperbolic phonon polaritons in low-dimensional heterostructures. Adv. Mater. **35**, 2300301 (2023). https://onlinelibrary.wiley.com/doi/abs/10.1002/adma.202300301
50. G. Álvarez-Pérez et al., Infrared permittivity of the biaxial van der Waals semiconductor α-MoO_3 from near- and far-field correlative studies. Adv. Mater. **32**, 1908176 (2020). https://onlinelibrary.wiley.com/doi/abs/10.1002/adma.201908176

Part III
Anomalous Optical Phenomena at the Nanoscale in Strongly Anisotropic Media

Reflection and refraction are arguably the two most fundamental optical phenomena. Reflection describes how light bounces off objects, and constitutes the basis for the development of mirrors and cavities. Refraction involves the bending of light as it passes through different media, enabling the design of lenses and prisms. Studying reflection and refraction at the nanoscale is crucial to understand how light interacts with matter, as well as to develop novel nano-optical devices based on these interactions, the aim of the field of nano-optics.

In this regard, hyperbolic polaritons in strongly anisotropic media are a very promising platform, in particular for the observation of anomalous instances of reflection and refraction. As introduced in the preceding chapters, the propagation of hyperbolic polaritons is highly counter-intuitive, to a large degree due to the misalignment between the direction of propagation and the direction of the wavefronts. As a result of this non-collinearity, hyperbolic polaritons can exhibit unconventional optical behaviors like anomalous refraction and negative reflection. Anomalous refraction occurs when light bends in unexpected directions, offering potential for superlenses that surpass the diffraction limit and enable high-resolution imaging. Negative reflection, on the other hand, opens the door to developing unconventional optical devices such as cloaking materials that can manipulate light flow to render objects invisible or redirect light around them. These phenomena challenge our conventional understanding of light propagation and present exciting opportunities in nanoscale optics and photonics, with applications in telecommunications, imaging, and biosensing.

In Chap. 5, we present the first visualization of negative reflection of nanoscale-confined polaritons and introduce the design of the first in-plane hyperbolic nanoresonator, substantially enhancing the intensity. Additionally, in Chap. 6, we discuss the first visualization of anomalous refraction of polaritons at the nanoscale and introduce the design of the first planar lens for polaritons, with a record-breaking focus size compared to incident light.

Chapter 5
Negative Reflection of Nanoscale-Confined Hyperbolic Polaritons

Abstract In this chapter, we discuss the direct visualization of nanoscale-confined polaritons negatively reflecting on subwavelength mirrors fabricated in the low-loss van der Waals crystal α-MoO_3. Our near-field nanoimaging results unveil unconventional and broad tunability of both the polaritonic wavelength and direction of propagation upon negative reflection. On the basis of these findings, we introduce a novel device in nano-optics: a hyperbolic nanoresonator, making possible to reflect hyperbolic polaritons with different momenta back to a common point source, enhancing the intensity. These results pave way to realize nanophotonics in low-loss natural media, providing an efficient route to control nanolight, key for future on-chip optical nanotechnologies.

The near-field experiments reported in this chapter were carried out in collaboration with Dr. Jiahua Duan (University of Oviedo, Spain). Sample fabrication was performed mainly by Javier Taboada Gutiérrez (University of Oviedo, Spain), Dr. Elizaveta Nikulina (CIC nanoGUNE; Spain) and Dr. Qingdong Ou (Monash University, Australia). The results were published in "Negative reflection of nanoscale-confined polaritons in a low-loss natural medium" by Gonzalo Álvarez-Pérez, Jiahua Duan et al., in Science Advances 8 (29), eabp8486 (2022).

Negative reflection is an uncommon optical phenomenon characterized by light being reflected towards the same side of the boundary's normal on which it is incident. Prior to the study reported in this chapter, published in Ref. [1] in 2022, this phenomenon had received little attention and had not been observed in natural materials, especially at the nanoscale, despite its significant implications for the control of interactions between light and matter at this scale. The only instances of negative reflection had been demonstrated in artificially engineered interfaces, such as metasurfaces [2–6] and chiral mirrors [7]. Unfortunately, these platforms do not allow for a direct visualization of the phenomenon in real space, which is crucial for a complete understanding of the process. Such visualization had been exceptionally difficult to accomplish experimentally, as not only low optical losses but also an adequate material platform are required to guarantee a clear real-space visualization. However, the visualization of negative reflection in real space has the

G. Álvarez Pérez, *Fundamentals of Nano-Optics in Hyperbolic van der Waals Materials*, Springer Theses, https://doi.org/10.1007/978-3-031-75767-9_5

potential to provide significant insights into this fundamental optical phenomenon, opening up new possibilities in the field of nano-optics.

The findings of hyperbolic phonon polaritons (HPhPs) in low-dimensional van der Waals (vdW) crystals provided low-loss natural platforms to explore exotic optical phenomena at the nanoscale. Owing to this, HPhPs have been proven to reflect anomalously in nanocones made of hexagonal boron nitride (hBN) [8] and to be edge steerable along counter-intuitive directions in microslits made of molybdenum trioxide [9]. By contrast, negative reflection of nanoscale-confined polaritons, which may provide another path toward manipulating light-matter excitations at the nanometer scale, remained unexplored before the findings discussed in this chapter were reported. Excitingly, and as presented in the preceding chapters, HPhPs in α-MoO_3 are ideal candidates for these optical studies, since they enable a direct experimental visualization along the surface due to their in-plane hyperbolic dispersion. In this chapter, we demonstrate the first real-space visualization of negative reflection of polaritons at the nanoscale and in a low-loss vdW crystal (α-MoO_3), unveiling a non-intuitive, unconventional, and broad tunability of the polaritonic wavelength and direction of propagation, providing fundamental insights into reflection of light and light-matter interactions at the nanoscale. To do so, we introduce an optical scheme that combines near-field interferometry measurements of HPhPs in α-MoO_3, using scattering-type near-field optical microscopy (s-SNOM, see Sect. 2.4) with the design and fabrication of tilted edges with subwavelength dimensions acting as mirrors.

5.1 Negative Reflection in Hyperbolic Media

Our finding relies on the exotic properties of HPhPs, which, as introduced in section "In-Plane Hyperbolic Phonon Polaritons in Alpha-Phase Molybdenum Trioxide" in Chap. 1, stem from the anisotropic crystal structure of the host medium, giving rise to different optical responses along different crystal directions. The anisotropy of HPhPs is captured by their hyperbolic isofrequency curve (IFC; Fig. 5.1a), a slice of the dispersion surface in wavevector—frequency ($\mathbf{k}\omega$) space—where $\mathbf{k} = (k_x, k_y)$ is the polaritonic in-plane wavevector—by a plane of a constant frequency ω. The IFC describes an open hyperbola (black curve). As a result, not all wavevectors are allowed in these media. In addition, the Poynting vector $\mathbf{S}$, which determines the direction of the polaritonic energy flux, and thus its propagation direction, is normal to the IFC.[1] Several consequences can be extracted from this.

Firstly, polaritons cannot propagate along all in-plane directions in real space. In fact, propagation is only allowed within the sectors limited by the asymptotes

[1] As commented in Sect. 1.3, the direction of the Poynting vector $\mathbf{S}$ is usually assumed to be parallel to that of the group velocity $\boldsymbol{v}_g$, and it is generally considered that both describe the propagation direction of the wave. We will also make these two assumptions along this thesis. In Ref. [10], we analytically prove that, for low losses, the direction of the imaginary part of the wavevector coincides with that of $\boldsymbol{v}_g$. However, the rigorous proof of the collinearity between $\mathbf{S}$ and $\boldsymbol{v}_g$ polaritons propagating along strongly anisotropic slabs is still an open problem.

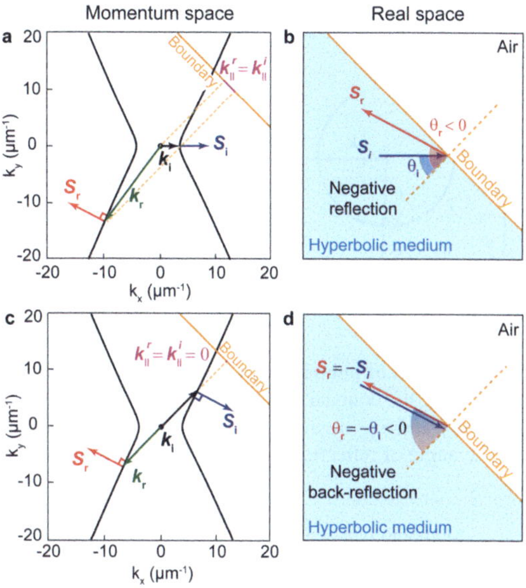

Fig. 5.1 Schematics of negative reflection of HPhPs. (**a**) Negative reflection of HPhPs upon momentum conservation at the boundary ($\Delta k_{\|} = k_{\|}^{r} - k_{\|}^{i} = 0$). The black solid line represents the IFC of HPhPs. $\mathbf{k}_{i,r}$ and $\mathbf{S}_{i,r}$, generally non-collinear, indicate the wavefronts and Poynting vectors of the incident/reflected HPhPs, respectively. (**b**) Real-space illustration of negative reflection of HPhPs. The angle of reflection θ_r is negative, and $\mathbf{S}_i$ and $\mathbf{S}_r$ are on the same side of the normal to the boundary. (**c**) Negative back-reflection of HPhPs ($k_{\|}^{r} = k_{\|}^{i} = 0$), yielding $\mathbf{k}_i = -\mathbf{k}_r$ non-collinear with $\mathbf{S}_i = -\mathbf{S}_r$. (**d**) Real-space illustration of negative back-reflection of HPhPs, yielding only a single beam off the normal

of the hyperbola in the plane, see also Eq. (3.12). Second, **k** and **S** are generally non-collinear in hyperbolic media, which has marked consequences in reflection phenomena. Specifically, upon momentum conservation at the boundary (the projection of the incident and reflected wavevectors on the boundary, $k_{\|}^{i,r}$, must be conserved, i.e., $\Delta k_{\|} = k_{\|}^{r} - k_{\|}^{i} = 0$), HPhPs can reflect at angles that are different to that of incidence, and the angle of reflection (θ_r; calculated as the angle that the reflected energy flux **S** forms with the normal to the boundary) can be even negative, i.e., $\mathbf{S}_i$ and $\mathbf{S}_r$ can be on the same side of the normal to the boundary (Fig. 5.1b). This is in stark contrast with the behavior in isotropic media, where the angle of reflection is positive and equal to that of incidence $\theta_r = \theta_i$ (specular reflection), as shown in Fig. 5.2. It is also in contrast with the behavior in elliptic media, where the angle of reflection can be already different to that of incidence, but is still positive (Fig. 5.3). In addition, the wavevector of the outcoming wave, $\mathbf{k}_r$, can be much greater in modulus than that of incidence $\mathbf{k}_i$, implying that the reflected wave can exhibit a much smaller wavelength, thereby providing a means to squeeze nanolight well beyond the diffraction limit by reflection.

The phenomenon of negative reflection can be clearly pictured under the condition of back-reflection. When the incident and reflected wavevectors are both normal to the reflecting boundary, i.e., $k_{\|}^{r} = k_{\|}^{i} = 0$ (Fig. 5.1c), the energy is

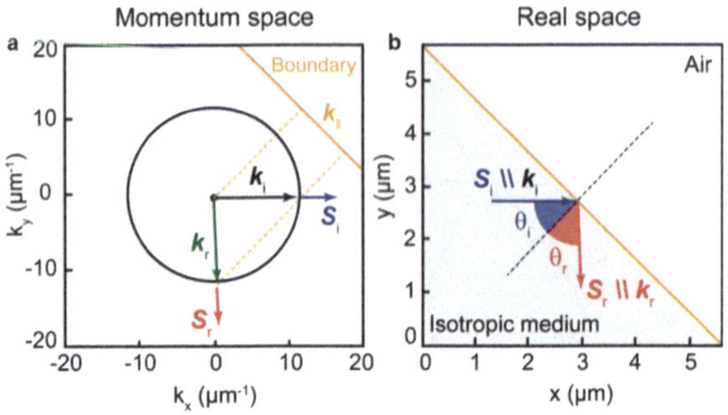

Fig. 5.2 Schematics of reflection of polaritons in isotropic media. (**a**) Reflection of polaritons upon momentum conservation at the boundary ($\Delta k_{\parallel} = k_{\parallel}^{r} - k_{\parallel}^{i} = 0$). The black solid line represents the polaritonic IFC, where $\mathbf{k}_{i,r}$ and $\mathbf{S}_{i,r}$ are always parallel. (**b**) Real-space illustration of reflection of isotropic polaritons. The angle of reflection θ_r is always equal to that of incidence θ_i (specular reflection)

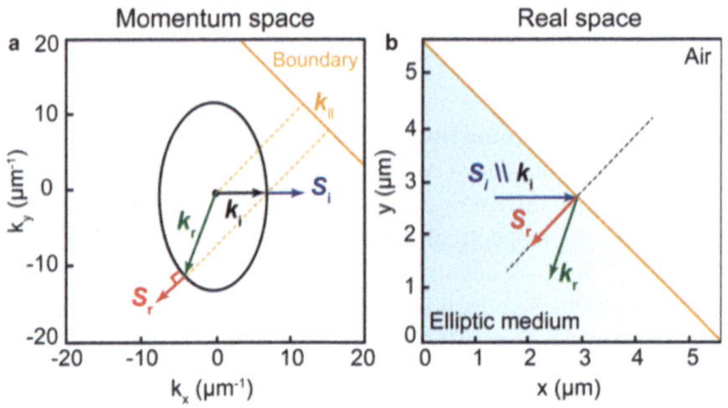

Fig. 5.3 Schematics of reflection of polaritons in elliptic media. (**a**) Reflection of polaritons upon momentum conservation at the boundary ($\Delta k_{\parallel} = k_{\parallel}^{r} - k_{\parallel}^{i} = 0$). The black solid line represents the polaritonic IFC, where $\mathbf{k}_{i,r}$ and $\mathbf{S}_{i,r}$ are generally non-collinear parallel. (**b**) Real-space illustration of reflection of elliptic polaritons. The angle of reflection θ_r is different to that of incidence θ_i

reflected back to the source, i.e., $\mathbf{S}_r = -\mathbf{S}_i$ and $\theta_r = -\theta_i < 0$ (Fig. 5.1d). Hence, the incident and reflected beams propagate along the same direction, thereby interfering and giving rise to a neat picture of negative reflection, with only a single interference beam off the normal.

5.2 Visualization of Negative Reflection of Nanoscale-Confined Polaritons in a Natural Medium

Crucially, and as commented in Sect. 2.4, back-reflection is the optical scheme used in s-SNOM polariton interferometry, where near-field images are obtained by raster scanning a metallic tip that acts both as source and detector of polaritonic fields. This is a usual constraint in s-SNOM experiments, as it does not allow directly imaging the wavefronts, but in this case serves as a notable advantage for our experiment, as it enables to selectively isolate a particular component of the wavevector and effectively image its back-reflection. This approach effectively circumvents the interference caused by the presence of other wavevectors, thereby yielding a more precise and visually coherent visualization.

Therefore, to experimentally demonstrate negative reflection at the nanoscale and in a natural medium, we performed s-SNOM polariton interferometry in α-MoO_3, which supports nanoscale-confined, low-loss in-plane HPhPs in the mid-infrared range. By taking advantage of the predictive capabilities enabled in the preceding chapters of this thesis, we study how HPhPs back-reflect at edges fabricated in thin slabs of α-MoO_3 by focused ion beam. Specifically, high-quality α-MoO_3 flakes on the SiO_2 (300 nm)/Si substrate were prepared via mechanical exfoliation of bulk crystals (see Sect. 2.1). The subwavelength mirrors in α-MoO_3 were then fabricated by focused ion beam etching, followed by a thermal annealing process [9]. The thicknesses of the α-MoO_3 flake is 250 nm. Sample fabrication was performed by Dr. Qingdong Ou and Prof. Qiaoliang Bao (Monash University, Australia). By carefully selecting the angle φ that the edges form with respect to the [100] crystal axis of α-MoO_3 and by fabricating them with subwavelength dimensions, we can unambiguously probe negative back-reflection of HPhPs along specific directions, avoiding the interference with HPhPs propagating along other directions. The use of large edges acting as mirrors would hinder the identification of the reflection direction due to interference effects (see Fig. 5.4a,b). In addition, the subwavelength edges are slightly rounded at their terminations to avoid the strong launching of HPhPs, which would also produce interferences that would distort our real-space visualization of negative reflection.

As a representative example, we fabricate edges tilted at an angle $\varphi = 38°$, thereby probing polaritonic states close to the asymptote of the hyperbolic IFC, where the non-collinearity between **k** and **S** is maximal (Fig. 5.5a). By s-SNOM imaging, we observe interferometric fringes revealing that HPhPs anomalously back-reflect at counterintuitive directions (Fig. 5.5b). In particular, the Poynting vector of the back-reflected wave (indicated by the direction of the maxima of the near-field amplitude, see Fig. 5.4c,d) is along the same side of the normal to the boundary as the incident one, unambiguously demonstrating negative reflection of HPhPs at an angle $\theta_r = -76°$. In addition, the back-reflected wavefronts appear tilted with respect to the direction of propagation and almost parallel to it (see Figs. 5.2 and 5.3), which is explained by **k** and **S** being almost perpendicular at some points of the hyperbolic IFC (Fig. 5.5a). To corroborate our experimental images and whether they demonstrate the negative reflection of PhPs, we perform full-wave simulations mimicking the experiment. As introduced in the previous chapters, to perform full-wave numerical simulations, we approximate the s-SNOM tip by a dipole source with a constant dipole moment (see Sect. 2.5). We assume that the polarizability of the dipole is weakly affected by the PhPs excited in the α-MoO_3 flake, and their back-action onto the tip can be thus neglected. Therefore,

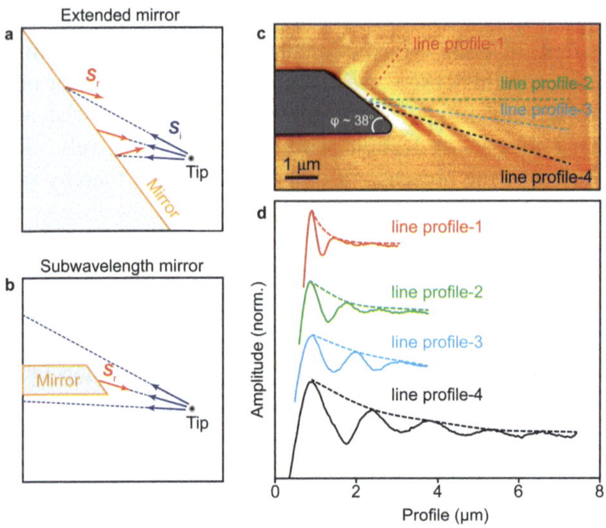

Fig. 5.4 Schematics of negative reflection of HPhPs upon extended and subwavelength mirrors and extraction of the direction of the Poynting vector from experimental near-field images. (**a**) HPhPs are launched by the tip along very different spatial directions (blue arrows). Upon reflection on an extended mirror, the outcoming HPhPs propagate along very different directions (red arrows), thereby generating complex interferences and obstructing clear visualization of the extreme case of negative reflection. (**b**) In contrast, a subwavelength mirror allows us to select waves propagating along one specific spatial direction (red arrow), enabling a clear visualization of negative reflection along a given direction by avoiding interferences with the rest of incoming waves (blue lines). Moreover, the subwavelength mirror guarantees a narrow beam of the reflected HPhPs, which is necessary to determine the direction of the polaritonic energy flux (i.e. the direction of Poynting vector, **S**) in the near-field images. (**c**) Experimental s-SNOM amplitude image of HPhPs back-reflecting on a mirror tilted at an angle $\varphi = 38°$ fabricated on α-MoO_3, at $\omega_0 = 881\,cm^{-1}$ (same data as those shown in Fig. 5.5b). We take several profiles (denoted as 1–4) along different in-plane directions. (**d**) By drawing the envelope to the line profiles, we find the direction of propagation of the Poynting vector as the direction along which the profile has the longest envelope (line profile-4)

we calculate the amplitude of the near field, $|E_z|$, as a function of the dipole position (x, y), using COMSOL Multiphysics. Figure 5.5c represents the absolute value of the z component of the electric field distribution, $|E_z|$, created by a vertical point dipole (representing the experimental tip; at a fixed lateral position, chosen along the expected direction of negatively reflected PhPs according to Fig. 5.5a). We can clearly recognize similar tilted interference fringes as those observed in the near-field image, thus unambiguously demonstrating the back-reflection of PhPs at the mirror.

For comparison, we also visualize in-plane elliptic and in-plane isotropic polaritons back-reflecting on mirrors fabricated in α-MoO_3 and isotopically enriched $h^{10}BN$ [11], respectively (Fig. 5.5d,g). The mirrors in hBN were fabricated by e-beam by Dr. Elizaveta Nikulina and Prof. Rainer Hillenbrand (CIC nanoGUNE,

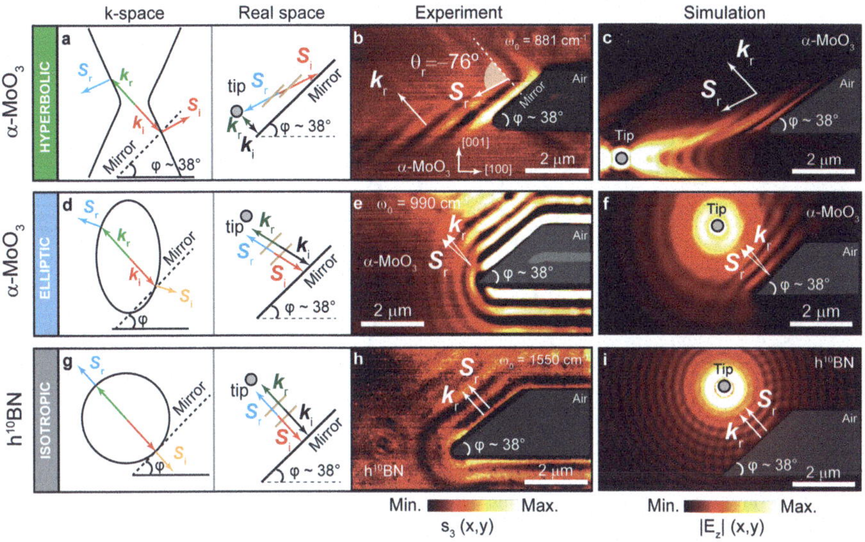

Fig. 5.5 Visualization of negative reflection of nanoscale-confined polaritons in natural media. Illustration of back-reflection of HPhPs in **k**-space and real space for (**a**) in-plane hyperbolic, (**d**) in-plane elliptic and (**g**) in-plane isotropic media. (**b**, **e** and **h**) Experimental s-SNOM amplitude image $s_3(x, y)$ of HPhPs back-reflecting on mirrors tilted at an angle of $\varphi = 38°$ fabricated on α-MoO_3, measured in in-plane hyperbolic (**b**) and in-plane elliptic (**e**) spectral bands, and in $h^{10}BN$ (**h**). $\mathbf{k}_{i,r}$ and $\mathbf{S}_{i,r}$ indicate the wavefronts and Poynting vectors of the incident/reflected HPhPs, respectively. (**c**, **f** and **i**) Full-wave numerical simulation $E_z(x, y)$ of back-reflected polaritons excited by a point dipole on α-MoO_3 in in-plane hyperbolic (**c**) and in-plane elliptic (**f**) spectral bands and in $h^{10}BN$ (**i**)

Spain). As expected, we observe interference fringes parallel to all the mirrors (Fig. 5.5e,h) and thus along all in-plane directions, in excellent agreement with our simulations (Fig. 5.5f,i). The only difference between these two cases lies in the fact that, in the in-plane elliptic scenario, the wavefronts deviate slightly from being perpendicular to the direction of propagation, whereas in the in-plane isotropic case they maintain a perpendicular orientation.

To gain further insights into the properties of negative reflection at the nanoscale, we study the dependence of the polaritonic wavelength λ_p and the angle of reflection θ_r as a function of the angle of the mirror φ and incident frequency ω_0. To do so, we fabricate mirrors in α-MoO_3 at different angles of $\varphi = 38°$, $60°$, and $90°$ and perform s-SNOM polariton interferometry of back-reflected HPhPs. Since, in anisotropic media, the wavevector is direction dependent, λ_p changes depending on the direction along which we map back-reflection and hence on φ. By selecting φ, we probe the behavior of HPhPs upon negative reflection for different **k** and **S**, i.e., different angles of incidence. This is illustrated in Fig. 5.6a,b and in Fig. 5.7 for $\omega_0 = 889\,\text{cm}^{-1}$. Note that Fig. 5.6a,b and the top panels in Fig. 5.7a,b show the same results for completeness.

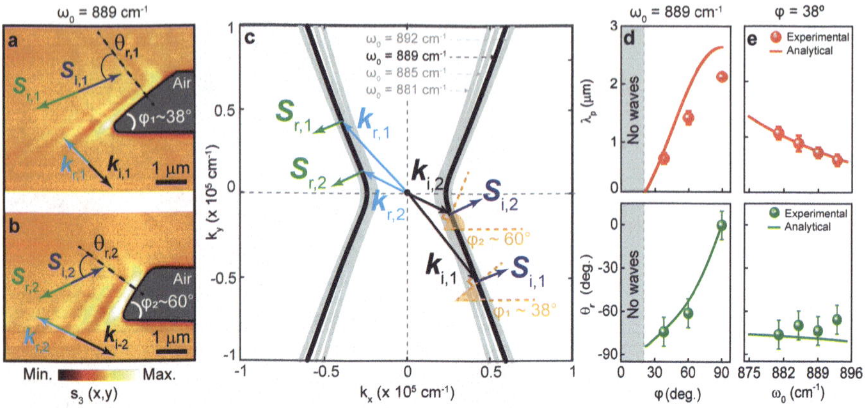

Fig. 5.6 Properties of negative reflection of HPhPs in α-MoO$_3$. (**a** and **b**) Experimental near-field images $s_3(x, y)$ of HPhPs back-reflecting on mirrors fabricated on an α-MoO$_3$ slab and tilted at $\varphi = 38°$ and 60°, respectively. The incident frequency is $\omega_0 = 889\,\text{cm}^{-1}$. (**c**) IFC of HPhPs showing the directions of $\mathbf{k}_{i,r}$ and $\mathbf{S}_{i,r}$ of the incident and back-reflected HPhPs on mirrors tilted at $\varphi = 38°$ and 60° at $\omega_0 = 889$ (black line), 892, 885, and 881 cm^{-1} (gray lines). (**d** and **e**) Dependence of the polaritonic wavelength λ_p (top) and angle of reflection θ_r (bottom) on φ ($\omega_0 = 889\,\text{cm}^{-1}$) and ω_0 ($\varphi = 38°$), respectively. The continuous lines in (**d**) and (**e**) are extracted from analytical calculations and the symbols from experimental s-SNOM images

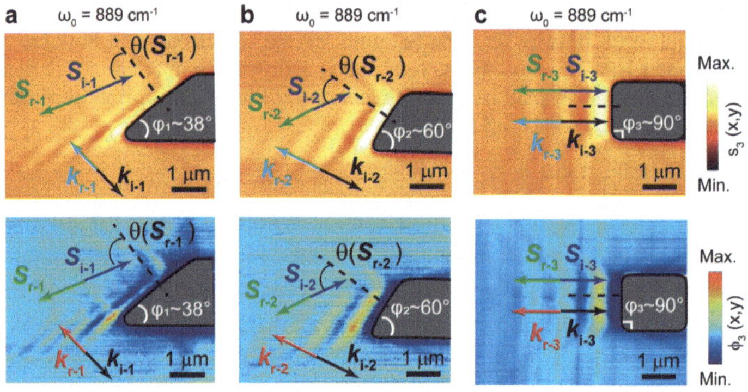

Fig. 5.7 Negative reflection of HPhPs in α-MoO$_3$ as a function of the angle of the boundary. Experimental near-field amplitude $s_3(x, y)$ (top row) and phase $\phi_3(x, y)$ (bottom row) of HPhPs back-reflecting on mirrors fabricated on an α-MoO$_3$ slab and tilted at (**a**) $\varphi = 38°$, (**b**) 60° and (**c**) 90°. The incident frequency is $\omega_0 = 889\,\text{cm}^{-1}$

Clearly, λ_p and θ_r change as a function of φ. Specifically, as a result of the hyperbolic-like shape of the IFC, when φ decreases (from 90° to 20°), $\mathbf{k}_i$ and $\mathbf{k}_r$ approach the asymptote (see black and cyan arrows in Fig. 5.6, a to c), and λ_p can be markedly squeezed (from 2.5 to 0 μm; Fig. 5.6d, top), while, on the other hand, θ_r takes values in a very broad angular sector (from 0° to −80°; Fig. 5.6d, bottom). Note that φ values smaller than 20° are forbidden, since the corresponding

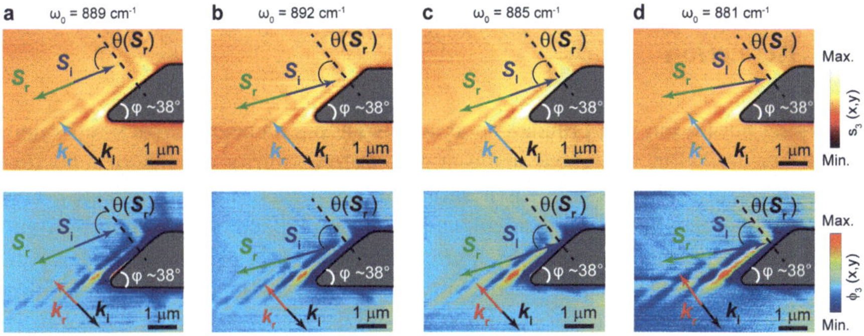

Fig. 5.8 Negative reflection of HPhPs in α-MoO_3 as a function of the incident frequency. Experimental near-field amplitude $s_3(x, y)$ (top row) and phase $\phi_3(x, y)$ (bottom row) of HPhPs back-reflecting on mirrors fabricated on an α-MoO_3 slab at incident frequencies (**a**) $\omega_0 = 889\,\mathrm{cm}^{-1}$, (**b**) $\omega_0 = 892\,\mathrm{cm}^{-1}$, (**c**) $\omega_0 = 885\,\mathrm{cm}^{-1}$ and (**d**) $\omega_0 = 881\,\mathrm{cm}^{-1}$. The angle of the boundary is $\varphi = 38°$

wavevectors are not allowed by the hyperbolic IFC (gray shaded area). The error bars in λ_p are the result of averaging five line profiles in the near-field images, while for θ_r, we take a conservative error of $\pm 10°$. The continuous lines show calculations based on the analytical IFC and momentum conservation at the mirror, in excellent agreement with the experiment. Note that the reasonable deviation in the top panel of Fig. 5.6d may be caused by the focused ion beam process involved in the fabrication of the reflectors, which may induce a gradient thickness in the α-MoO_3 slab. Together, these results show that both λ_p and θ_r exhibit unconventional and broad tunability as a function of φ, in stark contrast to back-reflected polaritons in isotropic media, where the same λ_p and θ_r is obtained for all φ. We note that polariton interferometry of back-reflected polaritons is the technique that allows us to probe these exotic aspects of negative reflection, yet the conclusions are broader and apply beyond the case of back-reflection and even of negative reflection, providing a strategy to squeeze and steer nanolight.

Given the strong dispersive nature of the back-reflected polaritons, we further study their dependence upon ω_0. Figure 5.8 shows near-field amplitude (top row) and phase (bottom row) of HPhPs negatively reflecting on a mirrors tilted at $\varphi = 38°$ as a function of ω_0, Note that, for completeness, the top panels in Fig. 5.7a,d display the same results as those shown in Fig. 5.6a and Fig. 5.5b. Data values of λ_p and θ_r extracted from these images are plotted in the top and bottom panels in Fig. 5.6e, together with analytical calculations (green continuous lines), showing an excellent agreement. The spectral dependence demonstrates that λ_p and θ_r can be tuned between 0.6 and 1.1 μm and $-76°$ and $-65°$, respectively, when ω_0 varies from 881 to $893\,\mathrm{cm}^{-1}$. Note that this dependence does not hold in isotropic media, where θ_r is always equal to zero under back-reflection regardless of ω_0.

5.3 A Hyperbolic Nanoresonator Based on Negative Reflection

Following our findings and fundamental understanding on negative back-reflection, we further derive the analytical shape of a polaritonic mirror capable of reflecting all the incident HPhPs back to their source, i.e., a hyperbolic retroreflector. This curve must be such that every incident wavevector is perpendicular to it. In isotropic media, this condition is satisfied by a circumference (inset in Fig. 5.9a). However, in hyperbolic media, because of the non-collinearity of **k** and **S**, the condition yields the following hyperbola [see Eq. (C.2) and the rest of Appendix C.1 for more insights into the analytical derivation]:

$$y(x) = \pm\sqrt{\frac{x^2|\varepsilon_y|}{|\varepsilon_x|} - a^2}, \tag{5.1}$$

where x and y are the directions along the main and conjugate axes of the hyperbola, ε_x and ε_y are the dielectric permittivities along x and y, and a is the

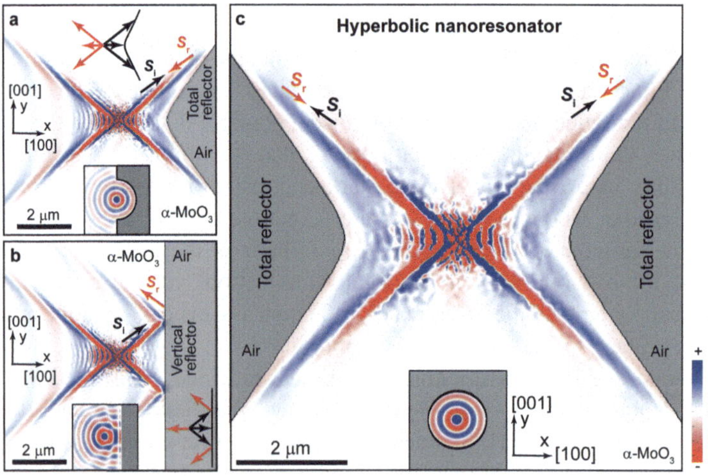

Fig. 5.9 Retroreflector and nanoresonator for HPhPs. Simulated near-field image Re[$E_z(x, y)$] of HPhPs reflecting on (**a**) a hyperbolic retroreflector, (**b**) a straight edge, and (**c**) a hyperbolic nanoresonator consisting of two retroreflectors on a 200-nm-thick α-MoO$_3$ slab at an incident laser frequency of $\omega_0 = 920\,\text{cm}^{-1}$. The source is a point dipole located at a distance equal to the HPhP wavelength of $\lambda_p = 1.7\,\mu\text{m}$ from the edge. The insets show Re[$E_z(x, y)$] of isotropic polaritons upon reflection on (**a**) a retroreflector in isotropic media (a semi-circumference), (**b**) a straight edge, and (**c**) a nanoresonator in isotropic media (a circumference)

distance between the center and the vertex of the hyperbola.[2] Figure 5.9a shows numerical simulations of the electric field $\mathrm{Re}[E_z(x, y)]$ on a α-MoO_3 slab using a retroreflector whose shape is given in Eq. (5.1) for HPhPs in α-MoO_3 at an incident frequency of $\omega_0 = 920\,\mathrm{cm}^{-1}$ and a distance a equal to $\lambda_p = 1.7$ μm, where λ_p is the analytical polariton wavelength along the x direction ([100] crystal direction). All the incident waves reflect to the location of the source. In order to determine whether there is any enhancement of the intensity ΔI_z produced under the dipole source by the total hyperbolic reflector, we calculate $\Delta I_z = (\Delta E_z)^2$, where ΔE_z is the electric field enhancement, given by

$$\Delta E_z = \frac{E_z^R}{E_z^0} - 1,$$

with E_z^R and E_z^0 as the z components of the electric field created by a dipole source placed on top of the system, with and without reflector, respectively. Both E_z^R and E_z^0 are probed at the location of the source, thereby providing a simple means to cancel the contribution from the dipole to the field enhancement. Pulling out the enhancement factor with further confidence with the strong dipole source is nontrivial and falls beyond the scope of this work. Evaluating ΔI_z for the structure shown in Fig. 5.9a, we get an enhancement of the intensity at the source of approximately +35%.

Figure 5.9b shows a comparison with a standard reflector, such as a straight edge, in which HPhPs propagate far away from the source (red arrows in Fig. 5.9b). By mirroring the hyperbolic retroreflector structure with respect to the location of the source, we design an optical nanoresonator for in-plane HPhPs (Fig. 5.9c), enabling all the hyperbolic waves launched by a dipole located at its center to continuously reflect back and forth between its boundaries, yielding a standing wave pattern which is perhaps more clearly observed by plotting the field amplitude (see Fig. 5.10). In this case, the intensity at the source is enhanced approximately +64%. We note that this value is very robust to displacements of the dipole of up to 1 μm inside the cavity, both along the x and y directions, which is vital for practical purposes. While this value could be further improved by making the cavity dimensions smaller, here, we introduce a first proof-of-concept hyperbolic nanoresonator that exploits both the strong directionality of in-plane HPhPs and their high momenta to enhance light-matter interactions and hence with outstanding prospects for nanophotonics.

We note here that polaritons are launched by a point dipole to study just the reflection phenomena, excluding any focusing effects due to far-field illumination.

[2] This analytical derivation was performed assuming an infinite hyperbolic medium along the z direction, thereby neglecting the dependence of the polaritonic wavelength on the thickness of the slab, which is then introduced ad hoc. A derivation that does consider the finite thickness of the slab is thus an interesting avenue for future work, and would provide a more accurate approach towards the experimental realization of the device.

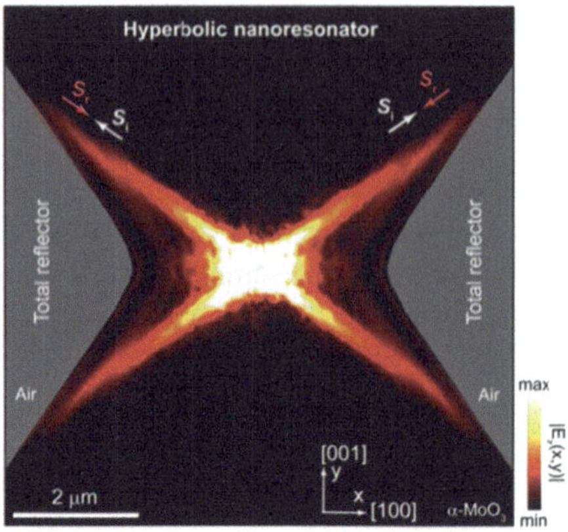

Fig. 5.10 **Nanoresonator for HPhPs.** Simulated near-field image $|E_z(x, y)|$ of HPhPs reflecting on a hyperbolic nanoresonator consisting of two retroreflectors on a 200-nm-thick α-MoO_3 slab at an incident laser frequency of $\omega_0 = 920\,cm^{-1}$ (the same structure as the one shown in Fig. 5.9c). The source is a point dipole located at a distance equal to the HPhP wavelength $\lambda_p = 1.7\,\mu m$ from the edge

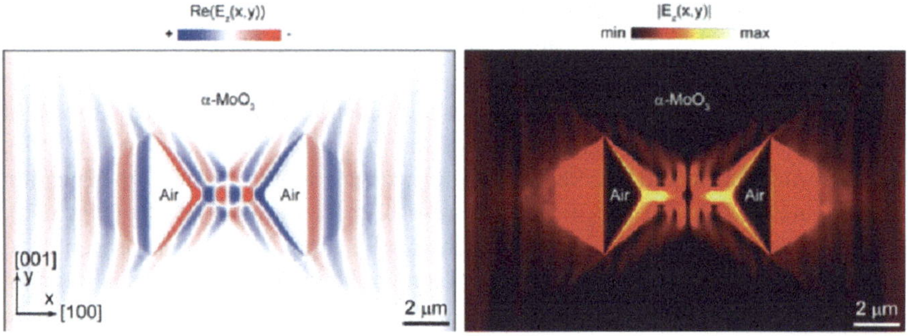

Fig. 5.11 **Nanoresonator for HPhPs under far-field illumination.** Simulated near-field image $Re[E_z(x, y)]$ (left) and $|E_z(x, y)|$ of HPhPs launched by the edges of a hyperbolic nanoresonator consisting of two retroreflectors, on a 200-nm-thick α-MoO_3 slab at an incident laser frequency of $\omega_0 = 920\,cm^{-1}$

To study the combination of both focusing and launching effects, we conducted a simulation of PhPs excited in a hyperbolic nanoresonator under far-field illumination. As shown in Fig. 5.11, the phases of PhP waves launched by the edges of each reflector element are opposite, which is analogous to the phase difference of the polariton waves launched by resonant antennas [12, 13]. As such, a destructive interference generates complex near-field patterns, leading to a decrease in intensity at the center of the structure. The shape of the total hyperbolic retroreflector shown in Fig. 5.9 is derived by assuming that all the waves arrive to origin with the same phase (see Appendix C.1), which is the case for dipole-launching PhPs. We note that the focusing effect of antenna-launching PhPs (i.e. under far-field illumination) arises from the interference between their high-momentum components, as studied in prior works [14–16]. Optimizing the geometry for both focusing and reflection effects is out of the scope of this discussion, but provides an interesting avenue for future work.

5.4 Conclusions

In conclusion, we have demonstrated the negative reflection of nanoscale-confined HPhPs in a low-loss natural medium (α-MoO_3). Our study unveils unconventional tunability of the direction of propagation and wavelength of the negatively reflected HPhPs as a function of the incident frequency ω_0 and the angle of the mirror φ, opening a plethora of possibilities to control nanolight by reflection in hyperbolic media. These fundamental insights allowed us to develop an optical nanodevice: a hyperbolic nanoresonator. Because of the high density of optical states in hyperbolic media, these cavities hold promises for controlling the decay rates of emitters. In addition, open cavities as our hyperbolic nanoresonator are more accessible from the outside and thus experimentally preferred to design a photonic circuit. Together, our findings provide a route for controlling and manipulating the flow of energy at the nanoscale by using hyperbolic waves and for squeezing light to nano- or subwavelength scales, thereby enabling to compress and enhance light-matter interactions to unprecedentedly small volumes, the general premise of nanophotonics. Our results thus hold promises for the development of applications in optoelectronics, biosensing, nanoimaging, and chemistry.

References

1. G. Álvarez-Pérez et al., Negative reflection of nanoscale-confined polaritons in a low-loss natural medium. Sci. Adv. **8**, eabp8486 (2022). https://www.science.org/doi/abs/10.1126/sciadv.abp8486
2. N. Yu et al., Light propagation with phase discontinuities: generalized laws of reflection and refraction. Science **334**, 333–337 (2011). https://www.science.org/doi/abs/10.1126/science.1210713
3. F. Aieta et al., Out-of-plane reflection and refraction of light by anisotropic optical antenna metasurfaces with phase discontinuities. Nano Lett. **12**, 1702–1706 (2012). https://doi.org/10.1021/nl300204s
4. B. Liu, W. Zhao, Y. Jiang, Apparent negative reflection with the gradient acoustic metasurface by integrating supercell periodicity into the generalized law of reflection. Sci. Rep. **6**, 38314 (2016). https://doi.org/10.1038/srep38314
5. S. Liu et al., Negative reflection and negative surface wave conversion from obliquely incident electromagnetic waves. Light Sci. Appl. **7**, 18008–18008 (2018). https://doi.org/10.1038/lsa.2018.8
6. B. Gérardin, J. Laurent, F. Legrand, C. Prada, A. Aubry, Negative reflection of elastic guided waves in chaotic and random scattering media. Sci. Rep. **9**, 2135 (2019). https://doi.org/10.1038/s41598-019-38480-3

7. C. Zhang, T.J. Cui, Negative reflections of electromagnetic waves in a strong chiral medium. Appl. Phys. Lett. **91**, 194101 (2007). https://doi.org/10.1063/1.2804124
8. A.J. Giles et al., Imaging of anomalous internal reflections of hyperbolic phonon-polaritons in hexagonal boron nitride. Nano Lett. **16**, 3858–3865 (2016). https://doi.org/10.1021/acs.nanolett.6b01341
9. Z. Dai et al., Edge-oriented and steerable hyperbolic polaritons in anisotropic van der Waals nanocavities. Nat. Commun. **11**, 6086 (2020). https://doi.org/10.1038/s41467-020-19913-4
10. K.V. Voronin, G. Álvarez-Pérez, C. Lanza, P. Alonso-González, A.Y. Nikitin, Fundamentals of polaritons in strongly anisotropic thin crystal layers. ACS Photonics **11**, 550–560 (2024). https://doi.org/10.1021/acsphotonics.3c01428
11. A.J. Giles et al., Ultralow-loss polaritons in isotopically pure boron nitride. Nat. Mater. **17**, 134–139 (2018). https://doi.org/10.1038/nmat5047
12. P. Pons-Valencia et al., Launching of hyperbolic phonon-polaritons in h-BN slabs by resonant metal plasmonic antennas. Nat. Commun. **10**, 3242 (2019). https://doi.org/10.1038/s41467-019-11143-7
13. P. Alonso-González et al., Controlling graphene plasmons with resonant metal antennas and spatial conductivity patterns. Science **344**, 1369–1373 (2014). https://www.science.org/doi/abs/10.1126/science.1253202
14. J. Martín-Sánchez et al., Focusing of in-plane hyperbolic polaritons in van der Waals crystals with tailored infrared nanoantennas. Sci. Adv. **7**, eabj0127 (2021). https://www.science.org/doi/abs/10.1126/sciadv.abj0127
15. Z. Zheng et al., Controlling and focusing in-plane hyperbolic phonon polaritons in α-MoO_3 with a curved plasmonic antenna. Adv. Mater. **34**, 2104164 (2022). https://onlinelibrary.wiley.com/doi/abs/10.1002/adma.202104164
16. Y. Qu et al., Tunable planar focusing based on hyperbolic phonon polaritons in α-MoO_3. Adv. Mater. **34**, 2105590 (2022). https://onlinelibrary.wiley.com/doi/abs/10.1002/adma.202105590

Chapter 6
Anomalous Refraction and Lensing of Nanoscale-Confined Hyperbolic Polaritons

Abstract Refraction between isotropic media is a well understood phenomenon, described by the simplest form of Snell's law and characterized by light bending towards the normal to the boundary when passing from a low- to a high-refractive-index medium. However, refraction between anisotropic media is a more exotic phenomenon which had remained barely investigated before the research presented in this chapter was carried out, particularly at the nanoscale. In this chapter, we visualize and comprehensively study the general case of refraction of electromagnetic waves between two strongly anisotropic (hyperbolic) media, and we do it by imaging nanoscale-confined polaritons in α-MoO_3. Upon refraction, these polaritons exhibit non-intuitive directions of propagation as they traverse planar nanoprisms, enabling to unveil an exotic optical effect: bending-free refraction. Furthermore, we develop an in-plane refractive hyperlens, yielding foci as small as $\lambda_p/6$, being λ_p the polariton wavelength ($\lambda_0/50$ compared to the wavelength of free-space light). These results set the grounds for planar nano-optics in strongly anisotropic media, with potential for effective control of the flow of energy at the nanoscale.

The near-field experiments reported in this chapter were carried out in collaboration with Dr. Jiahua Duan (University of Oviedo, Spain). Sample fabrication was performed mainly by Ana Isabel Fernández-Tresguerres Mata (University of Oviedo, Spain), Bingdong Chang and Prof. Sanshui Xiao (Technical University of Denmark, Denmark). The results were published in "Planar refraction and lensing of highly confined polaritons in anisotropic media" by Jiahua Duan, Gonzalo Álvarez-Pérez et al. in Nature Communications 12, 4325 (2021).

Refraction is one of the most fundamental optical phenomena, very well understood in isotropic media, and central in the development of many technologies, such as lenses or microscopes. When it comes to hyperbolic waves, refraction has traditionally been a much less explored phenomenon, despite the fundamental interest of this type of waves and their potential for the development of new optical applications, which have been extensively discussed in this thesis so far. Only very recently, and following the discoveries presented in this chapter, a

G. Álvarez Pérez, *Fundamentals of Nano-Optics in Hyperbolic van der Waals Materials*, Springer Theses, https://doi.org/10.1007/978-3-031-75767-9_6

couple of subsequent studies have documented the experimental observation of negative refraction in van der Waals (vdW) heterostructures. These structures were formed using hBN and α-MoO_3 [1], as well as graphene and α-MoO_3 [2]. In turn, before the findings presented in this chapter were reported, refraction of hyperbolic waves had only been studied for the case in which the incident beam comes from an isotropic medium, typically free space [3–5]. As such, the general case of refraction, involving hyperbolic waves in which both the incident and the refracted waves exhibit non-collinear wavevector **k** and energy flux **S** had remained experimentally unexplored, particularly at the nanoscale, where the specific case of negative refraction of highly confined polaritons had only been recently proposed in two theoretical studies [6, 7].

As already illustrated in the previous chapters, the discoveries of phonon polaritons (PhPs) in vdW crystals [8, 9] with hyperbolic dispersion, such as hBN [10–12], α-MoO_3 [13, 14], and α-V_2O_5 [15], provided unique material platforms to study exotic optical phenomena. In particular, we have shown in the preceding chapters that PhPs in α-MoO_3 feature in-plane hyperbolic propagation, ultra-low losses, and strong confinement, offering the possibility to visualize anomalous refraction directly on the crystal surface and at the nanoscale, thereby opening new routes in planar nano-optics [16].

In this chapter [19], we show, theoretically and experimentally, the general case of refraction at the interface between two strongly anisotropic (hyperbolic) media. Importantly, we do it at the nanoscale and in a low-loss natural medium by visualizing the propagation of polaritons as they traverse planar nanoprisms tailored on the surface of α-MoO_3. Our images show non-intuitive directions of propagation and strong confinement of the refracted waves, enabling to unveil an exotic optical effect: bending-free refraction, which extends the current capabilities to control the propagation of light at the nanoscale. Furthermore, we develop an in-plane refractive hyperlens, yielding foci as small as $\lambda_p/6$, being λ_p the polariton wavelength ($\lambda_0/50$ with respect to the wavelength of light in free space). Our findings add to the fundamental knowledge described in the preceding chapters, while also provide a novel effective strategy for the manipulation of polaritons in anisotropic media, paving the way for integrated flat subwavelength optics.

6.1 Anomalous Refraction Between Two Hyperbolic Media

The unique properties of polaritons in hyperbolic media can be understood by analyzing their isofrequency curve (IFC). The IFCs of polaritons in two different hyperbolic media are illustrated in Fig. 6.1a,b. For convenience, we consider these two media to be defined by the same hyperbolic slab (we select representative permittivity values $\varepsilon_x = -5$, $\varepsilon_y = 1$ and $\varepsilon_z = 5$) placed on two different dielectric substrates (with permittivities $\varepsilon_{sub} = 1$ and $\varepsilon_{sub} = 5$). As discussed in the previous chapter, the Poynting vector **S**—which determines the propagation direction of the

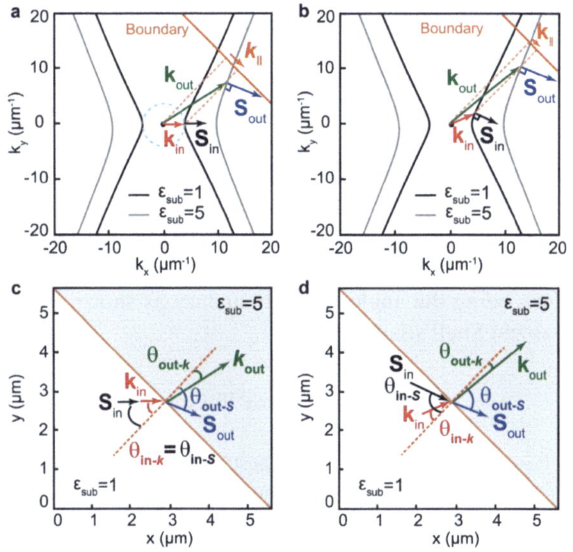

Fig. 6.1 Schematics of anomalous refraction of polaritons between two hyperbolic media. (**a**) Isofrequency curves of polaritons propagating in a hyperbolic slab (with $\varepsilon_x = -5; \varepsilon_y = 1; \varepsilon_z = 5$) placed on two different semi-infinite substrates with $\varepsilon_{sub} = 1$ (black curve) and $\varepsilon_{sub} = 5$ (gray curve) that define two different hyperbolic media (medium 1 and 2, respectively). The incident wave in medium 1 is characterized by collinear $\mathbf{k}_{\text{in}}$ and $\mathbf{S}_{\text{in}}$ (as in an isotropic medium, indicated by a dashed cyan circle). Upon refraction into medium 2, momentum conservation at the boundary (orange line), $k_{\|}$ ($k_{\|} = k_{\text{in}} \cdot \sin\varphi$), where φ is the angle of the boundary), is fulfilled by non-collinear $\mathbf{k}_{\text{out}}$ and $\mathbf{S}_{\text{out}}$. The dashed orange lines represent the normal to the boundary. (**b**) The general case of refraction between two hyperbolic media is represented by an incident wave from medium 1 with non-collinear $\mathbf{k}_{\text{in}}$ and $\mathbf{S}_{\text{in}}$ (normal to the isofrequency curve). When the wave refracts into medium 2, momentum conservation at the boundary (orange line) is fulfilled by non-collinear $\mathbf{k}_{\text{out}}$ and $\mathbf{S}_{\text{out}}$. The dashed orange lines represent the normal to the boundary. (**c**) Real-space illustration of refraction between two hyperbolic media shown in a where the incident wave exhibits collinear $\mathbf{k}_{\text{in}}$ and $\mathbf{S}_{\text{in}}$, i.e. $\theta_{\text{in-k}} = \theta_{\text{in}-\mathbf{S}}$, giving rise to non-collinear $\mathbf{k}_{\text{out}}$ and $\mathbf{S}_{\text{out}}$, i.e. $\theta_{\text{out}-\mathbf{S}} \neq \theta_{\text{out}-\mathbf{k}}$. (**d**) Real-space illustration of the general case of refraction between two hyperbolic media shown in (**b**) where both the incident and the outgoing wave exhibits non-collinear **k** and **S**, i.e. $\theta_{\text{in-k}} \neq \theta_{\text{in-S}}$ and $\theta_{\text{out-S}} \neq \theta_{\text{out-k}}$. The tangents parallel to both hyperbolas give rise to bending-free refraction, i.e. $\theta_{\text{in-S}} \approx \theta_{\text{out-S}}$. The orange dashed lines in (**c,d**) represent the normal to the boundary. The white and gray regions in (**c,d**) correspond to α-MoO_3/air and α-MoO_3/SiO_2, respectively

polariton and is normal to the IFC[1]—is in general not collinear with **k**, as indicated in Fig. 6.1a (they are collinear only along the x axis in Fig. 6.1a). As such, the

[1] As already pointed out in Sect. 1.3 and Chap. 5, the direction of the Poynting vector **S** is usually assumed to be parallel to that of the group velocity $\boldsymbol{v}_g$, and it is generally considered that both describe the propagation direction of the wave. We will also assume this here. In Ref. [17], we analytically prove that, for low losses, the direction of the imaginary part of the wavevector coincides with that of $\boldsymbol{v}_g$. However, the rigorous proof of the collinearity between **S** and $\boldsymbol{v}_g$ polaritons propagating along strongly anisotropic slabs is still an open problem.

properties of propagating polaritons in hyperbolic media are different to those in isotropic media, where the IFCs are circular (see dashed cyan curve in Fig. 6.1a) and polaritons, as is well-known, are allowed to propagate along all in-plane directions in real space with the same absolute value of the wavevector, which is always collinear to **S**. Importantly, such properties of polaritons propagating in hyperbolic media have a dramatic effect when they refract at a boundary between two different hyperbolic media. In analogy to the study of negative reflection of HPhPs in Chap. 5, momentum conservation at the boundary implies here that the projection of the incident and refracted wavevectors ($\mathbf{k}_{in}$ and $\mathbf{k}_{out}$, respectively) must be conserved (with $k_{\|} = k_{in}\sin\varphi$ being the angle of the boundary as shown in Fig. 6.1a), giving rise to the generalized Snell's law [18]:

$$k_{in}\sin(\theta_{in\text{-}k}) = k_{out}\sin(\theta_{out\text{-}\mathbf{k}}), \tag{6.1}$$

where $\theta_{in\text{-}k}$ and $\theta_{out\text{-}\mathbf{k}}$ are the angles that $\mathbf{k}_{in}$ and $\mathbf{k}_{out}$ form with the normal to the boundary, respectively (see Fig. 6.1c, d). The propagation directions of the incident and refracted waves are then given by $\mathbf{S}_{in}$ and $\mathbf{S}_{out}$, respectively, i.e. the directions normal to the hyperbolic IFCs for each case, which are in general non-collinear with $\mathbf{k}_{in}$ and $\mathbf{k}_{out}$ (Fig. 6.1b–d), and thus refraction can occur at angles $\theta_{out-\mathbf{S}}$ that can be different from $\theta_{out\text{-}\mathbf{k}}$. This behavior is in stark contrast to that in isotropic media, in which momentum conservation at the boundary implies that refraction occurs always at $\theta_{out\text{-}\mathbf{k}} = \theta_{out\text{-}\mathbf{S}}$. This case will be demonstrated experimentally in Sect. 6.2.

A particular case of refraction between hyperbolic media for an incident wave with collinear k_{in} and $\mathbf{S}_{in}$ is shown in Fig. 6.1a. We observe that the refracted wave ($\mathbf{S}_{out}$) bends away from the direction of $\mathbf{S}_{in}$ towards the boundary (Fig. 6.1c), in contrast to what is expected in isotropic media for a wave passing from a low refractive index to a high refractive index, where as is well known $\mathbf{S}_{out}$ bends towards the normal to the boundary. Also, the modulus of $\mathbf{k}_{out}$ is much larger than that of $\mathbf{k}_{in}$, showing a strong wavelength reduction.

The general case of refraction occurs when $\mathbf{k}_{in}$ and $\mathbf{S}_{in}$ of the incident wave are not collinear. This case will be tackled experimentally in this chapter for the first time, in Sect. 6.4, as well as in Ref. [19]. It is also sketched in Fig. 6.1d, showing a wave impinging at a tilting angle $\theta_{in\text{-}\mathbf{S}}$ on the boundary between two hyperbolic media. The boundary is also tilted a given angle with respect to the crystal axes. Interestingly, due to the similar shapes of the IFCs in the considered hyperbolic media (α-MoO_3/air and α-MoO_3/SiO_2, black and gray curves in Fig. 6.1b, respectively), the Poynting vectors of the incident and refracted waves are parallel for almost any $\mathbf{k}_{in}$ and boundary angle φ, especially in the region where the arms of both hyperbolic IFCs are straight. Hence, the refracted wave propagates almost parallel to the incident wave (i.e. $\theta_{in\text{-}\mathbf{S}} = \theta_{out\text{-}\mathbf{S}}$), as if the incident wave had been transmitted directly without any change in its propagation direction (black and blue arrows in Fig. 6.1d). This feature opens the door to the realization of bending-free refraction at arbitrary incident angles in anisotropic media, which is not possible in isotropic media. Note, however, that the modulus and direction of $\mathbf{k}_{out}$ is very different from the modulus and direction of $\mathbf{k}_{\text{in}}$, which opens the door to direct engineering of the wavelength and wavefront without modifying the direction of propagation of the wave.

6.2 Nanoimaging of Anomalous Refraction of Nanoscale-Confined In-Plane Hyperbolic Polaritons

In this section, we experimentally study the characteristics of refraction of polaritons propagating in hyperbolic media, providing the first demonstration of these effects at the nanoscale and/or in a natural medium. To do so, we design and fabricate planar prisms in a slab of the naturally hyperbolic vdW crystal α-MoO_3. To define the prisms, we etch away triangular regions on a silica (SiO_2) substrate on top of which we place a 160-nm-thick α-MoO_3 slab, thus forming a region α-MoO_3/air with a different refractive index, and thus different polaritonic dispersion, with respect to the region α-MoO_3/SiO_2. This is done by following a two-step process: (1) the triangular patterns were defined using a direct writing system (MLA 100, Heidelberg Instrument) equipped with a 365 nm LED light source on a 1.5-μm-thick positive tone photoresist (AZ MiR 701, MicroChemicals) previously deposited on the SiO_2/Si substrate; (2) part of the SiO_2 layer was etched away by a fluorine-based plasma using an inductively coupled plasma system (Advanced Oxide Etcher, SPTS), generating gaps with a depth of around 350 nm into the SiO_2 layer. To get rid of photoresist residues and contaminants, the etched substrates were cleaned with a plasma ashing system (PVA TePla 300, PVA TePla AG) before transferring the flakes. The fabrication was carried out in collaboration with Bingdong Chang and Prof. Sanshui Xiao (Technical University of Denmark, Denmark).

Once the flake is transferred on top, we visualize the propagation of hyperbolic phonon polaritons (HPhPs) passing through the prisms by polariton wavefront mapping using a scattering-type scanning near-field optical microscope (s-SNOM), as described in Sect. 2.4. The near-field image of Fig. 6.2a, taken at an incident wavelength $\lambda_0 = 11.3$ μm and showing the real part of the field, clearly corroborates the two different polaritonic dispersions in the regions α-MoO_3/air and α-MoO_3/SiO_2. Specifically, we observe HPhPs launched inside the prism (highlighted by white dashed lines) by the flake edge (see Fig. 6.3), which propagate with collinear $\mathbf{k}_{\text{in}}$ and $\mathbf{S}_{\text{in}}$ and longer wavelength λ_{in} (white arrow), i.e. smaller wavevector $|\mathbf{k}_{\text{in}}| = 2\pi/\lambda_{\text{in}}$ (black arrow), than outside the prism (λ_p, red arrow), thus revealing the lower refractive index of the prisms. We also observe in the same near-field image that when the HPhPs reach a boundary of the prism tilted at an angle $\theta_{\text{in}} \approx 55°$, they refract into the α-MoO_3/SiO_2 region (note that reflection is expected to be very weak) along a different direction ($\mathbf{S}_{\text{out-exp}}$, blue arrows) with respect to which their wavefronts are tilted ($\mathbf{k}_{\text{out-exp}}$, green arrows). Thus, the refracted energy flux and wavevector are not collinear, in agreement with our predictions for hyperbolic refraction in Fig. 6.1a. Importantly, while the

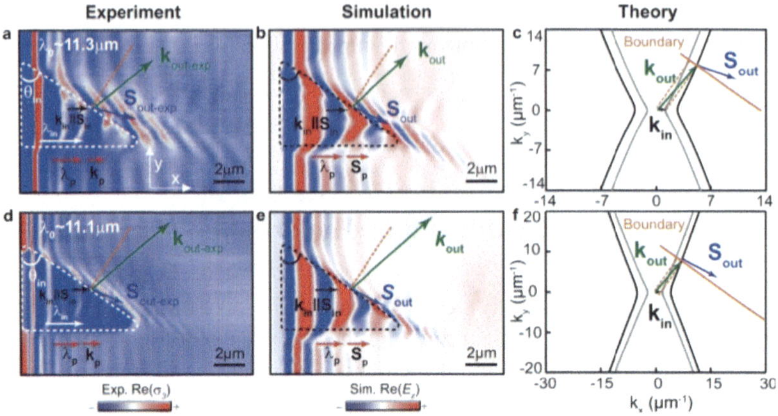

Fig. 6.2 Real-space visualization of refraction between two anisotropic media using highly confined polaritons with collinear incident k **and** S, $\mathbf{k}_{\text{in}} \parallel \mathbf{S}_{\text{in}}$. (**a**,**b**) Experimental $\text{Re}(\sigma_3(x, y))$ (**a**) and simulated $\text{Re}(E_z(x, y))$ (**b**) near-field images of HPhPs propagating in a 160-nm-thick α-MoO_3 flake at $\lambda_0 = 11.3\ \mu\text{m}$. The white and black dashed lines mark triangular prisms fabricated by etching an air cavity on the SiO_2 substrate below the α-MoO_3 flake. $\mathbf{S}_{\text{in}}$ and $\mathbf{k}_{\text{in}}$ display the direction of propagation and the wavevector of incident polaritons in α-MoO_3/air, respectively. Horizontal propagation of non-refracted polaritons is marked as $\mathbf{k}_p$ and $\mathbf{S}_p$. Upon refraction at a boundary of the prism with an angle $\theta_{\text{in}} \approx 55°$, HPhPs bend away from the normal, $\mathbf{S}_{\text{out-exp}}$ (blue arrow), with a tilted wavevector $\mathbf{k}_{\text{out-exp}}$ (green arrow). Compared to non-refracted HPhPs, indicated by $\mathbf{k}_p$, the refracted HPhPs are stronger confined (with a wavelength about 1.6 times shorter). (**c**) Analytic IFCs of α-MoO_3/SiO_2 (black hyperbolas) and α-MoO_3/air (gray hyperbolas) effective media in (**a**,**b**), and considering momentum conservation at the boundary (orange line), the extracted wavevector and direction of the refracted polaritons, $\mathbf{k}_{\text{out}}$ and $\mathbf{S}_{\text{out}}$, respectively, are in good agreement with both experiment and simulation. (**d**,**e**) Experimental $\text{Re}(\sigma_3(x, y))$ (**d**) and simulated $\text{Re}(E_z(x, y))$ (**e**) near-field images of HPhPs propagating in a 160-nm-thick α-MoO_3 flake at $\lambda_0 = 11.1\ \mu\text{m}$. The refracted HPhPs propagate almost parallel to the boundary with a wavelength 2.1 times smaller than λ_p. (**f**) Analytic IFCs of α-MoO_3/SiO_2 (black hyperbolas) and α-MoO_3/air (gray hyperbolas) effective media in (**d**,**e**). The orange dashed lines in a–f represent the normal to the boundary

wavevector, $\mathbf{k}_{\text{out-exp}}$, refracts towards the normal (note that polariton launching by the prism boundary can be ruled out, see the next section "Clarification of the Refractive Origin of the Polaritonic Waves in the Near-Field Images"), the energy flux, $\mathbf{S}_{\text{out-exp}}$, bends away from it. This is in stark contrast to what is expected for a wave propagating from a lower refractive index region to a higher refractive index region in isotropic media. In addition, the modulus of the refracted wavevector $\mathbf{k}_{\text{out-exp}}$ ($\approx 6.48\ \mu\text{m}^{-1}$) is considerably larger than that of the incident wavevector $\mathbf{k}_{\text{in}}$ ($\approx 2.09\ \mu\text{m}^{-1}$) and that of the polariton wavevector along the x direction outside the prism $\mathbf{k}_p$ ($\approx 4.08\ \mu\text{m}^{-1}$), revealing a strong wavelength reduction resulting from refraction in hyperbolic media. These findings are in perfect agreement with the case predicted in Figs. 6.1a–c for hyperbolic refraction considering collinear $\mathbf{k}_{\text{in}}$ and $\mathbf{S}_{\text{in}}$.

To unambiguously verify the features of refraction between hyperbolic media, full-wave numerical simulations mimicking the experiments were carried out,

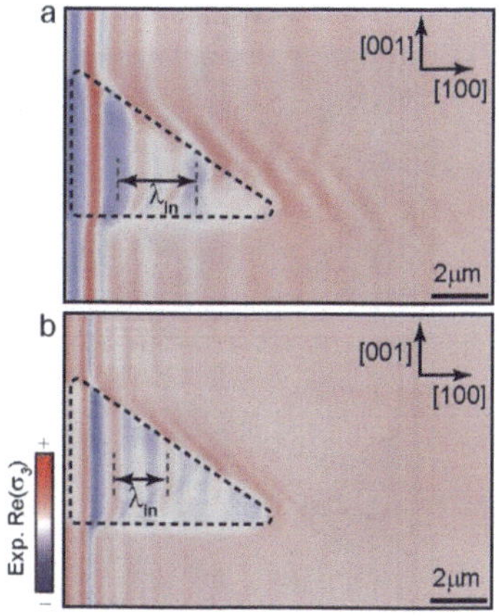

Fig. 6.3 Visualization of incident waves inside the prisms for refraction of polaritons in anisotropic media. (a–b) The same experimental near-field images as in Fig. 6.2a,b, but with different color saturation. With less saturation, fringes ascribed to incident waves are clearly observed inside the prisms. The wavelength of incident waves are marked by black arrows

as described in Sect. 2.5. The resulting spatial distribution of the out-of-plane component of the electric field, $\text{Re}(E_z(x, y))$, is plotted in Fig. 6.2b, clearly showing refraction of both the energy flux ($\mathbf{S}_{out}$) and the wavevector ($\mathbf{k}_{out}$), in excellent qualitative and quantitative agreement ($|\mathbf{k}_{out}| \approx 6.35\,\mu\text{m}^{-1}$) with the experimental image in Fig. 6.2a. In addition to numerical simulations, we also validate the experimental results by performing analytical calculations based on the dispersion derived in Chap. 3 and analogous to those shown in Fig. 6.1. Namely, we calculate the IFCs of HPhPs in α-MoO_3/air (gray curve), and α-MoO_3/SiO_2 (black curve) regions at $\lambda_0 = 11.3\,\mu\text{m}$ (Fig. 6.2c). Applying the condition of momentum (wavevector) conservation at the boundary (orange line), we extract the Poynting vector and wavevector of the refracted polaritons. Again, we observe refraction of the energy flux ($\mathbf{S}_{out}$, blue arrow) with a tilted wavevector ($\mathbf{k}_{out}$, green arrow) in excellent agreement with the experiment, as well as with the full-wave numerical simulations.

To further analyze refraction in hyperbolic media, we also perform experiments at a different illuminating wavelength λ_0 (different tilting angles of the prism boundary θ are also shown in Fig. 6.4), as shown in Fig. 6.2d–f for $\lambda_0 = 11.1\,\mu\text{m}$. Interestingly, we observe that, in this case, both the angular separation between $\mathbf{k}_{out\text{-}exp}$ and $\mathbf{S}_{out\text{-}exp}$ and the confinement effect are larger, being the refracted wave ($\mathbf{S}_{out\text{-}exp}$) almost parallel to the prism boundary and the modulus of $\mathbf{k}_{out\text{-}exp}$ ($\approx 12.47\,\mu\text{m}^{-1}$) about four times larger than that of $\mathbf{k}_{in}$ ($\approx 3.14\,\mu\text{m}^{-1}$) and 2.1 times larger than that of k_p ($\approx 5.92\,\mu\text{m}^{-1}$).

Altogether, these results demonstrate the efficient refractive nature of the planar prisms, enabling us to visualize in real-space three important features of highly confined polaritons refracted at the boundary between two hyperbolic media: (1) large tilting of their wavefronts (given by out-exp $\mathbf{k}_{out\text{-}exp}$) with respect to their propagation direction (given by out-exp $\mathbf{S}_{out\text{-}exp}$), (2) counter-intuitive directions of propagation, and (3) subwavelength confinement (with respect to polaritons along the same crystal axis in the same medium).

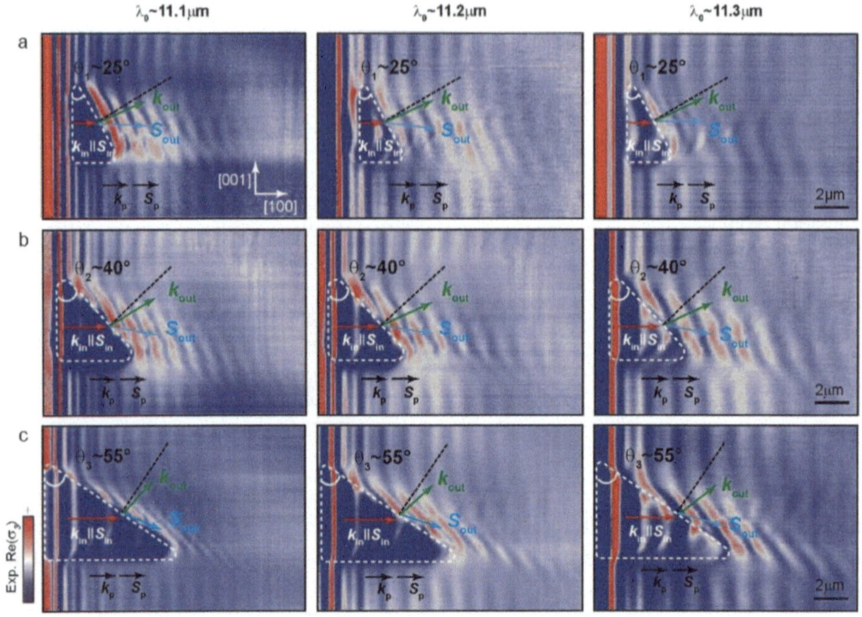

Fig. 6.4 Hyperbolic refraction of polaritons as a function of the angle of incidence and incident wavelength. (**a–c**) Experimental near-field images of polaritons propagating in a 160-nm-thick α-MoO_3 flake. White dashed lines marks triangular prisms fabricated by etching an air cavity on the SiO_2 substrate below the α-MoO_3 flake. The boundary angles are $\theta_1 \approx 25°$ (**a**), $\theta_2 \approx 40°$ (**b**), and $\theta_3 \approx 55°$ (**c**), respectively. The incident wavelength is 11.1 μm, 11.2 μm, and 11.3 μm in the left, middle and right panel, respectively. White arrows in the near-field image (top left) represent the crystal directions of α-MoO_3

Clarification of the Refractive Origin of the Polaritonic Waves in the Near-Field Images

Figure 6.2 describes refraction of hyperbolic polaritons in α-MoO_3 flakes as they pass through triangular prisms, which are defined by fabricating air cavities on the SiO_2 substrates below the α-MoO_3 flakes. To rule out any possible contribution to the near-field images stemming from the excitation of polaritons by the boundaries of the air cavities, we carried out numerical simulations using large metallic prism-like structures on α-MoO_3 flakes (with the same angles as the air-cavity prisms used in Fig. 6.2 of the main manuscript), which act as effective metal antennas to launch polaritons in the flakes. This design resembles a potential situation where polaritons are excited by the edges of the air prisms. However, as shown in Fig. 6.5, the simulated near-field images demonstrate that in the case of polaritons being launched by the metal antenna, the polaritonic waves always exhibit parallel wavefronts to the boundary of the antenna (black solid lines in Fig. 6.5), i.e. their wavevector (black arrows in Fig. 6.5) is perpendicular to the boundary. Such

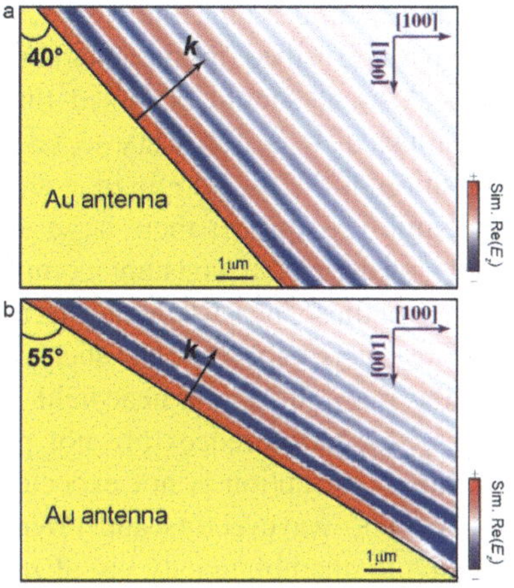

Fig. 6.5 Propagation of polaritons excited by gold antennas in hyperbolic media. (a–b) Simulated near-field images, Re(E_z), on gold antennas placed on top of a 160-nm- thick α-MoO_3 flake whose edge is oriented at an angle of 40° (**a**) and 55° (**b**) with respect to the [001] crystalline direction in α-MoO_3. Polaritons excited by the gold antennas, with wavevector **k** (black arrow), exhibit wavefronts which are parallel to the boundary of the antennas in stark contrast to what is obtained in Fig. 6.2 for refracted polaritons

observation of highly confined polaritons with parallel wavefronts with respect an extended source can be explained by the Huygens' principle, which states that every point on the boundary acts as a source of wavelets that interfere, thus forming the wavefront. As the momentum of polaritons in α-MoO_3 is much larger than the size of the source, such interference gives rise to patterns that are always parallel to the contour of the source.

The behavior of polaritons launched by the edge is therefore in stark contrast to what is shown in Fig. 6.2, where the outcoming waves always show tilted wavefronts with respect to the boundary. As such, their wavevector is not completely perpendicular to it, but tilted, which is expected by the general laws of refraction due to momentum conservation (see Figs. 6.1 and 6.2). Hence, this difference clearly confirms refraction as the origin of the propagating polaritonic waves upon passing through the air prisms (as shown in the near-field images in Fig. 6.2).

Another potential concern against the refractive origin of the polaritonic waves coming out from the prisms could be that there are hardly any visible waves inside the air-cavity prisms. Indeed, as expected in a situation in which the incident polaritons refract at each of the boundaries as they pass through the prism, the intensity of the waves inside the prisms should be much higher than that of the refracted waves (outside the prisms), due to the refraction loss at the second boundary. In this regard, no clear wave oscillations are observed inside the prisms in Fig. 6.2 due to the selected color saturation, which was chosen to highlight and improve the visibility of the weaker refracted waves.

Figure 6.3 shows the same results as the ones depicted in Fig. 6.2, but with lower saturation, where fringes ascribed to the incident waves can be clearly observed inside the prisms. Also, it should be noted that, regarding contrast in the experimental s-SNOM images, it is necessary to take into account the non-negligible near-field contribution of the tip-sample interaction (especially for metal tips, which are generally employed due to their large scattering cross-section). This contribution typically consists of a constant complex value added to the complex polaritonic signal. As in these measurements there are two effective media, α-MoO_3/air and α-MoO_3/SiO_2, it is expected that such contribution gives rise to small offset differences in the real part of the near-field plots in one region with respect to the other. In turn, these differences are not present in the numerical simulations. Nevertheless, this contribution is not expected to have any influence on the direction of the polaritonic wavevectors and Poynting vectors, neither on their relative contrast, allowing to unambiguously visualize refraction phenomena.

6.3 A Sub-diffractional Planar Lens Based on Anomalous Refraction of Hyperbolic Polaritons

The unique features that characterize refraction of polaritons in naturally in-plane hyperbolic media open the door to focus ultra-confined polaritons in a planar geometry. To demonstrate this possibility, we design and fabricate a planar lens in α-MoO_3 (Fig. 6.6a). As noted above (Fig. 6.2), when HPhPs with collinear $\mathbf{k}_{in}$ and $\mathbf{S}_{in}$ (red arrow in Fig. 6.6b) refract at a boundary with another hyperbolic medium with higher refractive index (such as when passing from α-MoO_3/air to α-MoO_3/SiO_2), they bend away from the normal to the boundary (blue arrow in Fig. 6.6b). This means that, in the case of considering a prism of triangular shape, as that shown in Fig. 6.6a, all the refracted polaritons (blue arrows) can converge into a single spot, and thus the prism acts as a focusing lens for highly confined polaritons (see Appendix C.2 for the analytical description of general shapes of lenses focusing hyperbolic waves).

More importantly, as such a lens is based on refraction of HPhPs, the refracted waves can potentially feature infinitely large wavevectors when the boundary is perpendicular to the asymptote of the hyperbolic IFC, which would yield deeply sub-diffractional foci sizes. However, as HPhPs decay exponentially, the intensity at a distant focus can be weak due to propagation losses, which would be more notable for large wavevectors approaching the asymptote of the hyperbolic IFC. Consequently, we designed our triangular lens looking for a compromise between a large refracted wavevector and a long propagation length of the refracted polaritons. According to our theoretical calculations (see Sect. C.3), this compromise is obtained for an angle of the in-plane wavevector of about 62°. Therefore, we fabricate a triangular prism with boundaries perpendicular to this angle (Fig. 6.6a).

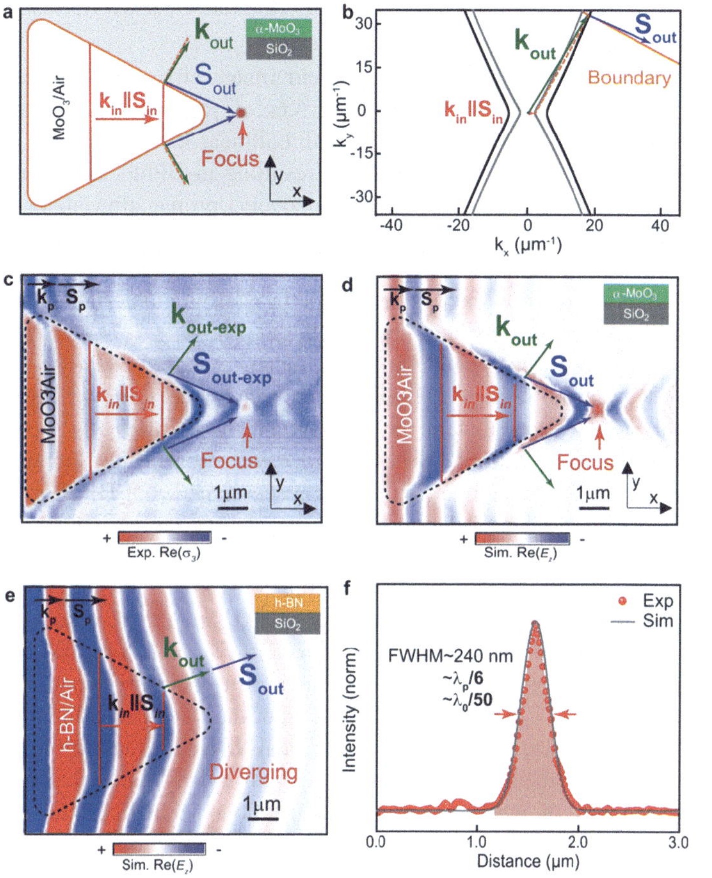

Fig. 6.6 Sub-diffractional planar lens based on refraction of HPhPs. (**a**) Schematics of a refractive hyperbolic lens fabricated by etching a triangular air cavity on the SiO_2 substrate below the α-MoO_3 flake. The top and bottom boundaries of the lens have the same slope as the boundary in (**b**) Upon refraction at the boundaries (orange contour line), polaritons bend far away from the normal, $\mathbf{S}_{out}$ (blue arrows), with a tilted wavevector $\mathbf{k}_{out}$ (green arrows), converging at a focal spot (red dot). $\mathbf{S}_{in}$ and $\mathbf{k}_{in}$ display the direction of propagation and the wavevector of incident polaritons in α-MoO_3/air, respectively. b Analytic isofrequency curves (IFCs) of polaritons propagating in α-MoO_3/SiO_2 (black hyperbola) and α-MoO_3/air (gray hyperbola). When the boundary (orange line) is nearly perpendicular to the asymptote of the open hyperbolic IFC, the refracted polaritons propagate ($\mathbf{S}_{out}$, blue arrows) almost parallel to the boundary with large non-collinear wavevector ($\mathbf{k}_{out}$, green arrows). (**c**) Experimental near-field image of the refractive planar hyperlens (black dashed line) for polaritons in a 170-nm-thick α-MoO_3 slab, at $\lambda_0 = 11.16\,\mu m$. The polaritons converge upon refraction at the triangular boundary. Compared to non-refracted polaritons, indicated by $\mathbf{k}_p$ and $\mathbf{S}_p$ (black arrows) the refracted polaritons, $\mathbf{S}_{out\text{-}exp}$ (blue arrows), propagate nearly parallel to the boundary. (**d**) Simulated near-field image of the refractive planar hyperlens (black dashed line) considered in b and visualized in (**c**). (**e**) Simulated near-field image of a refractive lens for in-plane isotropic polaritons in a 170-nm-thick h-BN slab, at $\lambda_0 = 6.5\,\mu m$. Upon refraction at the triangular boundary (black dashed line), polaritons bend towards the normal, $\mathbf{S}_{out}$ (blue arrow), with a collinear wavevector $\mathbf{k}_{out}$ (green arrow), yielding a diverging effect. Horizontal propagation of non-refracted polaritons is marked as $\mathbf{k}_p$ and $\mathbf{S}_p$. (**f**) Near-field intensity profiles (gray line and red dots) extracted through the focus spot along the vertical direction in (**d**) and (**c**), respectively. Both curves are normalized to the near-field intensity far away from the lens and flake edges. Confinement factors as large as $\approx \lambda_p/6$ and $\approx \lambda_0/50$ are obtained with respect to the polariton and free-space light wavelengths

The experimental and simulated near-field images for this lens design upon illumination at 11.16 μm are shown in Fig. 6.6c, d, respectively. In both images, we observe refraction of incident HPhPs with collinear $\mathbf{k}_{\mathrm{in}}$ and $\mathbf{S}_{\mathrm{in}}$ (red arrows) at the lens boundaries (black dashed contour) resulting in HPhPs with non-collinear $\mathbf{k}_{\mathrm{out}}$ and $\mathbf{S}_{\mathrm{out}}$ (green and blue arrows, respectively) propagating along directions almost parallel to the boundaries (blue arrows), which eventually converge, resulting in a focus. This result is in stark contrast to that observed in a similar lens based on refraction of highly confined polaritons in an isotropic material, such as h-BN, in which refracted polaritons bend towards the normal (Fig. 6.7), making all them to diverge (Fig. 6.6e). The near-field images of our hyperbolic lens reveal that the wavevector $\mathbf{k}_{\mathrm{out}}$ of the refracted HPhPs is much larger than the wavevector $\mathbf{k}_{\mathrm{in}}$ of the incident HPhPs (note that in the experimental image there is a contribution of tip-launched HPhPs with two different $\mathbf{k}_{\mathrm{in}}$ wavevectors). More importantly, the wavevector $\mathbf{k}_{\mathrm{out}}$ is also much larger than the wavevector $\mathbf{k}_p$ of HPhPs propagating along the x direction in the flake on top of SiO_2 (black arrows). In order to avoid the influence from the permittivity of substrate, we evaluate the focusing resolution by comparing the full-width at half-maximum (FWHM) with the wavelength of polaritons propagating in α-MoO_3/SiO_2 ($\lambda_p = 2\pi/|k_p|$). Since the focus shows a FWHM of ≈240 nm (red dots and gray curve in Fig. 6.6f, for experimental and simulated line profiles, respectively), we obtain a focus that is much smaller than the polariton wavelength (λ_p) along the x direction, or the free-space illumination (λ_0), namely of $\approx\lambda_p/6$, or $\lambda_0/50$. This resolution reveals that a diffraction-limited optical system in hyperbolic media can show a focus that is much smaller than the incident polaritonic wavelength, which again reflects the unique behavior of electromagnetic waves in hyperbolic media. Moreover, this result confirms our planar lens based on refraction of HPhPs as a nano-optical element that greatly exceeds the focusing resolution of any lens based on refraction of highly confined polaritons in isotropic media [20–23].

6.4 Visualization of the General Case of Refraction

So far, we have visualized refraction in hyperbolic media for the case in which collinear $\mathbf{k}_{\mathrm{in}}$ and $\mathbf{S}_{\mathrm{in}}$ refract into polaritons with non-collinear $\mathbf{k}_{\mathrm{out}}$ and $\mathbf{S}_{\mathrm{out}}$. However, the general phenomenon of refraction involves the incident polaritons exhibiting non-collinear $\mathbf{k}_{\mathrm{in}}$ and $\mathbf{S}_{\mathrm{in}}$ (as sketched in Fig. 6.1b, d). In the following, we study this fundamental phenomenon in hyperbolic media (Fig. 6.8). To do this, we again fabricate prisms in α-MoO_3 (following the same structure design as in Fig. 6.2) and visualize (by s-SNOM) the propagation of HPhPs refracting upon them. As shown in the near-field image of Fig. 6.8a, we observe that HPhPs launched by the edge of the flake (black arrows) refract at boundary-1 (Fig. 6.8c), and the outcoming HPhPs propagate with non-collinear $\mathbf{k}_R$ and $\mathbf{S}_R$ inside the prism (orange arrows). As such, these polaritons can now be used to visualize the general case of refraction at another boundary of the prism.

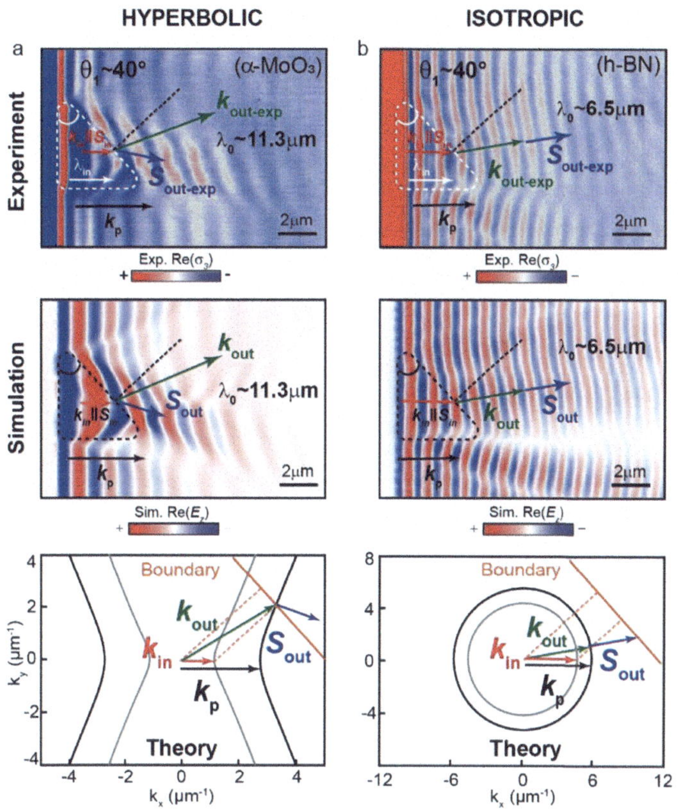

Fig. 6.7 Refraction of polaritons in hyperbolic and isotropic media. (**a**) Experimental (top row) and simulated (center row) near-field images of HPhPs propagating in a 160-nm-thick α-MoO_3 flake at $\lambda_0 = 11.3$ μm. Upon refraction at the right boundary of a triangular prism (dashed triangle) with an angle $\theta_1 \approx 40°$, HPhPs bend away from the normal, $\mathbf{S}_{\text{out-exp}}$ (blue arrow), with a tilted wavevector $\mathbf{k}_{\text{out-exp}}$ (green arrow). (**b**) Experimental (top row) and simulated (center row) near-field image of in-plane isotropic polaritons in a 93-nm-thick h-BN flake at $\lambda_0 = 6.5$ μm. Upon refraction at the right boundary of a triangular prism (dashed triangle) tilted an angle $\theta_1 \approx 40°$, in-plane isotropic polaritons in h-BN bend towards the normal, with collinear $\mathbf{k}_{\text{out-exp}}$ (green arrow) and $\mathbf{S}_{\text{out-exp}}$ (blue arrow). From the corresponding analytic IFCs (grey for the substrate as air while black for the substrate as SiO_2) and considering momentum conservation at the boundary (orange lines), the extracted $\mathbf{k}_{\text{out}}$ and $\mathbf{S}_{\text{out}}$ are in good agreement with experiment and simulation for both α-MoO_3 and h-BN cases

However, to carry out this experiment successfully, we need to ensure that: (1) the angle of the boundary allows refraction of polaritons according to momentum conservation, and (2) the HPhPs can reach this boundary within a reasonable propagation distance. We fulfill these conditions by considering a horizontal boundary (boundary-2) followed by a vertical boundary (boundary-3) in the triangular prism, as shown in the near-field image of Fig. 6.8a. In particular, we observe that HPhPs with non-collinear $\mathbf{k}_R$ and $\mathbf{S}_R$ are reflected at boundary-2 (Fig. 6.8d),

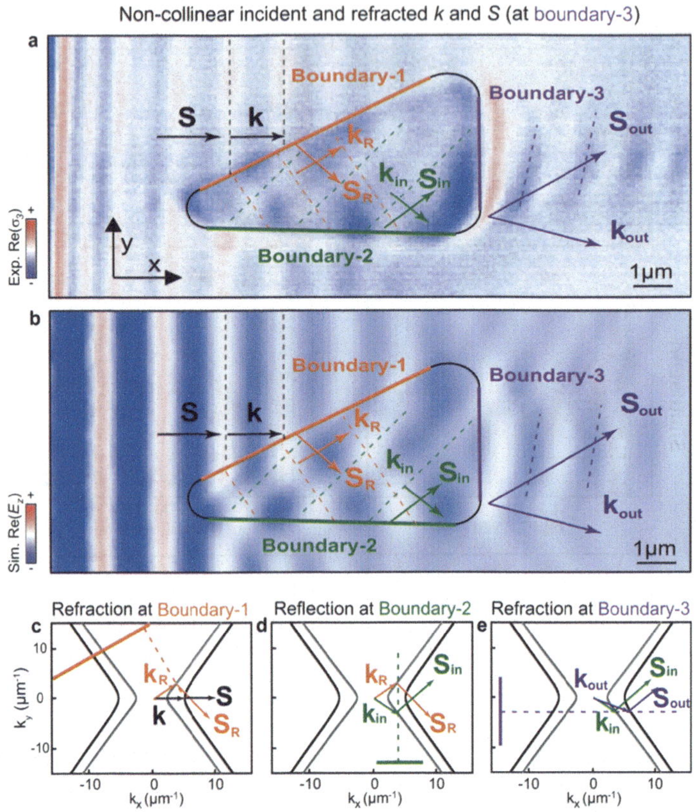

Fig. 6.8 Real-space visualization of the general case of refraction between two anisotropic media using nanoscale-confined HPhPs passing through a bending-free planar prism. (**a**) Experimental Re($\sigma_3(x, y)$) near-field images of polaritons propagating in a 231-nm-thick α-MoO_3 flake at $\lambda_0 = 11.0$ μm. The black contour line marks a triangular prism fabricated by etching an air cavity on the SiO_2 substrate below the α-MoO_3 flake. A first refraction takes place at boundary-1 (orange solid line) for incident polaritons with collinear **k** and **S** (black arrows), yielding refracted polaritons with non-collinear $\mathbf{k}_R$ and $\mathbf{S}_R$ (orange arrows). These polaritons then reflect at boundary-2 (green solid line), yielding polaritons with non-collinear $\mathbf{k}_{in}$ and $\mathbf{S}_{in}$ (green arrows). A second refraction at boundary-3 (violet solid line) yields polaritons with non-collinear $\mathbf{k}_{out}$ and $\mathbf{S}_{out}$ (violet arrows). (**b**) Simulated Re[$E_z(x, y)$] near-field images of HPhPs for the case shown in (**a**). Dashed lines in experimental and simulated near-field images indicate the wavefronts of polaritons as they pass through the prisms. (**c**–**e**) Analytic IFCs of HPhPs in α-MoO_3/air (gray curve) and α-MoO_3/SiO_2 (black curve), predicting the directions of refraction or reflection of HPhPs at boundary-1 (**c**), boundary-2 (**d**), and boundary-3 (**e**) based on momentum conservation. The orange (**c**), green (**d**), and purple (**e**) solid lines represent the boundary-1, boundary-2, and boundary-3, while the corresponding dashed lines represent the normal to the boundary

yielding polaritons with non-collinear $\mathbf{k}_{in}$ and $\mathbf{S}_{in}$ (green arrows) propagating directly towards boundary-3, and reaching it within a reasonably short distance. Consequently, these polaritons refract at boundary-3, which results in polaritons with non-collinear $\mathbf{k}_{out}$ and $\mathbf{S}_{out}$ (violet arrows), as predicted by momentum conservation (Fig. 6.8e), and in good agreement with numerical simulations mimicking the experiment (Fig. 6.8b). This result unambiguously demonstrates refraction of waves whose energy flux and wavevector directions are non-collinear, and thus constitutes the real-space visualization of the general case of refraction between two hyperbolic media, which, in addition, is demonstrated at the nanoscale and in a natural medium. Furthermore, we highlight that, since $\mathbf{k}_{in}$ and $\mathbf{k}_{out}$ are close to the asymptotes of the IFCs, where the tangents to both hyperbolas are parallel, $\mathbf{S}_{in}$ and $\mathbf{S}_{out}$ are almost parallel after refraction at boundary-3. As a result, polaritons refract upon boundary-3 behave as if they had been directly transmitted without any change in their direction of propagation, despite the wavevector does change upon this refraction phenomenon. Therefore, refraction upon this prism is bending-free, in excellent agreement with our theoretical prediction shown in Fig. 6.1d. This bending-free refractive prism could open possibilities to engineer polaritonic wavefronts at the nanoscale without the need of changing their direction of propagation.

6.5 Conclusions

In summary, the work discussed in this chapter explored the character of refraction between two strongly anisotropic media. Our observations of refraction of strongly confined polaritons in a hyperbolic biaxial vdW crystal reveal an exotic optical effect: bending-free refraction, which opens the door to on-demand steering of light at the nanoscale and in natural media. Furthermore, our demonstration of a subwavelength planar lens at the nanoscale based on hyperbolic refraction paves the way for the development of planar nano-optical elements in anisotropic media. Altogether, these results open new avenues for integrated flat optics, directional energy transfer, and heat management applications, as well as for mid-infrared (bio) sensing.

References

1. A.J. Sternbach et al., Negative refraction in hyperbolic hetero-bicrystals. Science **379**, 555–557 (2023). https://www.science.org/doi/abs/10.1126/science.adf1065
2. H. Hu et al., Gate-tunable negative refraction of mid-infrared polaritons. Science **379**, 558–561 (2023). https://www.science.org/doi/abs/10.1126/science.adf1251
3. T. Xu, A. Agrawal, M. Abashin, K.J. Chau, H.J. Lezec, All-angle negative refraction and active flat lensing of ultraviolet light. Nature **497**, 470–474 (2013). https://doi.org/10.1038/nature12158
4. A.J. Hoffman et al., Negative refraction in semiconductor metamaterials. Nat. Mater. **6**, 946–950 (2007). https://doi.org/10.1038/nmat2033
5. J. Yao et al., Optical negative refraction in bulk metamaterials of nanowires. Science **321**, 930–930 (2008). https://www.science.org/doi/abs/10.1126/science.1157566

6. X. Lin et al., All-angle negative refraction of highly squeezed plasmon and phonon polaritons in graphene–boron nitride heterostructures. Proc. Natl. Acad. Sci. **114**, 6717–6721 (2017). https://www.pnas.org/doi/abs/10.1073/pnas.1701830114
7. J. Jiang, X. Lin, B. Zhang, Broadband negative refraction of highly squeezed hyperbolic polaritons in 2D materials. Research **2018** (2018). https://spj.science.org/doi/abs/10.1155/2018/2532819
8. T. Low et al., Polaritons in layered two-dimensional materials. Nat. Mater. **16**, 182–194 (2017). https://www.nature.com/articles/nmat4792
9. D.N. Basov, M.M. Fogler, F.J.G. de Abajo, Polaritons in van der Waals materials. Science **354**, 195 (2016). https://science.sciencemag.org/content/354/6309/aag1992
10. S. Dai et al., Tunable phonon polaritons in atomically thin van der Waals crystals of boron nitride. Science **343**, 1125–1129 (2014). http://science.sciencemag.org/content/343/6175/1125
11. J.D. Caldwell et al., Photonics with hexagonal boron nitride. Nat. Rev. Mater. **4**, 552–567 (2019). https://www.nature.com/articles/s41578-019-0124-1
12. A.J. Giles et al., Ultralow-loss polaritons in isotopically pure boron nitride. Nat. Mater. **17**, 134–139 (2018). https://doi.org/10.1038/nmat5047
13. W. Ma et al., In-plane anisotropic and ultra-low-loss polaritons in a natural van der Waals crystal. Nature **562**, 557 (2018). https://www.nature.com/articles/s41586-018-0618-9
14. Z. Zheng et al., A mid-infrared biaxial hyperbolic van der Waals crystal. Sci. Adv. **5**, eaav8690 (2019). https://advances.sciencemag.org/content/5/5/eaav8690
15. J. Taboada-Gutiérrez et al., Broad spectral tuning of ultra-low-loss polaritons in a van der Waals crystal by intercalation. Nat. Mater. **19**, 964–968 (2020). https://doi.org/10.1038/s41563-020-0665-0
16. A.V. Kildishev, A. Boltasseva, V.M. Shalaev, Planar photonics with metasurfaces. Science **339**, 1289 (2013). https://science.sciencemag.org/content/339/6125/1232009
17. K.V. Voronin, G. Álvarez-Pérez, C. Lanza, P. Alonso-González, A.Y. Nikitin, Fundamentals of polaritons in strongly anisotropic thin crystal layers (2023). arXiv:2309.12935. https://arxiv.org/abs/2309.12935
18. N. Yu et al., Light propagation with phase discontinuities: generalized laws of reflection and refraction. Science **334**, 333–337 (2011). https://www.science.org/doi/abs/10.1126/science.1210713
19. J. Duan et al., Planar refraction and lensing of highly confined polaritons in anisotropic media. Nat. Commun. **12**, 4325 (2021). https://doi.org/10.1038/s41467-021-24599-3
20. P. Alonso-González et al., Controlling graphene plasmons with resonant metal antennas and spatial conductivity patterns. Science **344**, 1369–1373 (2014). https://www.science.org/doi/abs/10.1126/science.1253202
21. K. Chaudhary et al., Polariton nanophotonics using phase-change materials. Nat. Commun. **10**, 4487 (2019). https://doi.org/10.1038/s41467-019-12439-4
22. A.J. Huber, B. Deutsch, L. Novotny, R. Hillenbrand, Focusing of surface phonon polaritons. Appl. Phys. Lett. **92**, 203104 (2008). https://doi.org/10.1063/1.2930681
23. C. Zhao, Y. Liu, Y. Zhao, N. Fang, T. Jun Huang, A reconfigurable plasmofluidic lens. Nat. Commun. **4**, 2305 (2013). https://doi.org/10.1038/ncomms3305

Part IV
Controlling, Directing and Guiding Hyperbolic Polaritons at the Nanoscale

The rich plethora of optical phenomena exhibited by hyperbolic phonon polaritons (HPhPs) has been extensively explored during this thesis so far. These include: extremely directional propagation, concave wavefronts, negative reflection, anomalous refraction, and other intriguing effects. Moreover, we have illustrated how these phenomena can be leveraged for unprecedented manipulation of light at the nanoscale, exemplified through the design of a hyperbolic nanoresonator and the realization of an in-plane refractive hyperlens.

Nonetheless, the technological implementation of HPhPs, for instance in integrated circuits or on-chip devices, necessitates the ability to tune their properties on demand. Achieving this task poses significant challenges, given that the characteristics of HPhPs are inherently determined by the anisotropic crystal structure of the host material. Particularly, hyperbolicity in α-MoO_3, which has been extensively studied in this thesis, is deeply rooted in the orthorhombic nature of its crystal structure.

A drastic approach is to modify the crystalline structure of a polar dielectric by ion intercalation. As demonstrated in a work carried out during the course of this PhD thesis, α-MoO_3 can be intercalated with hydrogen in a non-volatile and recoverable way, allowing reversible switching of HPhPs while preserving their long lifetimes [1]. This can also enable spatially selective switching of PhPs in specific regions. An alternative approach, also demonstrated during this thesis, is the intercalation of Na atoms into α-V_2O_5, another orthorhombic vdW crystal [2], less-studied than α-MoO_3, allowing for a broad tuning of the *Reststrahlen* band. However, these two works will not be reviewed in this part, as the contribution from the candidate to both of them has not been central.

Instead, in this part, we will demonstrate that in-plane HPhPs can be directed, canalized, and guided at the nanoscale by different mechanisms. Chapter 7 shows that the propagation of HPhPs can be steered along forbidden directions through heterostructuring and hybridization with other polaritonic modes. Chapter 8 delves into the theoretical derivation and experimental demonstration of PhP canalization in twisted α-MoO_3 bi- and trilayers. These findings, along with other independent works from different research groups, have given birth to *twist-optics*, a new field in nanophotonics that seeks to control light at the nanoscale by stacking twisted slabs of strongly anisotropic materials. Furthermore, in the last two chapters we will devise a couple of strategies to actively govern their propagation using external stimuli. The strategy discussed in Chap. 9 is based on combining a gated graphene layer with α-MoO_3, allowing for electrical tuning and canalization of PhPs. Chapter 10 seeks a mechanical control through exerting force on twisted α-MoO_3 bilayers with the tip of an atomic force microscope. Altogether, these studies are crucial for the development of optoelectronic devices, such as infrared sensors or photodetectors, based on HPhPs with dynamically controllable properties.

References

1. Y. Wu et al., Chemical switching of low-loss phonon polaritons in α-MoO_3 by hydrogen intercalation. Nat. Commun. **11**, 2646 (2020). https://doi.org/10.1038/s41467-020-16459-3
2. J. Taboada-Gutiérrez et al., Broad spectral tuning of ultra-low-loss polaritons in a van der Waals crystal by intercalation. Nat. Mater. **19**, 964–968 (2020). https://doi.org/10.1038/s41563-020-0665-0

Chapter 7
Enabling Propagation of Hyperbolic Polaritons Along Forbidden Directions

Abstract As has been shown so far in this thesis, polaritons with in-plane hyperbolicity and ultralow losses in van der Waals (vdW) crystals promise unprecedented manipulation of light at the nanoscale. However, these polaritons present a crucial limitation: the directions along which they are allowed or not to propagate are intrinsically determined by the crystal structure of the host material. Therefore, there are forbidden directions of propagation which are rooted on the arrangement of the atoms of the host material, in principle hard to modify. In this chapter, we demonstrate that in-plane hyperbolic phonon polaritons in vdW crystals—in particular α-phase molybdenum trioxide—can be directed along forbidden directions by inducing an optical topological transition, which emerges when the slab is placed on a substrate with a given negative permittivity—in particular, we choose 4H-silicon carbide. By visualizing the transition in real space, we observe exotic polaritonic states between mutually orthogonal hyperbolic regimes, which unveil the topological origin of the transition: a gap opening in the dispersion. This work provides insights into optical topological transitions in vdW crystals, which introduce a route to direct light at the nanoscale.

The near-field experiments reported in this chapter were carried out in collaboration with Dr. Jiahua Duan (University of Oviedo, Spain). Sample fabrication was performed by Dr. Jiahua Duan (University of Oviedo, Spain) and Dr. Iván Prieto (Institute of Science and Technology, Austria). Some of the calculations were performed in collaboration with Kirill Voronin (Moscow Institute of Physics and Technology, Russia). The results were published in "Enabling propagation of anisotropic polaritons along forbidden directions via a topological transition" by Jiahua Duan, Gonzalo Álvarez-Pérez et al. Science Advances 7(14), eabf2690 (2021).

Low-loss in-plane HPhPs open the door to study extraordinary optical phenomena directly on the crystal surfaces, as has been discussed in detail in the preceding chapters of this thesis and exemplified with the direct observation of directional propagation, negative reflection and anomalous refraction, among others. Nevertheless, the direction of propagation of HPhPs in naturally hyperbolic media is difficult

G. Álvarez Pérez, *Fundamentals of Nano-Optics in Hyperbolic van der Waals Materials*, Springer Theses, https://doi.org/10.1007/978-3-031-75767-9_7

to manipulate, since it is predefined by the intrinsic anisotropy of the crystal, which forbids propagation along certain directions. This inherent limitation reduces the potential and versatility of HPhPs for applications requiring specific propagation directions of the electromagnetic energy flow at the nanoscale, such as coupling between quantum emitters [1, 2] or heat management [3]. Specifically, in-plane hyperbolic polaritons can only propagate along directions contained between the asymptotes of their IFC in the (k_x, k_y) plane, i.e., within the hyperbolic sectors $|\tan(k_x/k_y) < \sqrt{-\varepsilon_y/|\varepsilon_x|}$, where ε_x and ε_y are the in-plane permittivities of the hyperbolic medium and k_x and k_y are the in-plane components of the wavevector. Therefore, any engineering, manipulation, or reorientation of the IFC will obviously have a marked effect on the directional propagation of hyperbolic polaritons in real space. Attempts to manipulate this anisotropic propagation of polaritons in vdW crystals have been carried out, either by assembling crystal heterostructures [4] or by fabricating metasurfaces [5] and twisted bilayers [6–9]. The latter approach has even demonstrated that the propagation of HPhPs can be guided along one specific direction and will be discussed in detail in Chap. 8. However, despite this extraordinary progress, a solution to overcome the existence of intrinsically forbidden directions of propagation of HPhPs had remained elusive before the mechanism presented in this chapter was reported, even from a theoretical point of view. In this chapter [21], we theoretically predict and experimentally demonstrate that a reorientation of the IFC of hyperbolic polaritons in a biaxial slab can be engineered to allow propagation along forbidden directions. This is caused by a low-loss topological transition of the IFC, which is naturally induced when the biaxial slab is placed on a substrate with a given negative permittivity. In particular, we demonstrate that the direction of propagation of in-plane HPhPs in an α-MoO_3 slab can be rotated by 90° (and thus occur along intrinsically forbidden directions in the studied spectral range) by placing the slab on a 4H-SiC substrate. The low-loss character of the topological transition allows us to experimentally reveal exotic polaritonic intermediate states, in which HPhPs propagate along mutually orthogonal directions. By theoretically analyzing these states, we unveil the exotic topological origin of this transition, characterized by a gap opening in the polaritonic dispersion.

7.1 Propagation of Hyperbolic Polaritons Along Forbidden Directions: Theoretical Prediction

To elaborate on the theoretical conditions upon which this reorientation of the directional propagation of HPhPs can take place, let us start analyzing the general case of a biaxial slab embedded between two semi-infinite isotropic media with dielectric permittivities ε_1 and ε_3. Without loss of generality, we assume the representative dielectric functions $\mathrm{Re}(\varepsilon_x) < 0$, $\mathrm{Re}(\varepsilon_y), \mathrm{Re}(\varepsilon_z) > 0$, in analogy with the convention for orthorhombic vdW materials in which in-plane HPhPs have been reported, such as α-MoO_3 [10–12] and α-V_2O_5 [13]. Propagation of in-plane HPhPs occurs within hyperbolic sectors centered along the direction with negative permittivity, i.e., the x axis. The dispersion relation of polaritons in a biaxial slab (considering the layers stacked along the z axis) was derived in Chap. 3 and Ref. [14]. We reproduce it here for completeness:

$$q = \frac{\rho}{k_0 d}\left[\arctan\left(\frac{\varepsilon_1\rho}{\varepsilon_z}\right) + \arctan\left(\frac{\varepsilon_3\rho}{\varepsilon_z}\right) + \pi l\right], \quad l = 0, 1, 2... \tag{7.1}$$

where $q = k_t/k_0$ is the normalized in-plane momentum ($k_t^2 = k_x^2 + k_y^2$), ε_1 and ε_3 are the permittivities of the superstrate and substrate, respectively, d is the thickness of the α-MoO_3 flake, l is the mode index, and $k_0 = \omega/c$ is the wavevector in free space and

$$\rho = i\sqrt{\frac{\varepsilon_z q^2}{\varepsilon_x q_x^2 + \varepsilon_y q_y^2}} = i\sqrt{\frac{\varepsilon_z}{\varepsilon_x \cos^2\alpha + \varepsilon_y \sin^2\alpha}},$$

with α being the angle between the x axis and the in-plane component vector. Setting $k_y = 0$ (or $\alpha = 0$) in Eq. (7.1), we find that conventional propagation of in-plane hyperbolic polaritons centered along the x axis is allowed when $\mathrm{Re}(\varepsilon_1)+\mathrm{Re}(\varepsilon_3) > 0$ (see details in Appendix D.1). We also find that propagation of hyperbolic polaritons is allowed along the dielectric y axis—i.e., physical solutions of Eq. (7.1) for $\alpha = \pi/2$—, initially a non-polaritonic direction [10–12], when the condition $\mathrm{Re}(\varepsilon_1) + \mathrm{Re}(\varepsilon_3) < 0$ is fulfilled (see Appendix D.1). Therefore, these results predict a 90° reorientation of the hyperbolic IFC of polaritons in a biaxial slab in the neighborhood of $\mathrm{Re}(\varepsilon_1) + \mathrm{Re}(\varepsilon_3) = 0$. We note that although propagation of polaritons along any given in-plane direction is found for both α and $\alpha + \pi$, for simplicity, we show the dispersion only for α, which corresponds to waves propagating with positive Poynting vector $\mathbf{S}$.

This prediction can be addressed using natural materials, such as the biaxial vdW semiconductor α-MoO_3 placed on top of the polar crystal 4H-SiC. As has been shown throughout this thesis, α-MoO_3 supports low-loss in-plane HPhPs in its RB2 (821–963 cm^{-1}) [12], along which 4H-SiC has negative permittivity ($\mathrm{Re}(\varepsilon_3) < 0$ from 797 to 970 cm^{-1}) [15] (see Appendix D.2). The permittivity of 4H-SiC attains a value $\mathrm{Re}(\varepsilon_3) = -1$ at the surface optical (SO) phonon frequency $\omega_{SO} = 943\,cm^{-1}$, where it thus approximately fulfills the critical condition derived for the rotation of the IFC (see Appendix D.2), $\mathrm{Re}(\varepsilon_1) + \mathrm{Re}(\varepsilon_3) = 0$, when air is considered as superstrate ($\varepsilon_1 = 1$). To explore this possibility, we compare the analytical dispersion of HPhPs in a slab of α-MoO_3 when it is placed on top of a 4H-SiC substrate and on top of a commonly used dielectric substrate, such as SiO_2, where the condition $\mathrm{Re}(\varepsilon_1) + \mathrm{Re}(\varepsilon_3) > 0$ is always fulfilled in the studied spectral range (see Appendix D.2). Figure 7.1a shows the dispersion of propagating HPhPs, calculated analytically from Eq. (7.1) (solid line) and by full-wave numerical simulations (dots), along both in-plane crystal directions ([100]

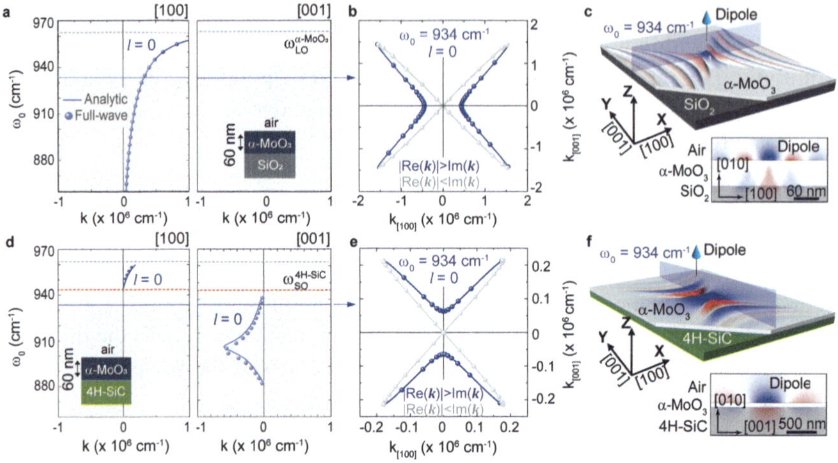

Fig. 7.1 Rotation of the directional propagation of in-plane HPhPs in an α-MoO_3 biaxial slab. (**a**) Dispersion of HPhPs in a 60-nm-thick α-MoO_3 slab on top of SiO_2 along both in-plane directions. The dashed gray line shows the LO phonon of α-MoO_3, while the continuous blue line marks $\omega_0 = 934\,\text{cm}^{-1}$. (**b**) IFC of HPhPs propagating in the α-MoO_3/SiO_2 heterostructure at an incident frequency $\omega_0 = 934\,\text{cm}^{-1}$: The major axis of the hyperbola-like curve lays along $k_{[100]}$. (**c**) Schematics of HPhPs propagating in the α-MoO_3/SiO_2 heterostructure, launched by a dipole source at $\omega_0 = 934\,\text{cm}^{-1}$. Propagation occurs within hyperbolas centered along the [100] crystal direction of α-MoO_3. The inset shows a cross section of the simulated electric field distribution $\text{Re}(E_z)$, revealing that HPhPs in α-MoO_3/SiO_2 are volume-confined. (**d**) Dispersion of HPhPs in a 60-nm-thick α-MoO_3 slab on top of 4H-SiC along both in-plane directions. The dashed gray line shows the LO phonon of α-MoO_3, while the shadowed area indicates the range below the SO phonon of 4H-SiC (dashed red line), also shown in (**a**) for the sake of comparison. The continuous blue line marks $\omega_0 = 934\,\text{cm}^{-1}$. (**e**) IFCs of HPhPs propagating in the α-MoO_3/4H-SiC heterostructure at an incident frequency $\omega_0 = 934\,\text{cm}^{-1}$, showing available wavevectors along $k_{[001]}$. (**f**) Schematics of PhPs propagating in the α-MoO_3/4H-SiC heterostructure, launched by a dipole source at $\omega_0 = 934\,\text{cm}^{-1}$. Propagation occurs within hyperbolas centered along the [001] crystal direction of α-MoO_3, in stark contrast with that in (**c**). The inset shows a cross section of the simulated electric field distribution $\text{Re}(E_z)$, revealing that HPhPs in α-MoO_3/4H-SiC are surface-confined (see Fig. 7.2)

and [001], corresponding to the x and y axes) of a 60-nm-thick α-MoO_3 slab placed on top of a SiO_2 substrate. In this *Reststrahlen* band of α-MoO_3, below the longitudinal optical (LO) phonon frequency [12], $\omega_{LO} = 963\,\text{cm}^{-1}$ (Fig. 7.1a), the α-MoO_3/SiO_2 heterostructure supports the lowest-order polaritonic mode ($l = 0$) along the [100] crystal direction of α-MoO_3, i.e., along the crystal direction with negative permittivity, while there are no polaritonic modes along the dielectric [001] direction, which therefore constitutes a forbidden propagation direction in this spectral range. Hence, the $l = 0$ polaritonic mode exhibits in-plane hyperbolic dispersion, consistently with what has been reported in prior works [10, 11]. This hyperbolic dispersion is more clearly visualized by plotting the IFC at a representative incident frequency, $\omega_0 = 934\,\text{cm}^{-1}$ (Fig. 7.1b). In this case, the IFC is a closed bow tie-shaped curve composed of a finite hyperbola whose major axis

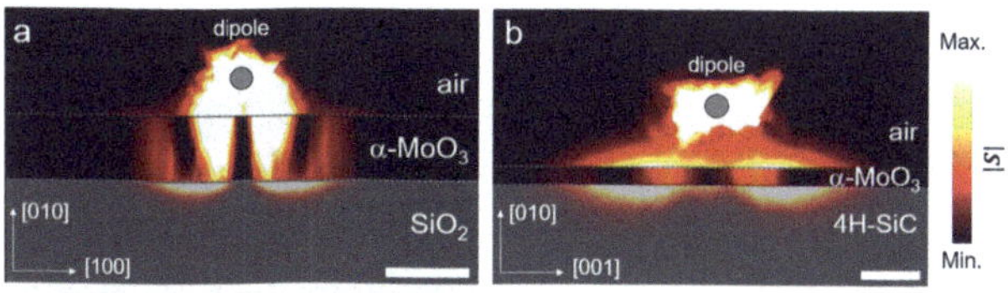

Fig. 7.2 Simulated |**S**| for PhPs propagating on (**a**) a 360-nm-thick α-MoO_3 flake on top of SiO_2 along the [100] crystal direction and (**b**) a 61-nm-thick α-MoO_3 flake on top of 4H-SiC along the [001] crystal direction

lays along $k_{[100]}$ and a crossing through the origin. While the hyperbola describes propagating modes for which $\text{Re}(\mathbf{k}) > \text{Im}(\mathbf{k})$ (blue lines in Fig. 7.1), the crossing represents overdamped modes which fulfill $\text{Re}(\mathbf{k}) < \text{Im}(\mathbf{k})$ (gray lines in Fig. 7.1b), the latter regime often overlooked in the literature. Blue and gray dots shown in Fig. 7.1b are obtained from full-wave numerical simulations, showing an excellent agreement with both parts of the analytical IFC. Hence, the IFC gives rise to propagation of HPhPs within hyperbolic sectors centered along the [100] crystal direction (Fig. 7.1c). These modes present a slab- or volume-confined propagation, as shown by extracting cross sections of both the near-field distribution $\text{Re}(E_z)$ (inset in Fig. 7.1c) and the norm of the Poynting vector |**S**| across the blue-shaded section in Fig. 7.1c (see Fig. 7.2).

In contrast, when the α-MoO_3 slab is placed on top of 4H-SiC (with air as superstrate, $\varepsilon_1 = 1$), the polaritonic dispersion changes markedly (Fig. 7.1d). The lowest-order hyperbolic mode $l = 0$ shows now dispersion along either the [100] or the [001] α-MoO_3 crystal directions (blue line and dots on the left and right panels of Fig. 7.1d, respectively), depending on the excitation frequency. In particular, the $l = 0$ mode is observed along the [100] direction between the SO phonon frequency of 4H-SiC, $\omega_{SO} = 943\,\text{cm}^{-1}$, and the LO phonon frequency of α-MoO_3, $\omega_{LO} = 963\,\text{cm}^{-1}$, i.e., when $\text{Re}(\varepsilon_3) > -1$. On the other hand, the $l = 0$ mode emerges along the [001] direction below ω_{SO}, i.e., when $\text{Re}(\varepsilon_3) > -1$, where it shows negative phase velocity, $v_p = \omega/k < 0$. Along the [001] direction, the α-MoO_3 slab has positive permittivity and, thus, is optically equivalent to a dielectric, whereas the 4H-SiC substrate has negative permittivity and, thus, behaves (optically) as a metal. Hence, surface polaritons exist at the interface α-MoO_3/4H-SiC, and their dispersion tends asymptotically to the frequency of the SO phonons at the interfaces α-MoO_3/4H-SiC and air/4H-SiC at large and small k, respectively (see Fig. 7.3) [17]. The dispersion relation for surface waves at the α-MoO_3/4H-SiC interface can be calculated from Eq. (3.46) in Chap. 3. For waves propagating along the y axis, it can be significantly simplified, leading to the following expression:

$$k = k_0\sqrt{\varepsilon_1\varepsilon_z\frac{\varepsilon_1 - \varepsilon_y}{\varepsilon_1^2 + \varepsilon_y\varepsilon_z}}.$$

On the other hand, along the [100] direction, no polaritonic mode is supported in this spectral range (see Appendix D.1), although both 4H-SiC and α-MoO_3

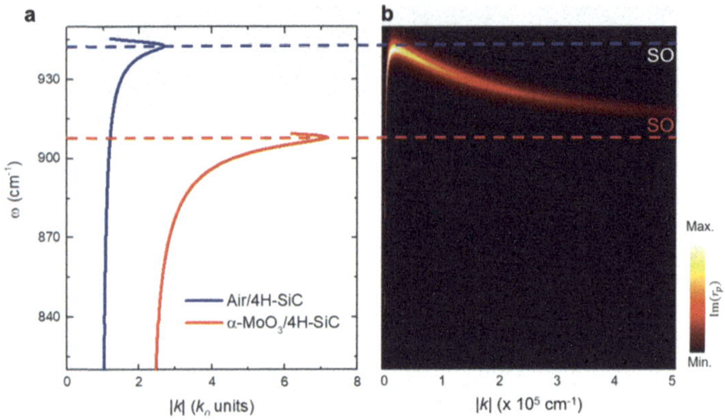

Fig. 7.3 (**a**) Dispersion of surface polaritons at the interfaces air/4H-SiC and α-MoO_3/4H-SiC along the [001] direction. Dashed lines mark the corresponding surface optical (SO) phonon frequencies. (**b**) False color-plot of the imaginary part of Fresnel's reflection coefficient, $\text{Im}(r_p(\omega, \mathbf{k}))$ at the interface air/α-MoO_3/4H-SiC along the [001] direction, obtained from transfer-matrix calculations [16]. The polaritonic dispersion corresponds to the maxima of $\text{Im}(r_p(\omega, \mathbf{k}))$

individually support polaritons in this frequency range. Therefore, as a consequence of the α-MoO_3 biaxial slab being embedded in an environment, which fulfills the condition $\text{Re}(\varepsilon_1) + \text{Re}(\varepsilon_3) < 0$, the corresponding IFC of the $l = 0$ mode at $\omega_0 = 934\,\text{cm}^{-1}$ in the α-MoO_3/4H-SiC heterostructure is a closed bow tie-shaped curve composed of a finite hyperbola-like curve for propagating modes with its major axis laying now along $k_{[001]}$. We highlight that this anomalous effect is theoretically derived when considering the heterostructure as a whole, not only for the substrate or the slab as separate entities. The emergence of available wavevectors in the mode dispersion along the $k_{[001]}$ axis leads to propagation of HPhPs within hyperbolic sectors centered along the previously non-polaritonic [001] α-MoO_3 crystal direction (Fig. 7.1e). This reorientation of the IFC clearly illustrates a 90° rotation of the propagation direction of HPhPs in the α-MoO_3/4H-SiC heterostructure. HPhPs along the [001] α-MoO_3 crystal direction show surface-confined propagation, as observed by extracting cross sections of the near-field distribution $\text{Re}(E_z)$ (inset of Fig. 7.1f) and the norm of the Poynting vector $|\mathbf{S}|$ across the blue-shaded section in Fig. 7.1f (see Fig. 7.2).

7.2 Propagation of Hyperbolic Polaritons Along Forbidden Directions: Experimental Demonstration

To experimentally demonstrate this unique rotation of the directional propagation of HPhPs, we assembled both α-MoO_3/4H-SiC and α-MoO_3/SiO_2 heterostructures and compared the propagation of polaritons supported by each of them. Specifically,

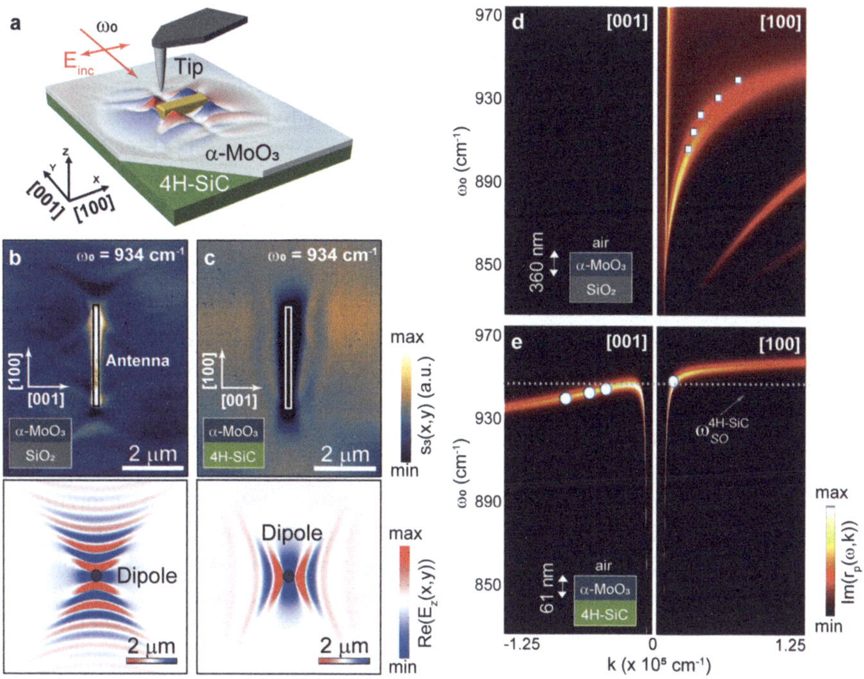

Fig. 7.4 Visualization of the rotation of the directional propagation of in-plane HPhPs in an α-MoO_3/4H-SiC heterostructure. (**a**) Illustration of antenna-launched HPhPs in an α-MoO_3/4H-SiC heterostructure. The spatial distribution of the near field (shown in red and blue) is adapted from the simulation of the near-field distribution, Re(Ez(x, y)). (**b**) Top: Experimental near-field amplitude images $s_3(x, y)$ of antenna-launched HPhPs on a 360-nm-thick α-MoO_3 flake on top of SiO_2 at an incident frequency $\omega_0 = 934\,\text{cm}^{-1}$. Propagation occurs within hyperbolic sectors centered along the [100] α-MoO_3 crystal direction. Bottom: Simulated near-field distribution, $\text{Re}(E_z(x, y))$, excited by a point dipole on a 360-nm-thick α-MoO_3 flake on top of SiO_2 at $\omega_0 = 934\,\text{cm}^{-1}$. a.u., arbitrary units. (**c**) Top: Experimental near-field amplitude images $s_3(x, y)$ of antenna-launched HPhPs on a 61-nm-thick α-MoO_3 flake on top of 4H-SiC at $\omega_0 = 934\,\text{cm}^{-1}$. Propagation occurs within hyperbolic sectors centered along the [001] α-MoO_3 crystal direction. Bottom: Simulated near-field distribution, $\text{Re}(E_z(x, y))$, excited by a point dipole on a 61-nm-thick α-MoO_3 flake on top of 4H-SiC at $\omega_0 = 934\,\text{cm}^{-1}$. (**d**) Transfer-matrix calculations (false color plot) of HPhPs propagating on a 360-nm-thick α-MoO_3 flake on top of SiO_2 along the [001] (left) and [100] (right) crystal directions. White dots indicate experimental data from monochromatic s-SNOM imaging. (**e**) Transfer-matrix calculations (false color plot) of HPhPs propagating on a 61-nm-thick α-MoO_3 flake on top of 4H-SiC along the [001] (left) and [100] (right) crystal directions. White dots indicate experimental data from monochromatic s-SNOM imaging. The dashed white line marks the SO phonon of 4H-SiC at $\omega_{SO} = 943\,\text{cm}^{-1}$

we performed real-space infrared nanoimaging (using s-SNOM, see Sect. 2.4) of polaritons that are efficiently launched by a metallic antenna [18] fabricated on top of the heterostructures (see schematics in Fig. 7.4a). The fabrication of the antennas was performed by Dr. Iván Prieto (Institute of Science and Technology, Austria). We acquired near-field amplitude images at an incident laser frequency $\omega_0 = 934\,\mathrm{cm}^{-1}$, since our theoretical calculations predict that, at this frequency, HPhPs will propagate along orthogonal in-plane directions in these material heterostructures. For the α-MoO_3/SiO_2 system, our images show HPhPs propagating with concave wavefronts within hyperbolic sectors centered along the [100] direction, consistently with prior reports [10, 11] (Fig. 7.4b, top). However, in the α-MoO_3/4H-SiC heterostructure (Fig. 7.4c), we observe HPhPs propagating with concave wavefronts within hyperbolic sectors centered along the [001] direction, which has thus far been an intrinsically forbidden polaritonic direction in single slabs of α-MoO_3 [10, 11] in the spectral band of 821.4–963.0 cm^{-1}. To further validate these experimental findings, we perform full-wave numerical simulations, where, instead of a metallic antenna, a vertically oriented electric dipole placed on top of the heterostructures acts as the polaritonic source. The obtained near-field simulations show HPhPs propagating along the [100] in-plane direction in the α-MoO_3/SiO_2 heterostructure (Fig. 7.4b, bottom) and along its [001] orthogonal direction in the α-MoO_3/4H-SiC heterostructure (Fig. 7.4c, bottom). Together, our results unambiguously uncover a 90° rotation of the major axis of the polaritonic IFC, corroborating that the directional propagation of in-plane HPhPs in a biaxial slab can be effectively engineered along previously forbidden directions, as predicted by our analytical study. Note that we choose α-MoO_3 flakes of different thicknesses in both heterostructures (360 and 61 nm for α-MoO_3/SiO_2 and α-MoO_3/4H-SiC, respectively) so that the wavelengths of HPhPs supported by them are similar, thus facilitating the comparison of the IFCs.

We can gain further insights into the rotation of the IFC by extracting the HPhPs dispersion in each of the heterostructures from near-field measurements at different incident frequencies ω_0, and comparing them to a false color plot of the imaginary part of Fresnel's reflection coefficient, $\mathrm{Im}(r_p(\omega, \mathbf{k}))$, obtained from transfer-matrix calculations [16] (Fig. 7.4, d and e). As expected, the calculated polaritonic dispersion in α-MoO_3/SiO_2 (visible as bright maxima in the color plot) appears only along the [100] crystal direction (Fig. 7.4d) [10–12]. In turn, the α-MoO_3/4H-SiC heterostructure shows polaritonic branches along both the [100] crystal direction (Fig. 7.4e), above ω_{SO}, and [001] crystal direction, below ω_{SO}, i.e., when the permittivity of 4H-SiC takes values below -1, in agreement with our analytical theory (Fig. 7.1). Besides, as illustrated in Fig. 7.1d, polaritons exhibit negative phase velocity $v_p = \omega/k$, and a very flat dispersion curve [17, 19]. To characterize the propagation of HPhPs in an α-MoO_3/4H-SiC heterostructure, we calculate their group velocity along the [001] crystal direction by taking the derivative of the analytical dispersion, $\partial\omega/\partial k$, in Eq. (7.1) for a 60-nm-thick α-MoO_3 flake, as shown in Fig. 7.5. The result yields ultra-low group velocities, as low as $10^{-5}c$.

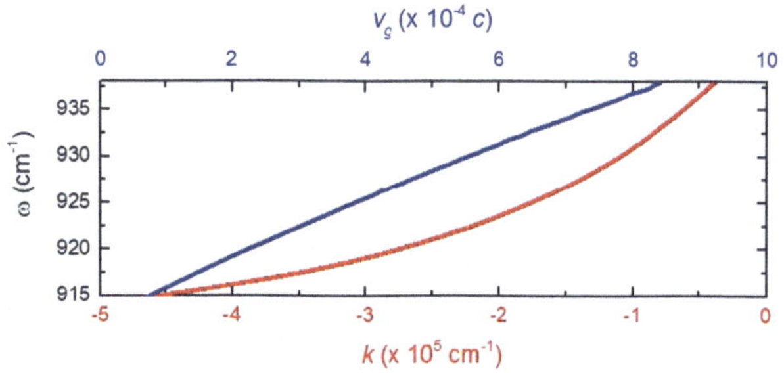

Fig. 7.5 Analytical dispersion (red) and group velocity (blue) of HPhPs propagating in a 60-nm-thick α-MoO_3 flake on top of 4H-SiC, along the [001] crystal direction of α-MoO_3

7.3 Topological Transition Between Mutually Rotated Hyperbolic Propagation Regimes

To explore in detail the transition occurring between these mutually rotated propagation regimes, we image antenna-launched polaritons in the α-MoO_3/4H-SiC heterostructure at several intermediate frequencies between $\omega_0 = 948\,\text{cm}^{-1}$ and $\omega_0 = 937\,\text{cm}^{-1}$, i.e., in the frequency range in which the rotation takes place.

At the highest frequency, $\omega_0 = 948\,\text{cm}^{-1}$, we observe HPhPs propagating within a hyperbolic sector whose major axis lays along the [100] α-MoO_3 direction (Fig. 7.6a). However, when lowering the frequency to $\omega_0 = 943\,\text{cm}^{-1}$, propagation of HPhPs takes place also within a hyperbolic sector whose major axis lays along the [001] direction (Fig. 7.6b). This behavior becomes even more evident at a slightly lower frequency, $\omega_0 = 941\,\text{cm}^{-1}$ (Fig. 7.6c), where exotic propagation of polaritons within hyperbolic sectors whose major axes lay along both orthogonal directions is clearly seen. Last, by further lowering the frequency to $\omega_0 = 937\,\text{cm}^{-1}$, propagation of polaritons is only observed within a hyperbolic sector whose major axis lays along the [001] crystal direction (Fig. 7.6d), thus completing a 90° rotation from the initial directional propagation of HPhPs at $\omega_0 = 948\,\text{cm}^{-1}$.

We perform additional theoretical analyses to understand this exotic behavior of propagating polaritons. Namely, we numerically calculate the near-field distributions $\text{Re}(E_z(x, y))$ generated at the aforementioned frequencies by a localized source placed above a 61-nm-thick α-MoO_3 flake on top of 4H-SiC, and subsequently, we Fourier-transform (FT) them (left of Fig. 7.6, e–h). The resulting FTs clearly show the rotation of the major axes of the hyperbolic IFCs from the [100] to the [001] α-MoO_3 crystal directions when comparing the furthermost cases at $\omega_0 = 948$ and $937\,\text{cm}^{-1}$. However, apart from the rotation effect, the FTs also show nontrivial features appearing at intermediate frequencies, $\omega_0 = 943$ and $941\,\text{cm}^{-1}$, and relatively low momenta (Fig. 7.6, f and g).

These features suggest a complicated multimode excitation in this frequency range. Therefore, we perform a careful analysis of the generated field patterns based on the dyadic Green's function (DGF, see Appendix E) and simplified dispersion

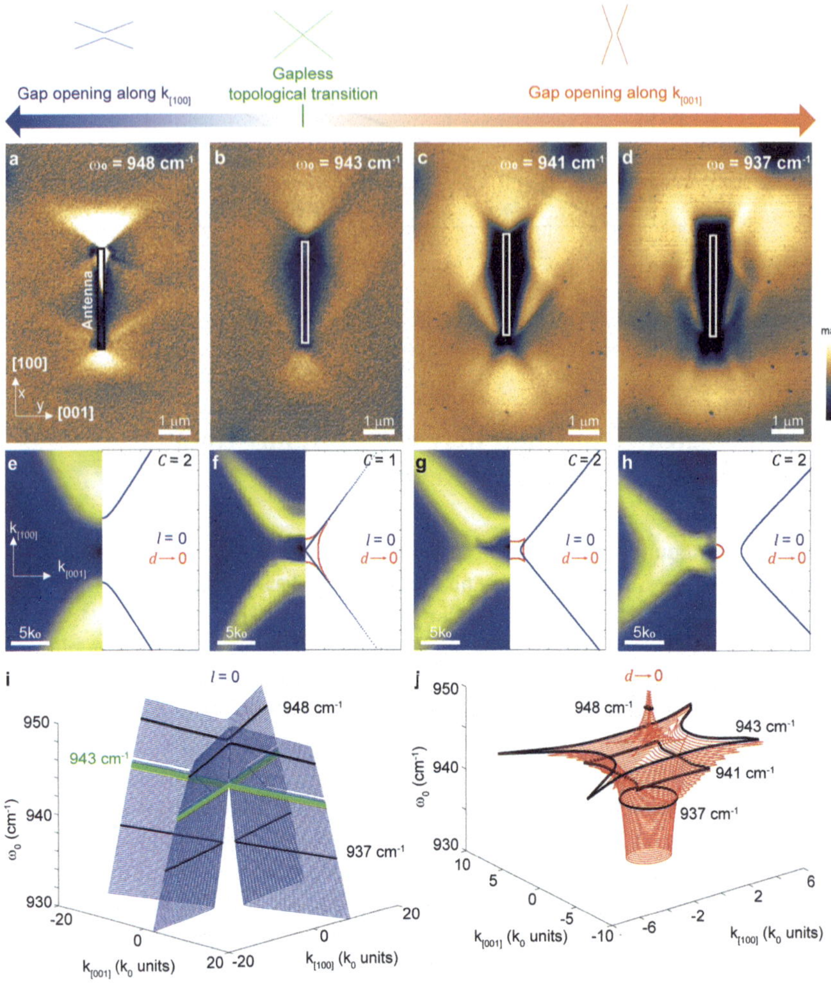

Fig. 7.6 Topological transition of the polaritonic IFC between orthogonal hyperbolic regimes in an α-MoO_3/4H-SiC heterostructure. (**a–d**) Experimental near-field amplitude images $s_3(x, y)$ of antenna-launched hyperbolic PhPs on a 61-nm-thick α-MoO_3 flake on top of 4H-SiC at different illuminating frequencies. (**e–h, left**) Simulated IFCs, calculated by Fourier-transforming simulated near-field distributions, $\mathrm{Re}(E_z(x, y))$ created by a point dipole on a 61-nm-thick α-MoO_3 flake on top of 4H-SiC. (**e–h, right**) Analytical IFCs for the fundamental mode $l = 0$ (blue) and the mode $d \to 0$ (red). (**i,j**) 3D visualization of the topological transition of the fundamental mode $l = 0$ and of the mode $d \to 0$, in blue and red, respectively. Black lines show the IFCs at the frequencies at which the near-field images were acquired. The green line depicts the IFC at the topological transition

relations. Following this approach, we find that only two modes excited in the heterostructure present a major contribution to the fields launched by a localized source: the lowest-order mode—the $l = 0$, according to Eq. (7.1)—and the mode

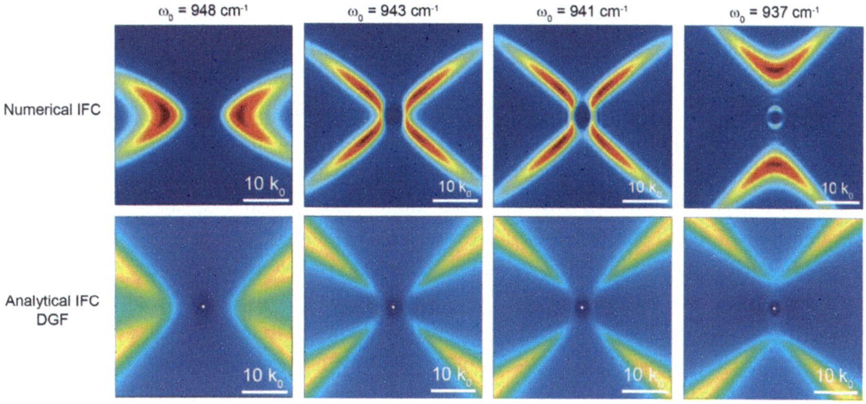

Fig. 7.7 (**Top row**) Simulated IFCs, calculated by Fourier-transforming simulated real-space near-field distributions, Re$[E_z(x, y)]$ created by a point dipole on a 61-nm-thick α-MoO_3 flake on top of 4H-SiC. (**Bottom row**) Analytical IFCs calculated from the distribution of the absolute value of the z component of the DGF

arising under the limit of a vanishing slab thickness $d \rightarrow 0$, referred to as the *ultrathin film limit* in Chap. 3, which has been extensively studied throughout this thesis. The dispersion relation of the mode $d \rightarrow 0$, written for biaxial slabs of small but nonzero thickness, was derived in Sect. 3.4, see Eq. (3.60). This equation does not require the assumption of high k, but it gives only zero-order modes with different polarizations, in contrast to Eq. (7.1), which describes only one polarization (transverse magnetic).

The domination of these two modes is shown by the good agreement between the numerical IFCs, where all modes are excited, and those calculated using the DGF in Fig. 7.7, which only takes into account the fundamental modes. Hence, we can describe properly the main nontrivial features of the FTs in this frequency range in terms of the real part of the analytical IFCs of these two modes (Fig. 7.7). Namely, the branches of the hyperbolic IFC of the mode $l = 0$ linearize and intersect in one point at one specific frequency (blue curve in the right of Fig. 7.6f), and wavevectors with small moduli become available along all in-plane directions for the $d \rightarrow 0$ mode (red curve in the right of Fig. 7.6f).

It is worth having a look at the path connectedness, i.e., the number of components in which a path can be drawn between any two points of the curve within it, of the analytical real-part IFCs for the $l = 0$ mode, as it is one of the main properties that are used to distinguish different topological spaces. Unlike size or shape, path connectedness remains unchanged under homeomorphisms, and thus, based on it, we can distinguish when two curves are not topologically equivalent, i.e., not homeomorphic. At $\omega_0 = 948$, 941, and $937\,\text{cm}^{-1}$ the number of path-connected components C of the IFC of the $l = 0$ mode is 2. However, at $\omega_0 = 943\,\text{cm}^{-1}$, the gap in the IFC notably closes, and the number of path-connected components of the IFC evolves from $C = 2$ to $C = 1$. This IFC gap opening/closing of the IFC corresponds to cutting/attaching the polaritonic dispersion in momentum space and alters the topology of the system [20], as it represents a discontinuous deformation of the polaritonic mode $l = 0$. This unveils one of the hallmark features of the polaritonic topological transition that emerges for the IFC of the lowest-order polaritonic mode in the frequency range between both mutually rotated regimes in the α-MoO_3/4H-SiC heterostructure. This transition can thus be more clearly visualized in a 3D plot of the in-plane dispersion of the IFC as a function of frequency ω, i.e., in the three-dimensional (3D) space (k_x, k_y, ω) (Fig. 7.6i). The resulting surface is a hyperbolic cylinder composed of two disconnected sheets with their in-plane major axes along the $k_{[001]}$ direction at low frequencies and along the $k_{[100]}$ direction at high frequencies, respectively. These sheets become connected at the transition frequency $\omega_0 = 943\,\text{cm}^{-1}$, forming a cross-like curve (depicted in green) and clearly illustrating the topological transition of the IFC. The crossing point constitutes a degeneracy between topologically inequivalent IFCs [20]. In Fig. 7.6j, we show a 3D plot of the mode $d \to 0$ as a function of frequency ω, a surface composed of closed IFCs with low momentum. The cross sections of this surface markedly change as a function of frequency (from a circle-like to a four-vertex star-like IFC) although preserving the same topology up to $\omega_0 = 943\,\text{cm}^{-1}$, where the surface virtually collapses into a point. Thus, the mode $l = 0$ undergoes a topological transition in this frequency range, while the mode $d \to 0$ does not. However, we attribute the peculiarities observed in near-field images in Fig. 7.6 (a to d) to the strong frequency dependence of the propagation direction of polaritons, manifesting itself as a robust change of the IFCs of both $l = 0$ and $d \to 0$ modes, particularly in the vicinity of the topological transition frequency.

7.4 Conclusions

To sum up, we report on the engineering of the directional propagation of low-loss HPhPs in a natural medium along intrinsically forbidden directions, driven by an optical topological transition that emerges when the crystal lays on a substrate with a specific negative permittivity. Contrary to other expected polaritonic topological transitions, as those at the spectral limits between *Reststrahlen* bands, and thereby close to transverse optical phonons where losses in the medium are high, the topological transition here reported when α-MoO_3 is placed on 4H-SiC takes place at $\omega_{SO} = 943\,\text{cm}^{-1}$, where losses are low and thus optimal visualization of its unique features can be carried out in real space. In turn, our results reveal two different regimes in which HPhPs propagate along mutually orthogonal directions, including the naturally forbidden [001] crystal direction for polaritonic propagation in bare α-MoO_3 in the studied spectral range, and exotic intermediate states that unveil the topological origin of the transition, where HPhPs propagate within hyperbolas centered along both the [001] and [100] α-MoO_3 directions. This work opens up stimulating horizons for the investigation of topology-related features of polaritons in vdW materials. In addition, extensive and effective control of the propagation of hyperbolic polaritons at the nanoscale is potentially possible via hybridization with plasmon polaritons in various polaritonic heterostructures, thanks to the growing palette of polaritonic vdW materials that sum up to other conventional polaritonic media, such as SiC, aluminum nitride (AlN), or quartz. Furthermore, the introduction of new degrees of freedom—such as the twist angle (as will be shown in the next chapter)—and controlled external stimuli—such as elastic strain, electric doping, or photoinjection of carriers—can enable further external tuning of the propagation direction and the control of optical topological transitions naturally emerging in polaritonic heterostructures. In this context, our findings promise new opportunities for developing mid-infrared and terahertz optical devices and for manipulating the flow of light at the nanoscale with unprecedented degree of control. Moreover, this study provides fundamentally relevant insights into the emergence of optical topological transitions in low-loss vdW crystals, as well as a pathway through which to translate recent developments in topological engineering in low-dimensional electronic materials to photonics and polaritonics.

References

1. E. Bermúdez-Ureña et al., Coupling of individual quantum emitters to channel plasmons. Nat. Commun. **6**, 7883 (2015). https://doi.org/10.1038/ncomms8883
2. C.L. Cortes, Z. Jacob, Super-Coulombic atom–atom interactions in hyperbolic media. Nat. Commun. **8**, 14144 (2017). https://doi.org/10.1038/ncomms14144
3. L. Tranchant et al., Two-dimensional phonon polariton heat transport. Nano Lett. **19**, 6924–6930 (2019). https://doi.org/10.1021/acs.nanolett.9b02214
4. K. Chaudhary et al., Polariton nanophotonics using phase-change materials. Nat. Commun. **10**, 4487 (2019). https://doi.org/10.1038/s41467-019-12439-4
5. P. Li et al., Collective near-field coupling and nonlocal phenomena in infrared-phononic metasurfaces for nano-light canalization. Nat. Commun. **11**, 3663 (2020). https://doi.org/10.1038/s41467-020-17425-9
6. G. Hu et al., Topological polaritons and photonic magic angles in twisted α-MoO_3 bilayers. Nature **582**, 209–213 (2020). https://doi.org/10.1038/s41586-020-2359-9
7. J. Duan et al., Twisted nano-optics: manipulating light at the nanoscale with twisted phonon polaritonic slabs. Nano Lett. **20**, 5323–5329 (2020). https://doi.org/10.1021/acs.nanolett.0c01673
8. Z. Zheng et al., Phonon polaritons in twisted double-layers of hyperbolic van der Waals crystals. Nano Lett. **20**, 5301–5308 (2020). https://doi.org/10.1021/acs.nanolett.0c01627
9. M. Chen et al., Configurable phonon polaritons in twisted α-MoO_3. Nat. Mater. **19**, 1307–1311 (2020). https://doi.org/10.1038/s41563-020-0732-6
10. W. Ma et al., In-plane anisotropic and ultra-low-loss polaritons in a natural van der Waals crystal. Nature **562**, 557 (2018). https://www.nature.com/articles/s41586-018-0618-9
11. Z. Zheng et al., A mid-infrared biaxial hyperbolic van der Waals crystal. Sci. Adv. **5**, eaav8690 (2019). https://advances.sciencemag.org/content/5/5/eaav8690
12. G. Álvarez-Pérez et al., Infrared permittivity of the biaxial van der Waals semiconductor α-MoO_3 from near- and far-field correlative studies. Adv. Mater. **32**, 1908176 (2020). https://onlinelibrary.wiley.com/doi/abs/10.1002/adma.201908176
13. J. Taboada-Gutiérrez et al., Broad spectral tuning of ultra-low-loss polaritons in a van der Waals crystal by intercalation. Nat. Mater. **19**, 964–968 (2020). https://doi.org/10.1038/s41563-020-0665-0

14. G. Álvarez-Pérez, K.V. Voronin, V.S. Volkov, P. Alonso-GonzáLlez, A.Y. Nikitin, Analytical approximations for the dispersion of electromagnetic modes in slabs of biaxial crystals. Phys. Rev. B **100**, 235408 (2019). https://link.aps.org/doi/10.1103/PhysRevB.100.235408
15. T.E. Tiwald et al., Carrier concentration and lattice absorption in bulk and epitaxial silicon carbide determined using infrared ellipsometry. Phys. Rev. B **60**, 11464–11474 (1999). https://link.aps.org/doi/10.1103/PhysRevB.60.11464
16. N.C. Passler, A. Paarmann, Generalized 4×4 matrix formalism for light propagation in anisotropic stratified media: study of surface phonon polaritons in polar dielectric heterostructures. J. Opt. Soc. Am. B **34**, 2128–2139 (2017). https://opg.optica.org/josab/abstract.cfm?URI=josab-34-10-2128
17. A. Karalis, E. Lidorikis, M. Ibanescu, J.D. Joannopoulos, M. Soljačić, Surface-plasmon-assisted guiding of broadband slow and subwavelength light in air. Phys. Rev. Lett. **95**, 063901 (2005). https://link.aps.org/doi/10.1103/PhysRevLett.95.063901
18. P. Pons-Valencia et al., Launching of hyperbolic phonon-polaritons in h-BN slabs by resonant metal plasmonic antennas. Nat. Commun. **10**, 3242 (2019). https://doi.org/10.1038/s41467-019-11143-7
19. P. Li et al., Reversible optical switching of highly confined phonon–polaritons with an ultrathin phase-change material. Nat. Mater. **15**, 870–875 (2016). https://doi.org/10.1038/nmat4649
20. M. Kim, Z. Jacob, J. Rho, Recent advances in 2D, 3D and higher-order topological photonics. Light Sci. Appl. **9**, 130 (2020). https://doi.org/10.1038/s41377-020-0331-y
21. J. Duan et al., Enabling propagation of anisotropic polaritons along forbidden directions via a topological transition. Sci. Adv. **7**, eabf2690 (2021). https://www.science.org/doi/abs/10.1126/sciadv.abf2690

Chapter 8
Twist-Optics: Controlling the Propagation of Phonon Polaritons with Twisted van der Waals Stacks

Abstract The discovery of flat-band unconventional superconductivity and Mott insulating behavior in twisted graphene bilayers in 2018 was a groundbreaking and highly impactful development in condensed matter physics. It generated significant attention and excitement, sparking intense research into emergent quantum phenomena, challenging theoretical frameworks, and potentially revolutionizing superconducting materials. This finding spurred intense research on aiming to uncover novel physical phenomena by twisting layers of vdW materials, initially in the realm of electronics, thereby initiating a field that received the name of *twistronics*.

The first section of this chapter extends the concepts of twistronics to the world of nano-optics and polaritonics. Namely, we show that the superposition of two twisted thin slabs of highly anisotropic materials provides a neat platform for controlling and steering polariton propagation. Through near-field nanoimaging on twisted bilayers of α-MoO_3, the in-plane polaritonic dispersion is shown to undergo a transition from elliptic to hyperbolic as a function of both frequency and twist angle. Interestingly, at the critical twist angle between the two regimes, the so-called *magic angle*, the two arms of the polaritonic dispersion become flat and parallel, thereby allowing a single direction for propagation that is free of geometrical spreading. This leads to a guiding, collimation effect achievable across the entire *Reststrahlen* band by tuning the twist angle that has received the name of *canalization*. The twisting of layers to tailor the transition frequency at which the phonon polaritons become canalized provides an innovative avenue to steer and guide light-matter interactions that has since been coined as *twist-optics*.

Building on this discovery, the effect of a third layer is studied in the second section of the chapter, to explore novel capabilities for controlling the propagation of PhPs. Repeatedly reassembling a third layer offers a novel strategy for reconfiguring the twist angle in stacked structures. In fact, the addition of a third layer gives rise to unique features that are not present in twisted bilayers and single layers, and provides a major difference with respect to polariton canalization, such as all-angle tunable, robust and broadband canalization. Moreover, IFCs in trilayers can have highly non-intuitive asymmetric shapes due to the breaking of

G. Álvarez Pérez, *Fundamentals of Nano-Optics in Hyperbolic van der Waals Materials*, Springer Theses, https://doi.org/10.1007/978-3-031-75767-9_8

mirror symmetries, similar to those found in crystals with lower symmetry than orthorhombic, such as monoclinic.

The near-field experiments reported in this chapter were carried out by Dr. Jiahua Duan (University of Oviedo, Spain). Sample fabrication was performed by Dr. Jiahua Duan (University of Oviedo, Spain). Dr. Iván Prieto (Institute of Science and Technology, Austria) contributed to sample fabrication in the first section of this chapter, while Ana Isabel Fernández-Tresguerres Mata (University of Oviedo, Spain) contributed to sample fabrication in the second section. The derivations in the first section were performed together with Nathaniel Capote-Robayna (Donostia International Physics Center, Spain), while the calculations in the second section were performed in collaboration with Christian Lanza (University of Oviedo, Spain), Kirill Voronin and Nathaniel Capote-Robayna (Donostia International Physics Center, Spain).

The results reported in the first section of this chapter were published in "Twisted Nano-Optics: Manipulating Light at the Nanoscale with Twisted Phonon Polaritonic Slabs" by Jiahua Duan et al. Nano Letters 2020, 20, 5323–5329 (2020). The results reported in the second section of the chapter were published in "Multiple and spectrally robust photonic magic angles in reconfigurable α-MoO_3 trilayers" by Jiahua Duan, Gonzalo Álvarez-Pérez, Christian Lanza, Kirill Voronin et al. Nature Materials 22, 867–872 (2023).

The discovery of flat-band superconductivity [1–3], correlated insulators [4] and ferromagnetism [5, 6] in twisted bilayer graphene [7] boosted an intense research in twisted layered structures composed of van der Waals (vdW) crystals, aiming to uncover novel physical phenomena as a function of a relative twist angle θ between the in-plane crystalline axes of individual layers. In particular, twisted layered structures very soon impacted optics and photonics with the demonstrations of atomic photonic crystals in twisted bilayer graphene [8], Moiré excitons in stacked structures composed of transition-metal dichalcogenides [9–13], and plasmons in graphene moiré superlattices [14] and interband collective excitations in twisted bilayer graphene [15].

Interestingly, two theoretical reports predicted in 2019 and 2020 that a rich plethora of novel polaritonic phenomena, including topological transitions of polariton dispersions and broadband field canalization, also arises within twisted hyperbolic metasurfaces via near-field coupling of polaritons [16, 17]. In this regard, mid-infrared PhPs with in-plane anisotropic propagation, ultralow losses and subdiffractional confinement in α-MoO_3 [18, 19] provides an optimal test bench to study near-field mediated coupling effects on the propagation of PhPs in twisted layered structures.

8.1 Manipulating Light at the Nanoscale with Twisted Phonon Polaritonic Bilayers

In this first section of the chapter [20], we demonstrate that the propagation of PhPs can be effectively manipulated at the nanoscale by making use of vdW twisted structures. To do so, we stack two α-MoO_3 slabs with a relative rotation angle between their in-plane optical axes and study the propagation of PhPs along them by infrared nanoimaging. Remarkably, we uncover a topological transition of the polaritonic IFC from an open hyperbola-like curve to a closed circumference-like one, finding a critical angle where the IFC flattens, giving rise to strongly directional and diffraction-less propagation of PhPs—canalization regime—along one specific in-plane direction. We corroborate our experimental results by full-wave simulations as well as by analytical modeling, specifically developed in Appendix F, as well as in Ref. [20], to represent the IFCs of twisted bilayer polaritonic structures.

Theory of Polariton Propagation in Twisted vdW Bilayers

To study the propagation of PhPs in twisted bilayers of α-MoO_3, we first perform full-wave numerical simulations, in which we model two superimposed slabs of α-MoO_3 with identical thickness $d = 200$ nm and whose in-plane crystallographic axes are mutually rotated by an angle θ (Fig. 8.1a). To excite PhPs in the stacked structure, we perform the same strategy that has been carried out throughout this thesis: namely, we place a vertically oriented electric dipole (at an illuminating wavelength $\lambda_0 = 11.0$ μm) close to the top surface, generating highly confined electromagnetic fields that provide large momenta, as was shown in Sect. 2.4. We start our numerical study for the case $\theta = 0°$, that is, when the respective crystalline axes of both slabs are aligned. Figure 8.1b (left) shows the real part of the electric field along the z direction, $\mathrm{Re}[E_z(x, y)]$, revealing the fundamental PhPs mode in the biaxial slab [21] propagating with concave wavefronts centered along the α-MoO_3 [100] crystal direction. The corresponding IFC in **k**-space (with $\mathbf{k} = (k_x, k_y)$ the two-dimensional (2D) in-plane wavevector), is visually represented by Fourier transforming (FT) the field $E_z(x, y)$. The IFC presents an open hyperbola-like curve with the transverse and conjugate axes coinciding with the [100] and [001] crystal axes, respectively, showing a range of forbidden directions with no available **k** along them. Note that, to better correlate the real-space plots and IFCs, it is helpful to consider the Poynting vector **S**, as it determines the energy propagation direction of PhPs, being normal to the IFC (see arrows in the figure). This in-plane hyperbolic propagation of PhPs is qualitatively similar to that of single slabs of α-MoO_3 as previously reported [18, 19]. Therefore, as expected, a stacked structure at $\theta = 0°$ does not introduce any change in the propagation direction of PhPs

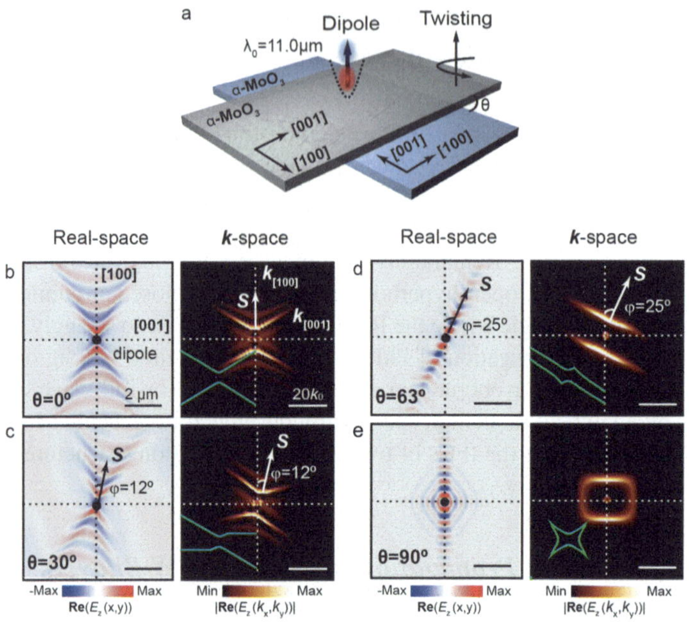

Fig. 8.1 **PhPs in hyperbolic twisted vdW structures.** (**a**) Schematics of two stacked 200 nm thick α-MoO_3 slabs with their in-plane crystalline directions rotated by a twist angle θ. A vertically oriented electric dipole, placed on top of the structure, originates highly concentrated electric fields that allow for launching PhPs at an incident wavelength $\lambda_0 = 11.0\ \mu m$. (**b–e**) Simulated real-space field distribution of PhPs propagating in the stacked structure and corresponding polaritonic IFC for twist angles $\theta = 0°$ (**b**), 30° (**c**), 63°, (**d**) and 90° (**e**), showing a topological transition from an open hyperbola (**b**) to a closed curve (**e**) and canalization of PhPs at the critical angle (**d**) (insets). The green solid lines represent corresponding analytical calculations for IFCs of PhPs in stacked α-MoO_3 slabs

in α-MoO_3, since the crystallographic axes (and thus the optical axes) are the same. At $\theta = 30°$ we observe again PhPs propagating with concave wavefronts (Fig. 8.1c); however, they appear now centered along a direction that is tilted at an angle $\varphi = 12°$ with respect to the [100] crystal axis of the top α-MoO_3 layer. The tilting effect is also visualized in the corresponding IFC, where an open hyperbola whose transverse axis forms an angle φ with the [100] crystal axis is clearly seen. The change of the PhPs propagation angles by φ is attributed to polaritonic coupling between the slabs in the twisted stack, giving rise to novel polaritonic propagation regimes, in comparison to single slabs. A particularly remarkable propagation regime occurs at $\theta = 63°$ (Fig. 8.1d), where PhPs seem to propagate only along one specific in-plane direction at $\varphi = 25°$. Accordingly, the IFC appears flattened and shows a unique direction for **S** at $\varphi = 25°$, uncovering the so-called canalization regime in the stacked structure, which holds great promise for guiding and manipulating the energy flow of electromagnetic fields at the nanoscale [22]. Intriguingly, when we further increase the twist angle to $\theta = 90°$, PhPs are found to propagate along all in-plane directions (Fig. 8.1e), as described by a closed IFC. Thus, our simulations show that the topology of the polaritonic IFC can thus be tuned in twisted slabs from an open hyperbola to a closed shape, when the twist angle θ varies from 0° to 90°. In particular, they reveal a polaritonic topological transition at a critical angle $\theta = 63°$, where the IFC flattens. These results point out that the propagation of PhPs can be effectively engineered by considering stacks of twisted slabs of α-MoO_3. Together with the FTs of the simulated electric field distribution, we also plot the analytical IFCs (green curves in Fig. 8.1b–e), calculated upon a simplified model based on the small thickness of both layers, thus not requiring computation of the fields inside the slabs. This approximation has already been described in Sect. 3.4 and avoids the calculation of the fields inside the slab [23]. It has been proven to be valid for: graphene, as shown in Sect. 1.3 and Ref. [24], transition layer polaritons [25], single layers of biaxial crystals, as shown in Sect. 3.4 and Refs. [18, 21], polaritonic crystals of hexagonal boron nitride [26], and Moiré hyperbolic metasurfaces [17] with a layer thickness much smaller than the polariton wavelength ($d \gg \lambda_p$). In the model, the anisotropic conducting layers of thickness d are rotated a relative angle θ and are separated a distance $2h$, with dielectric permittivity tensors $\hat{\varepsilon}_t$ (top layer) and $\hat{\varepsilon}_b$ (bottom layer), respectively. The effective conductivity for each uniaxial layer is given by:

$$\hat{\sigma}_i = \frac{cd}{2i\lambda_0}\hat{\varepsilon}_i, \qquad i \in \{t, b\}$$

where c is the speed of light in vacuum, λ_0 its wavelength and $\hat{\varepsilon}$ the in-plane uniaxial permittivity of the layer (both $\hat{\sigma}_i$ and $\hat{\varepsilon}_i$ are 2×2 matrices). Note that $\hat{\sigma}_i$ scale linearly with d, thus considering the effect of the small slab thickness (see Appendix F for further details on the derivation). This approximation is similar to the approximations developed in Refs. [16, 17] but valid for two twisted biaxial layers of a finite thickness and separated by an arbitrary distance. The analytical model qualitatively agrees with the numerically calculated IFCs, and it successfully accounts for their open or closed character. Therefore, we can use this model for a quick qualitative estimation of the IFC. Nevertheless, such agreement between the analytical and the simulated IFCs can be substantially improved. For instance, the analytical IFC in the right panel of Fig. 8.1e describes the closed character of the numerical IFC properly, but the shape is not perfectly reproduced. These discrepancies elucidate that polariton coupling in bilayers, which is intimately related to the finite thickness of the slabs, is not accurately described by our 2D approximation model. This effect is even more noticeable when adding more layers to the structure. For this reason, in Sect. 8.2 we will develop an analytical model that does consider the finite thickness of the slabs.

Experiments on the Polariton Propagation in Twisted vdW Bilayers

On the basis of our theoretical findings, in the following, we experimentally demonstrate the modification of the IFCs and the emergence of a canalization regime for PhPs in twisted α-MoO_3 slabs. To that end, we fabricate stacks of two α-MoO_3 slabs with different relative twist angles θ and visualize the propagation of PhPs launched along them by metal antennas, acting as localized sources, by infrared nanoimaging employing s-SNOM, as explained in Sect. 2.4. As schematized in Fig. 8.2a, the s-SNOM tip and sample are illuminated with s-polarized mid-infrared light—with the electric field parallel to the longitudinal direction of the metal antenna—spanning wavelengths from 10.8 to 11.2 μm within the hyperbolic *Reststrahlen* band RB2 of α-MoO_3 (10.4–12.2 μm). Importantly, because of the much larger launching

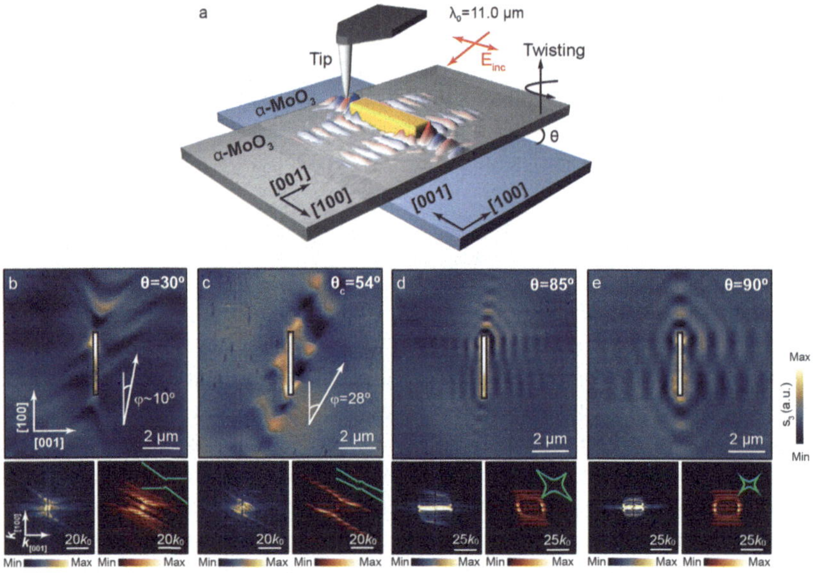

Fig. 8.2 Real-space imaging of PhPs propagating in twisted α-MoO_3 structures as a function of twist angle θ. (**a**) Schematics of the s-SNOM experimental scheme used to image PhPs on twisted α-MoO_3 structures. A gold antenna (yellow) confines infrared light (λ_0 = 11.0 μm) that allows to launch PhPs, which are probed by a metallized AFM tip. (**b–e**) Near-field amplitude images s_3 of twisted α-MoO_3 structures at an illuminating wavelength λ_0 = 11.0 μm for different rotation angles $\theta = 30°$ (**b**), 54° y, 85°, (**d**) and 90° (**e**). The bottom panels show experimental (**left**) and simulated (**right**) IFCs for the near-field images in (**b–e**) and numerical simulations mimicking the experiment, respectively. A clear topological transition from an open hyperbola (**b**) to a closed curve (**e**) as a function of the twist angle is observed, and canalization of PhPs at the critical angle $\theta_c = 54°$ (**c**). The thickness of the top/bottom α-MoO_3 layers are 240/180 nm, 180/280 nm, 90/120 nm, and 200/200 nm in (**b**–e), respectively. The corresponding analytical IFCs are plotted as green solid lines. The white arrows in (**b**, **c**) represent the Poynting vector, indicating the energy propagating direction of PhPs

efficiency of the resonant gold antenna compared to the s-SNOM tip under such illumination scheme, direct visualization of the PhPs wavefronts is possible. For a twist angle $\theta = 30°$, we observe that the polaritonic wavefronts are concave and propagate along a direction $\varphi \approx 10°$ with respect to the [100] crystal axis of the top α-MoO_3 slab (Fig. 8.2b). The experimental IFC (left bottom panel), extracted from the FT of the near-field image, shows an open hyperbola with its transverse axis tilted at an angle φ with respect to the [100] crystal axes of the top α-MoO_3 layer, in excellent agreement with our numerical simulations mimicking the experiment (the FT of the simulated electric field distribution is shown in the right bottom panel). Excitingly, at $\theta = 54°$, the propagation of PhPs appears to be strongly guided along one specific direction $\varphi = 28°$ (Fig. 8.2c), yielding a flattened IFC (left bottom panel) corresponding to a canalization regime, again in excellent agreement with numerical simulations (right bottom panel). We note that there is, however, a small difference between the critical angles from the experiment ($\theta_c = 54°$) and the prediction in Fig. 8.1, ($\theta_c = 63°$), which we attribute to a slightly different thickness for the top and bottom α-MoO_3 layers in our fabricated stacks, as it is challenging to exfoliate several homogeneous α-MoO_3 flakes with exactly the same thickness. When $\theta = 85°$, we observe PhPs propagating along all in-plane directions, yielding a closed IFC (Fig. 8.2d), which is also observed for $\theta = 90°$ (Fig. 8.2e). The IFCs of these two cases are, however, slightly different, the IFC of PhPs for $\theta = 90°$ being more squared-like, pointing out that even a small difference of 5° on the twist angle significantly alters the propagation of PhPs, which evidence twisting as a sensitive method for the manipulation of the polaritons propagation. In agreement with the predictions shown in Fig. 8.1, our experimental results demonstrate that (1) a topological transition from an open hyperbola (Fig. 8.2b) to a closed IFC (Fig. 8.2e) takes place as a function of the twist angle θ, and (2) canalization of PhPs occurs at specific twist (critical) angles θ_c.

Given the strong dispersive nature of PhPs in α-MoO_3, that has been extensively discussed throughout this thesis so far, we also explore their propagation in a twisted structure (at $\theta = 90°$) as a function of incident wavelength λ_0 from 10.8 to 11.1 μm (Fig. 8.3). For $\lambda_0 = 10.8$ μm, near-field imaging shows PhPs propagating with concave wavefronts and within hyperbolic sectors whose transverse axes lay along both the [100] and the [001] crystal directions (Fig. 8.3a), appearing to be less pronounced along the latter. We explain this difference by interference of PhPs launched at different antenna locations along the [100] direction, which makes the polaritonic pattern flatten. The corresponding IFC consists of a hyperbola with its transverse axis along the [100] crystal axis of the top α-MoO_3 layer and two bright spots along the [001] crystal axis (bottom left panel). This IFC agrees well with our numerical simulations (bottom right panel), showing an open hyperbola along the two in-plane crystal directions. Interestingly, these two hyperbolas do not intersect, so that the topology of the IFC in each slab is not affected by the stacking, revealing the absence of polaritonic coupling between the slabs, and therefore, the propagation of PhPs is independent in each of them. For $\lambda_0 = 10.9$ μm, we observe again PhPs propagating within hyperbolic sectors centered along both in-plane crystal axes; however, intriguingly, we also observe PhPs along previously forbidden directions

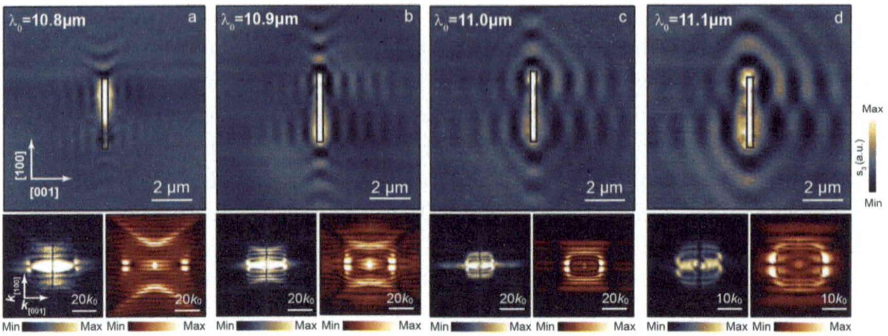

Fig. 8.3 Real-space imaging of PhPs propagating in 90°-twisted α-MoO_3 stacked structure as a function of incident wavelength λ_0. (**a–d**) Near-field amplitude images s_3 of α-MoO_3 twisted structures for λ_0 = 10.8 μm (**a**), λ_0 = 10.9 μm (**b**), λ_0 = 11.0 μm (**c**), and λ_0 = 11.1 μm (**d**). The bottom panels show the corresponding experimental (**left**) and simulated (**right**) IFCs for the near-field images in (**a–d**) and numerical simulations mimicking the experiment, respectively. A clear topological transition from an open hyperbola (**a**) to a closed curve (**d**) as a function of the incident wavelength is observed. The thicknesses of the top/bottom α-MoO_3 layers are 200/200 nm

(Fig. 8.3b), which indicates that, at this wavelength, due to coupling of the PhPs in the upper and lower slabs, the IFC topology changes, giving rise to a closed IFC curve, both in experiment and theory (bottom panels). This coupling effect is even more remarkable for λ_0 = 11.0 μm, where we clearly visualize PhPs propagating along all in-plane directions (Fig. 8.3c), and thus the polaritonic IFC forms an entirely square-like closed shape in both experiment and theory (bottom panels). Similarly, PhPs propagate along all in-plane directions for λ_0 = 11.1 μm (Fig. 8.3d), yet in a slightly different manner, as the IFC evolves from a square-like to a circumference-like closed shape, which uncovers nearly isotropic propagation of PhPs. These near-field measurements demonstrate that the propagation of PhPs in a twisted polaritonic structure can also be conveniently tuned by varying the incident wavelength, as a result of wavelength-dependent polaritonic coupling between the slabs, that, excitingly, also induces a topological transition of polaritonic IFCs from open to closed curves.

Polariton Canalization in Twisted vdW Bilayers

Canalization of PhPs—that is, collimated propagation of PhPs along one specific direction φ—offers unique possibilities for nanoscale guiding of electromagnetic fields and can occur in our twisted structures at certain critical twist angles θ_c, as shown in Fig. 8.2c. To get more insights into the propagation of PhPs in the canalization regime of our stacked α-MoO_3 slabs, we extract a line profile from the near-field amplitude image taken at λ_0 = 11 μm and $\theta_c = 54°$ (dashed green line in Fig. 8.4a) and fit it to an exponentially decaying oscillating function that takes into

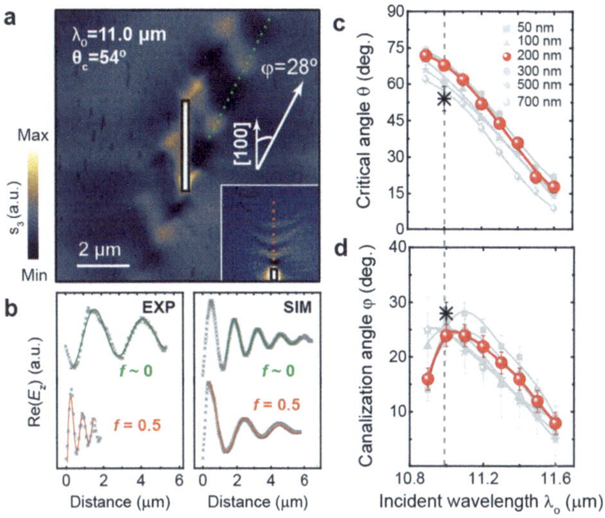

Fig. 8.4 **Canalization of PhPs in twisted α-MoO_3 stacked structures.** (**a**) Near-field amplitude image $s_3(x, y)$ of a α-MoO_3 stacked structure at an illuminating wavelength $\lambda_0 = 11.0\ \mu m$ and a twist angle $\theta = 54°$. The thicknesses of the top/bottom α-MoO_3 layers are 180/280 nm, respectively (**inset**). For comparison, a near-field amplitude image of PhPs propagating on a single slab of α-MoO_3 with a thickness of 100 nm. (**b**) (**left**) Line profiles (gray symbols) along the near-field experimental images in (**a**) (green and orange dashed lines for the stacked structure and the single slab (bottom), respectively), together with fitted curves following Eq. (8.1) (green and orange solid lines for the stacked structure and the single slab, respectively) (**right**). Line profiles (gray symbols) along near-field simulations (not shown) for canalization in a stacked structure (**top**) and hyperbolic propagation in a single slab (**bottom**), together with fitted curves following Eq. (8.1) (green and orange solid lines for the stacked structure and the single slab, respectively). (**c**, **d**) Dependence of the critical angle θ_c and the propagation direction φ on the incident wavelength λ_0. The thicknesses of the top/bottom α-MoO_3 layers in the simulations are set to 50/50 nm, 100 nm/100 nm, 200 nm/200 nm, 300 nm/300 nm, 500 nm/500 nm, and 700 nm/700 nm. Red dots indicate the case for a typical thickness of 200 nm, while gray points represent other corresponding thicknesses (interpolated curves are also shown for each case). Black symbols represent the experimental results

account both the polaritonic intrinsic damping and its geometrical decay, having the following form [27] (see also Eq. (4.4) in Chap. 4):

$$\xi(x) = \frac{A}{x^f} e^{ik_x x} \tag{8.1}$$

where k_x is the complex polariton wavevector, and f is the geometrical decay factor. Interestingly, the resulting curve (green line in the left panel of Fig. 8.4b) yields a value of zero for f, meaning that the geometrical decay in the canalization regime is negligible, and thus the PhPs propagate diffraction-less, in stark contrast to the case of a single α-MoO_3 slab (orange curve), where f takes a value of 0.5, in agreement with previous reports [18, 19, 28]. This result is further corroborated by numerical simulations mimicking the experiments (right panel in Fig. 8.4b). To better exploit the potential application of such diffraction-less propagation of PhPs in our stacked structures at critical angles, we numerically study the expected values of θ_c and propagation angles φ as a function of the incident wavelength λ_0 (Fig. 8.4c,d, respectively, where stacked structures consisting of two rotated α-MoO_3 slabs with the same thickness ranging from 50 to 700 nm are considered). Interestingly, in all cases we observe a large variation of θ_c from $-10°$ to $-70°$ (symbols in Fig. 8.4c. Red dots indicate the case for a typical thickness of 200 nm), in the rather narrow spectral range from 10.8 to 11.6 μm, which can be explained by the strongly dispersive nature of in-plane hyperbolic PhPs in α-MoO_3. More importantly, the propagation angle φ of *canalized* PhPs also shows in all cases a variation from $-5°$ to $-30°$ (symbols in Fig. 8.4d. Red dots indicate the case for a typical thickness of 200 nm) for the same spectral range of incident wavelengths. The similar values obtained for both θ_c and φ, despite the large variation of slab thicknesses considered, indicate the robustness of the interlayer coupling in the twisted structures. Note that, in this study, we only considered twisting angles from 0° to 90°, as angles from 0° to $-90°$ would result in complementary values ($-\theta_c$,$-\varphi$) which, however, would extend the range of expected angles by a factor of 2. Together with these numerical simulations, we plot our experimental results (stars in Fig. 8.4c,d), which show a good agreement, the small discrepancy between them being attributed to the slightly different thicknesses of the two slabs in the fabricated stack. Such tunability of φ, which in the future could be extended by considering a controlled difference of thicknesses of the slabs, has thus great potential for routing flows of energy at the nanoscale.

Conclusions

Our theoretical analysis and experimental observations on twisted bilayers of α-MoO_3 prove that the propagation of PhPs can be unprecedentedly controlled in twisted vdW stacks. The topological transition from hyperbolic to elliptic IFCs is directly demonstrated in this tunable system, as a function of both the twist angle and the incident wavelength. Notably, at the critical twist angle, a flat-IFC state emerges, resulting in canalization of PhPs with diffraction-less propagation. The clean process of fabricating twisted α-MoO_3 devices, achievable through well-established 2D material transfer techniques, makes this approach very appealing. By adjusting the relative alignment of two flakes and stacking them together, the crystalline order of α-MoO_3 is preserved, preserving the long-lived character of PhPs in α-MoO_3, which feature some of the longest lifetimes observed so far.

These findings thus extend the principles of *twistronics* to control polaritons at the nanoscale using twisted 2D materials, establishing the field of *twist-optics*. Here, hyperbolic twisted α-MoO_3 structures play a pivotal role in governing the flow of light at the nanoscale—a long-pursued goal in nanophotonics. Analogous to the revolutionary impact of natural hyperbolic materials on dispersion material research, this new *twist-optics* toolkit promises to inspire a multitude of optical effects, as well as to boost the design of optical components [29].

8.2 Multiple and Spectrally Robust Photonic Magic Angles in Reconfigurable α-MoO_3 Trilayers

As shown in the previous section, polariton canalization, as well as the development of twist-optics, have great potential not only for a deeper understanding of fundamental topological and optical phenomena at the nanoscale but also for applications such as the directional coupling of quantum emitters, waveguides, photonic circuits and infrared sensors. However, before the work presented in this section was carried out [30], twisted α-MoO_3 bilayers remained the only known system in which the low-loss propagation of canalized polaritons at a photonic magic angle had been demonstrated. Importantly, twisted α-MoO_3 bilayers present a fundamental limitation: there exists only one photonic magic angle for each frequency ω_c. This limitation has important technological implications since for a determined twisted structure, both the propagation direction and wavelength of the canalized PhPs are intrinsically fixed by θ_c and the thickness of the α-MoO_3 layers. In addition, the accessible canalization directions in twisted bilayers span a rather narrow in-plane angular sector even for different combinations of thicknesses, as shown in Fig. 8.4, indicating that canalization is far from being all-angle tunable. Indeed, the propagation of PhPs is forbidden in a wide range of directions around the [001] crystal direction in α-MoO_3 in its RB2, as was shown in Chap. 7.

In this section, we overcome these fundamental limitations of twistoptics by demonstrating the existence of multiple spectrally robust photonic magic angles in twisted α-MoO_3 trilayers, thereby enabling on-demand canalization of PhPs along any spatial direction and within a wide spectral range that covers up to half (from 870 to 940 cm^{-1}) of the hyperbolic *Reststrahlen* bands under study. To this end, we design and fabricate reconfigurable devices by stacking three single α-MoO_3 slabs, precisely controlling the twist angles between them (Fig. 8.5a,b shows the schematics). Here θ_{1-2} and θ_{1-3} are the twist angles between the [001] crystal direction of the bottom layer and the [001] crystal direction of the middle and top layers, respectively (all the angles are defined as positive counterclockwise). Figure 8.5c shows the optical image of a representative twisted α-MoO_3 trilayer device in which the top, middle and bottom layers are marked by white, yellow and black dashed rectangles, respectively. To excite PhPs, we fabricate an Au rod antenna between the top and middle layers. To visualize the propagation of PhPs, we perform near-field infrared nanoimaging employing s-SNOM. As shown in Fig. 8.5b, the tip of s-SNOM and the sample are illuminated with s-polarized mid-infrared light spanning the hyperbolic *Reststrahlen* band of α-MoO_3 under study.

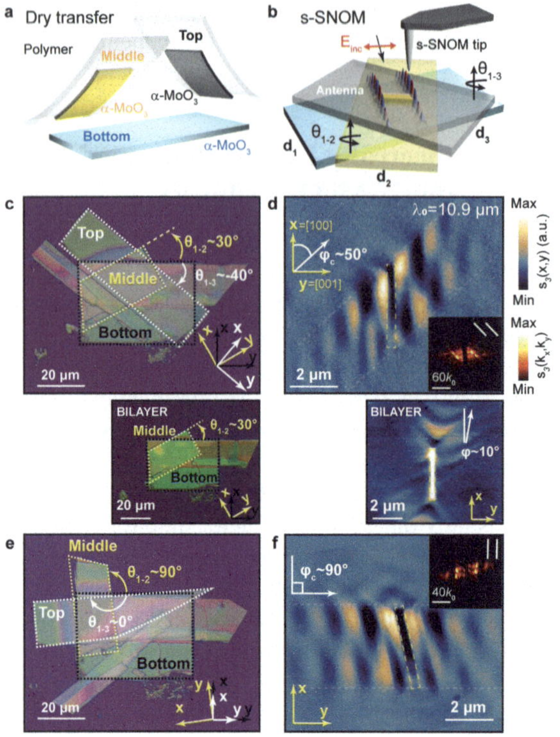

Fig. 8.5 Photonic magic angles and polariton canalization in reconfigurable trilayers. (**a**) Schematic of the dry transfer fabrication of twisted trilayers, which can be reconfigured multiple times by disassembling and reassembling the layers at different twist angles. (**b**) Schematic of the s-SNOM experimental scheme used to image PhPs in twisted α-MoO_3 trilayers. An Au antenna aligned along the [100] crystal direction of the middle α-MoO_3 layer launches the PhPs, whose wavefronts and propagation are probed with the s-SNOM tip. Here θ_{1-2} and θ_{1-3} are the twist angles between the [001] crystal direction of the bottom layer and the [001] crystal direction of the middle and top layers, respectively. The thickness of the top/middle/bottom layer is denoted by $d_3/d_2/d_1$. (**c**) Optical image of a twisted α-MoO_3 trilayer ($d_3 = 250\,\text{nm}$, $d_2 = 240\,\text{nm}$, $d_1 = 180\,\text{nm}$) with $\theta_{1-2} \approx 30°$ and $\theta_{1-3} \approx 40°$ (**top**). The contour of the top, middle and bottom layer is marked by white, yellow and black dashed rectangles, respectively. The x and y axes are defined along the [100] and [001] crystal directions of the α-MoO_3 layers, respectively. Optical image of the bilayer device ($\theta_{1-2} \approx 30°$, $d_1 = 240\,\text{nm}$, $d_2 = 180\,\text{nm}$) intermediate between (**c**) and (**e**) (**bottom**). (**d**) Near-field amplitude image of the twisted trilayers in (**c**) at an illuminating wavelength of $\lambda_0 = 10.9\,\mu\text{m}$ (top). The PhPs launched by both extremities of the antenna are constrained within a narrow spatial sector along the direction of $\varphi_c = 50°$ (white dashed lines). The experimental IFC (FFT of the near-field image) is shown in the inset. The white solid lines are guides to the eye. Near-field amplitude image of the bilayer device in the bottom panel of (**c**) at an illuminating wavelength of $\lambda_0 = 10.9\,\mu\text{m}$ (bottom). (**e**) Optical image of the same α-MoO_3 layers as in (**c**), but twisted at $\theta_{1-2} = 90°$ and $\theta_{1-3} = 0°$. (**f**) Near-field amplitude images of the twisted trilayers in (**e**) at $\lambda_0 = 10.9\,\mu\text{m}$. The experimental IFC (**inset**) verifies polariton canalization along the direction of $\varphi_c = 90°$. The detected near-field signal is demodulated at the third harmonic, with s_3 representing the near-field amplitude

Reconfigurable Twisted α-MoO_3 Trilayers

As shown in the preceding section, α-MoO_3 bilayers exhibit canalization of PhPs when they are twisted at the photonic magic angle. Note that although an individual α-MoO_3 slab can exhibit a flat IFC close to the TO or LO phonon frequencies, this does not usually result in the propagation of canalized PhPs due to high absorption at the TO phonon frequency and near-zero permittivity at the LO phonon frequency. For the case of twisted bilayers, at any other twist angle (for example, the bilayer with $\theta_{1-2} = 30°$; Fig. 8.5c, bottom), PhP propagation is not canalized, ranging from elliptic to strongly anisotropic, such as the hyperbolic propagation centered along $\varphi \approx 10°$ (that is, the angle between the PhPs propagation direction and the [100] crystal direction of the bottom α-MoO_3 slab) shown in the near-field image in Fig. 8.5d (bottom) taken at an incident wavelength of $\lambda_0 = 10.9$ μm.

Excitingly, if we add a third layer of α-MoO_3 (Fig. 8.5c) on top of this bilayer device (at $\theta_{1-3} \approx 40°$), we observe that the propagation of PhPs changes dramatically, now showing a strongly directional regime along the in-plane angle $\varphi_c \approx 50°$ (Fig. 8.5d). This effect is better revealed by performing a fast Fourier transform (FFT) on the near-field image to extract the PhP IFC (Fig. 8.5d, inset), which exhibits two flat branches. Since the propagation direction is in general normal to the IFC, these flat bands unambiguously demonstrate the appearance of a canalization regime. Therefore, the addition of a third layer clearly extends the possibilities for canalization beyond those found in twisted bilayers. Interestingly, by disassembling the trilayer device (Fig. 8.5c) into individual blocks and reassembling them into a new trilayer configuration with different twist angles, namely, $\theta_{1-2} = 90°$ and $\theta_{1-3} = 0°$ (Fig. 8.5e), we again observe PhP canalization, which, however, emerges along a different direction, that is, $\varphi_c \approx 90°$ (Fig. 8.5f). In particular, such a canalization direction corresponds to the [001] direction of the middle α-MoO_3 layer, which is forbidden for propagating PhPs in twisted bilayers and individual slabs at this frequency. These findings indicate that multiple photonic magic angles can exist in twisted trilayers, and consequently, PhP canalization can be tuned in such structures, opening the door to reconfigure and steer polaritons along any in-plane direction.

All-Angle Tunable Polariton Canalization

To explore the extent to which PhP canalization can be steered in twisted trilayers, we calculate—both analytically and numerically—the IFCs of PhPs in α-MoO_3 trilayers twisted at representative combinations of twist angles θ_{1-2} and θ_{1-3}, and an illuminating wavelength of $\lambda_0 = 10.9$ μm. Note that the approximation of finite-thickness slabs by two-dimensional layers commonly used in previous theoretical models, and discussed in the previous section, does not accurately describe all the features of polariton propagation in twisted stacks. As a result, it does not allow the efficient design of multilayer structures supporting canalized polaritons. Therefore, we analytically derive the dispersion of anisotropic polaritons in twisted structures considering the finite thickness of the constituent slabs. The derivation follows the same methodology as the one presented in Chap. 3 but for three biaxial slabs instead of one. The permittivity tensor and electric fields are written down for each of the slabs, and then the boundary conditions from Maxwell's equations are applied, yielding a system of equations. The dispersion relation is obtained by equaling the determinant of the system to zero, which implies looking for nontrivial solutions to the system. The expression is greatly simplified in the limit of high polaritonic momenta $k_p \gg k_0$. Due to length constrains, we will not show the detailed derivation here, but the reader is referred to the Supplementary Materials of Ref. [30]. The improved model allows us to fully explain our experimental results in twisted trilayers, as shown below, as well as to realistically address the complexity and thus the potential versatility of these structures. Figure 8.6a shows the resulting numerical IFCs for $\theta_{1-2} = 0°, 30°, 60°$ and $90°$ and $\theta_{1-3} = 0°, \approx 40°, \approx 60°$ and

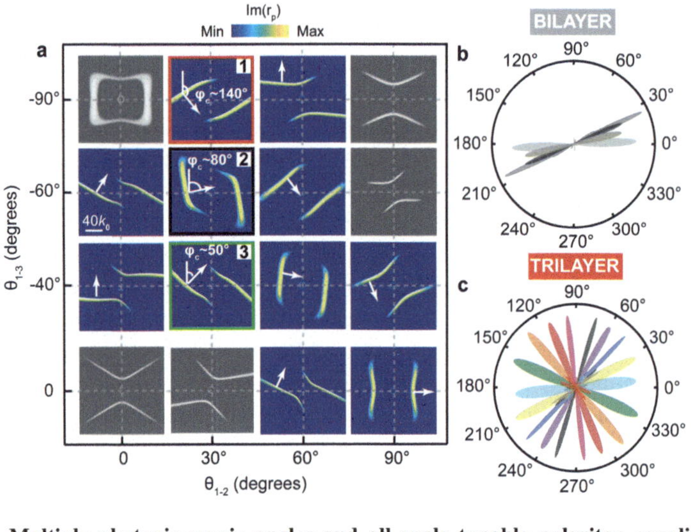

Fig. 8.6 Multiple photonic magic angles and all-angle tunable polariton canalization in twisted trilayers. (**a**) Simulated IFCs for PhPs in a twisted α-MoO_3 trilayer with twist angles $\theta_{1-2} = 0°, 30°, 60°$ and $90°$ and $\theta_{1-3} = 0°, \approx 40°, \approx 60°$ and $\approx 90°$ (Fig. 8.7 shows variation of all angles from $\approx 90°$ to $90°$) at an illuminating wavelength of $\lambda_0 = 10.9$ μm. The panels with coloured boxes (see e.g., **1**, **2** and **3**) represent PhP canalization regimes (that is, flattened IFCs) along different directions (white arrows). The grey-coloured plots represent non-canalized propagation regimes. (**b**) Polar plot of the field intensity in twisted bilayers. The field intensity ($I = |E_z|^2$) is calculated as a function of angle φ (all-angle variation of in-plane angle in the range from 0° to 360°), for all the possible twist angles, at a distance from the source equal to the propagation length ($L_p = 1/\mathrm{Im}(k)$). The grey lobes indicate that the allowed PhP canalization direction in twisted bilayers spans a narrow in-plane angular sector (from 0° to 30°). (**c**) Polar plot of the field intensity in twisted trilayers, showing all-angle tunable canalization. The coloured lobes indicate all-angle tunable polariton canalization in twisted trilayers

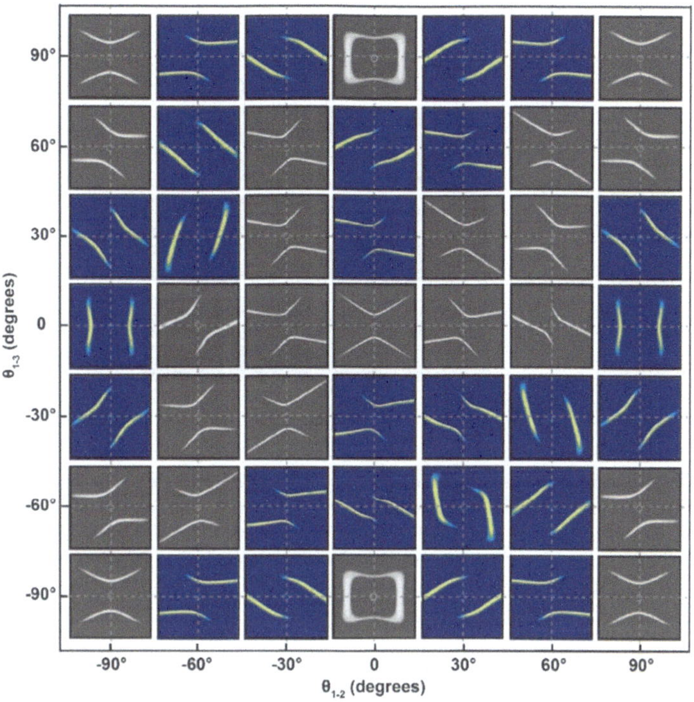

Fig. 8.7 **Numerical IFCs in twisted α-MoO_3 trilayers with all-angle variation of θ_{1-2} and θ_{1-3} from $-90°$ to $90°$**. Numerical PhP IFCs are calculated based on transfer-matrix calculations of twisted trilayers, in which the thicknesses of the top, middle and bottom α-MoO_3 layers are $d_3 = 250$ nm, $d_2 = 240$ nm, and $d_1 = 180$ nm, respectively. All calculations are performed at an incident wavelength $\lambda_0 = 10.9$ μm. In strong contrast to one flattened IFC along a specific direction in each twisted bilayer, there exists flattened IFCs along different directions in twisted trilayers. Moreover, some IFCs even exhibit strong asymmetries (i.e. $\theta_{1-2} \approx 60°$ and $\theta_{1-3} \approx 60°$, $\theta_{1-2} \approx -30°$ and $\theta_{1-3} \approx -30°$, $\theta_{1-2} \approx 60°$ and $\theta_{1-3} \approx 60°$)

$\approx 90°$ (Fig. 8.7 shows variations of all angles from $\approx 90°$ to $90°$ and the analytical results), in excellent agreement with the analytical results (Fig. 8.8). In particular, when θ_{1-3} varies from 0° to $\approx 90°$ for a fixed $\theta_{1-2} = 0°$, we observe a transition from open to closed IFCs (Fig. 8.6a, first column) passing through a flattened IFC that indicates the existence of a photonic magic angle. This *open-to-closed* topological transition is, thus, similar to the one shown in the previous section for twisted bilayers of α-MoO_3. On the other hand, when θ_{1-3} varies from 0° to $\approx 90°$ with a fixed $\theta_{1-2} = 30°$ (Fig. 8.6a, second column) or a fixed $\theta_{1-2} = 60°$ (Fig. 8.6a, third column), we observe that the IFCs are always flattened (except in the case of $\theta_{1-2} = 30°$ and $\theta_{1-3} = 0°$, where the IFCs are open hyperbolas), but, interestingly, with different slopes (that is, PhP canalization along different directions). This finding reveals the existence of multiple photonic magic angles at the same incident frequency in twisted trilayers. As a result, polaritons can be canalized along multiple

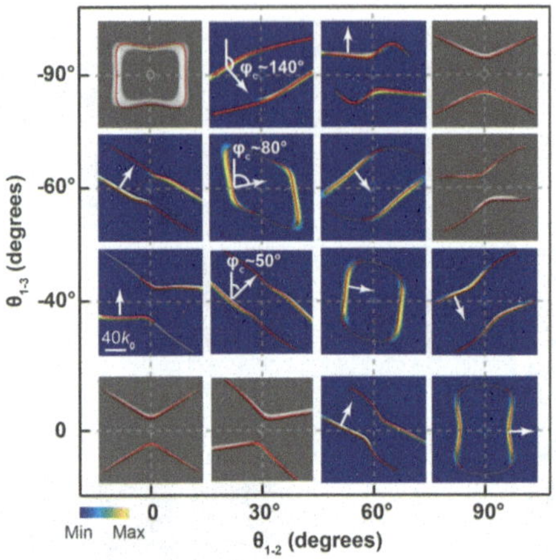

Fig. 8.8 Agreement between analytical IFCs and numerical IFCs in twisted $\alpha-MoO_3$ trilayers. Analytical PhP IFCs in twisted α-MoO_3 trilayers with representative twist angles $\theta_{1-2} = 0°, 30°, 60°, 90°$ and $\theta_{1-3} = 0°, -40°, -60°, -90°$. The numerical IFCs (background colorplot) based on transfer-matrix calculations perfectly match the analytical results (red solid lines), where the grey parts represent high propagation loss in the analytical calculation. Grey color plots represent non-canalized propagation regimes

directions φ_c (Fig. 8.6a, white arrows) at the same frequency, depending on the mutual orientation of the twisted layers in the structure. For example, $\varphi_c \approx 140°$ at $\theta_{1-2} = 30°$ and $\theta_{1-3} \approx 90°$ (Fig. 8.6a,1); $\varphi_c \approx 80°$ at $\theta_{1-2} = 30°$ and $\theta_{1-3} \approx 0°$ (Fig. 8.6a, bottom panel of the fourth column), canalization of PhPs can be achieved along the direction of $\varphi_c \approx 90°$ and thus along the [001] main crystal axis (this particular case is experimentally shown in Fig. 8.5f). Note that for a single α-MoO_3 layer, no PhP can propagate along this direction due to their hyperbolic dispersion. To compare the features of polariton canalization in twisted bilayers (Fig. 8.6b) and in twisted trilayers (Fig. 8.6c), we introduce a rigorous criterion to identify the canalization regime (see Appendix G). Specifically, based on the dyadic Green's function, we calculate the field intensity emitted by the dipole in real space as a function of the in-plane polar angle (φ) at a distance from the source equal to the PhP propagation length (Fig. 8.6b,c). These polar plots reveal that, in the canalization regime, the field intensity reaches its maximum along the in-plane canalization direction (φ_c) and PhPs propagate without geometrical spreading (see Appendix G), commonly referred to as diffractionless propagation. For the case of twisted bilayers, by varying all the possible twist angles, the canalization direction takes place within an angular sector of 0–30° (Fig. 8.6b, grey lobes). In contrast, in the case of twisted trilayers, the allowed canalization directions cover the whole angular sector from 0° to 180° (Fig. 8.6c, coloured lobes), giving rise to all-angle tunable canalization.

Our theoretical calculations (Figs. 8.6 and 8.7) predict that PhPs in twisted α-MoO_3 trilayers can be designed to exhibit IFCs with much more diverse shapes

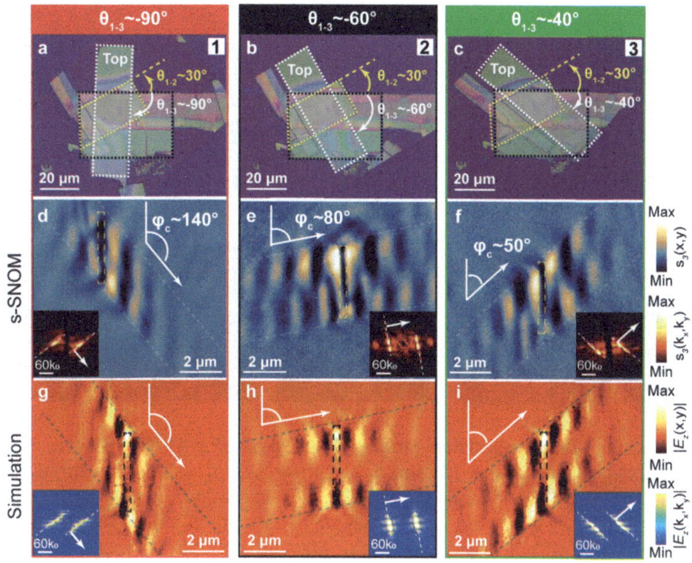

Fig. 8.9 **Experimental demonstration of multiple photonic magic angles and tunable polariton canalization in twisted α-MoO_3 trilayers.** (**a–c**) Optical images of twisted trilayers for $\theta_{1-2} = 30°$ and $\theta_{1-3} \approx 90°$ (**a**), $\approx 60°$ (**b**) and $\approx 40°$ (**c**). All the samples are composed of the same three α-MoO_3 layers ($d_3 = 250$ nm, $d_2 = 240$ nm, $d_1 = 180$ nm). (**d–f**) Near-field amplitude images of the trilayer in (**a–c**), respectively, at $\lambda_0 = 10.9$ μm. The propagation directions of PhPs are marked by white arrows, pointing along $\varphi_c = 140°$ (**d**), $\varphi_c = 80°$ (**e**) and $\varphi_c = 50°$ (**f**). The insets show the corresponding experimental IFCs, calculated through the FFT of the near-field images in (**d–f**). (**g–i**) Simulated near-field images of the trilayer in (**a–c**), respectively, at $\lambda_0 = 10.9$ μm. The propagation directions of PhPs are marked by white arrows, pointing along $\varphi_c = 140°$ (**g**) $\varphi_c = 80°$ (**h**) and $\varphi_c = 50°$ (**i**). The insets show the corresponding simulated IFCs, calculated through the FFT of the simulated images in (**g–i**). The white and black dashed lines are guides to the eye for the Au antennas. Images in (**d–f**) correspond to the IFCs shown in Fig. 8.6a(1–3), respectively. Figure 8.10 shows the canalization along other directions (all-angle tunability)

compared with twisted bilayers of the same vdW crystal: (1) flat IFCs with different slopes, resulting in canalized PhPs along any direction in the plane (Fig. 8.10); (2) strongly asymmetric IFCs in analogy to hyperbolic shear polaritons recently reported in low-symmetry crystals [31, 32].

To experimentally demonstrate the tunability of PhP canalization, we performed near-field imaging of reconfigurable twisted α-MoO_3 trilayers in which we fixed the twist angle $\theta_{1-2} = 30°$ and varied the angle θ_{1-3} multiple times [at $\approx 90°$ (Fig. 8.9a), $\approx 60°$ (Fig. 8.9b) and $\approx 40°$ (Fig. 8.9c)]. When $\theta_{1-3} \approx 90°$, the resulting near-field image shows that the propagation of PhPs is strongly guided along a specific in-plane direction $\varphi_c = 140°$ (Fig. 8.9d). This is further verified by the experimental IFC (FFT of the near-field image), showing a flat contour formed by two parallel lines (Fig. 8.9d, inset). In addition, we also performed full-wave numerical simulations mimicking the experiment (Fig. 8.9g) and extracted the resulting IFC as the FFT of the simulated electric-field distribution (Fig. 8.9g, inset),

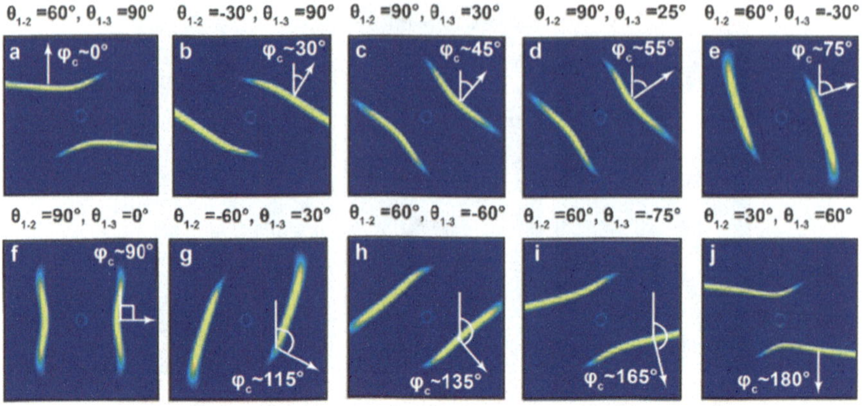

Fig. 8.10 All-angle tunable polariton canalization in twisted α-MoO_3 trilayers. (**a–j**) Numerical IFCs in twisted trilayers ($d_3 = 250$ nm, $d_2 = 240$ nm and $d_1 = 180$ nm) with twist angles of $\theta_{1-2} \approx 60°$ and $\theta_{1-3} \approx 90°$ (**a**), $\theta_{1-2} \approx 30°$ and $\theta_{1-3} \approx 90°$ (**b**), $\theta_{1-2} \approx 90°$ and $\theta_{1-3} \approx 30°$ (**c**), $\theta_{1-2} \approx 90°$ and $\theta_{1-3} \approx 25°$ (**d**), $\theta_{1-2} \approx 60°$ and $\theta_{1-3} \approx -30°$ (**e**), $\theta_{1-2} \approx 90°$ and $\theta_{1-3} \approx 0°$ (**f**), $\theta_{1-2} \approx -60°$ and $\theta_{1-3} \approx 30°$ (**g**), $\theta_{1-2} \approx 60°$ and $\theta_{1-3} \approx -60°$ (**h**), $\theta_{1-2} \approx 60°$ and $\theta_{1-3} \approx -75°$ (**i**), $\theta_{1-2} \approx 30°$ and $\theta_{1-3} \approx 60°$ (**j**), respectively. Flattened IFCs represent various canalization regimes where canalized PhPs travel in different directions (white arrows): $\varphi_c = 0°$ (**a**) $\varphi_c = 30°$ (**b**) $\varphi_c = 45°$ (**c**) $\varphi_c = 55°$ (**d**) $\varphi_c = 75°$ (**e**) $\varphi_c = 90°$ (f), $\varphi_c = 115°$ (**g**) $\varphi_c = 135°$ (**h**) $\varphi_c = 165°$ (**i**) $\varphi_c = 180°$ (**j**). Note that our reconfigurable twisted trilayer devices enable all-angle tunable polariton canalization, which is inaccessible in twisted α-MoO_3 bilayers

which are in excellent agreement with the experimental IFC (Fig. 8.9d, inset). These results show that polaritons are canalized along the direction given by $\varphi_c = 140°$. Interestingly, by changing the twist angle θ_{1-3} up to $\approx 60°$, we observe that the canalization angle changes to $\varphi_c = 80°$ (Fig. 8.9e), yielding a flat IFC (Fig. 8.9e, inset), in excellent agreement with the numerical simulations (Fig. 8.9h). Further varying θ_{1-3} to $\approx 40°$, we observe polariton canalization along another in-plane direction, namely, $\varphi_c = 50°$ (Fig. 8.9f,i). These experimental results, together with our theoretical and numerical calculations, unequivocally demonstrate the existence of multiple photonic magic angles in twisted trilayers, which provide a platform to steer canalized polaritons along any direction in the plane (Fig. 8.10 shows other possible canalization angles). Note that canalization in twisted trilayers is diffractionless (Fig. 8.11). This is independent of the thickness of each layer. In fact, canalization in twisted trilayers is also robust to thickness variations (Fig. 8.12), in clear contrast with the case of twisted bilayers. Therefore, the wide tunability and steering capability in reconfigurable trilayers offer unprecedented possibilities for guiding mid-infrared electromagnetic fields at will at the nanoscale.

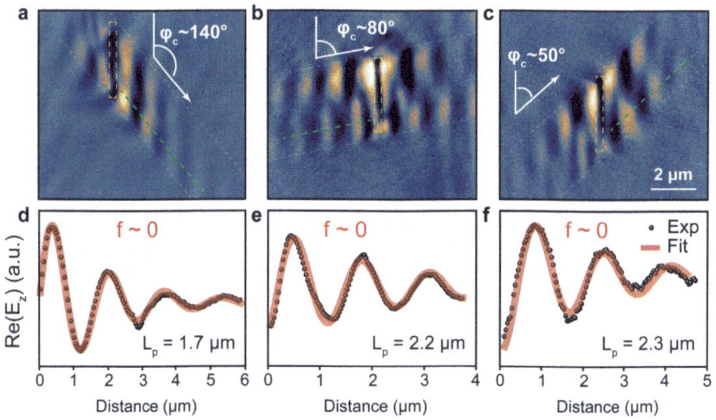

Fig. 8.11 Extraction of propagation length of canalized PhPs in twisted α-MoO_3 trilayers. **(a-c)** Near-field amplitude images of the trilayer in Fig. 8.9d ($\theta_{1-2} = 30°$ and $\theta_{1-3} = -90°$), Fig. 8.9e ($\theta_{1-2} = 30°$ and $\theta_{1-3} = -60°$), and Fig. 8.9f ($\theta_{1-2} = 30°$ and $\theta_{1-3} = -40°$) at $\lambda_0 = 10.9$ μm, respectively. **(d–f)** Line profiles (black symbols) along the green dashed lines in the near-field experimental images in **(a–c)** together with fitted curves (red solid lines) following Eq. (8.1). The resulting curves yield a value of 0 for f, indicating that the geometrical decay in the canalization regime is negligible. Note that large propagation length ($L_p = 1/\mathrm{Im}(k_x)$) barely varies when we change the twist angles, $\mathrm{Im}(k_x)$ giving rise to low-loss and diffractionless propagation of light at the nanoscale in the twisted trilayers

Spectrally Robust Photonic Magic Angles

To fully achieve the goal of steering nanolight on demand, there was still another obstacle that had not been addressed in twistoptics before the completion of the work presented in this chapter: the photonic magic angle exists for a specific frequency (ω_c) for each twisted bilayer device. Therefore, polariton canalization lacks spectral robustness in twisted bilayers, thereby limiting their potential for technological applications requiring canalization at multiple frequencies, such as broadband guiding of polaritons.

To study the spectral dependence of the photonic magic angles (and thus of canalization) in twisted trilayers, we performed near-field infrared nanoimaging of a twisted α-MoO_3 trilayer at different incident frequencies (ω_0) for a given combination of twist angles ($\theta_{1-2} \approx 30°$ and $\theta_{1-3} \approx 40°$). Excitingly, the obtained near-field images (Fig. 8.13a–d for ω_0 values of 930, 917, 909 and 901 cm^{-1}, respectively) show that the propagation of PhPs in the twisted trilayer is strongly collimated along a specific direction ($\varphi_c = 50°$) in all cases, that is, for a wide range of ω_0. This result is further verified by performing the FFT of the near-field images. The resulting IFCs show strongly flattened regions in all the cases (Fig. 8.13a–d, insets), thus unambiguously demonstrating the robustness of the photonic magic angle in twisted trilayers, giving rise to canalized polaritons along a specific direction in a broadband frequency range. To theoretically corroborate

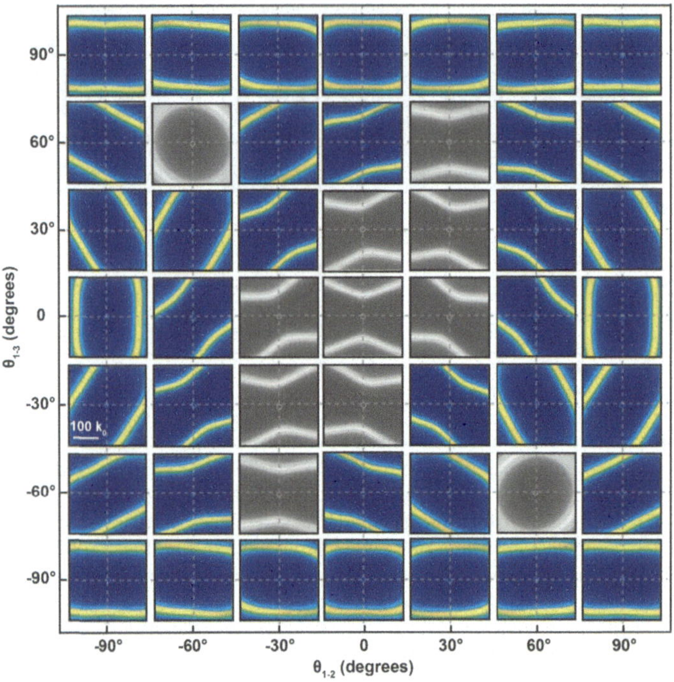

Fig. 8.12 Thickness independence on tunable polariton canalization in twisted trilayers. The color plot represents numerical PhP IFCs in twisted α-MoO_3 trilayers with small thicknesses ($d_3 = d_2 = d_1 = 25$ nm) and all-angle variation of θ_{1-2} and θ_{1-3} from $-90°$ to $90°$. All calculations are performed at an incident wavelength $\lambda_0 = 10.9$ μm. Analogous to the thick trilayer in Fig. 8.6 ($d_3 = 250$ nm, $d_2 = 240$ nm, $d_1 = 180$ nm), there exists flattened IFCs along different directions in thin twisted trilayers. The results demonstrate that multiple photonic magic angles and all-angle tunable polariton canalization in twisted trilayers are independent on the thickness of each layer

our experimental findings, we calculate the analytic IFCs as a function of incident frequency (ω_0) for the twisted trilayer shown in Fig. 8.13a–d (Fig. 8.13e). In the three-dimensional view and contour plot (Fig. 8.13e, inset), the analytic IFCs show a topological transition from a closed elongated ellipse to an open curve with a rather complex contour. Importantly, at this transition, the IFCs remain flat along the same direction within a wide frequency range (870–940 cm^{-1}), covering up to half of the hyperbolic *Reststrahlen* band of α-MoO_3 under study and thus revealing the emergence of a photonic magic angle with spectral robustness in twisted trilayers. This is better observed by extracting two-dimensional cross sections at the four representative frequencies shown in Fig. 8.13a–d [red, green, purple and black squares (Fig. 8.13e, inset) for ω_0 values of 930, 917, 909 and 901 cm^{-1}, respectively]. In all these cases, the cross sections show IFCs that are flat along the same direction, namely, $\varphi_c = 50°$ (Fig. 8.13e, black arrows). Our analytical calculations are, thus, in excellent agreement with our experimental results, corroborating the existence of robust photonic magic angles in twisted α-MoO_3 trilayers. Furthermore, to

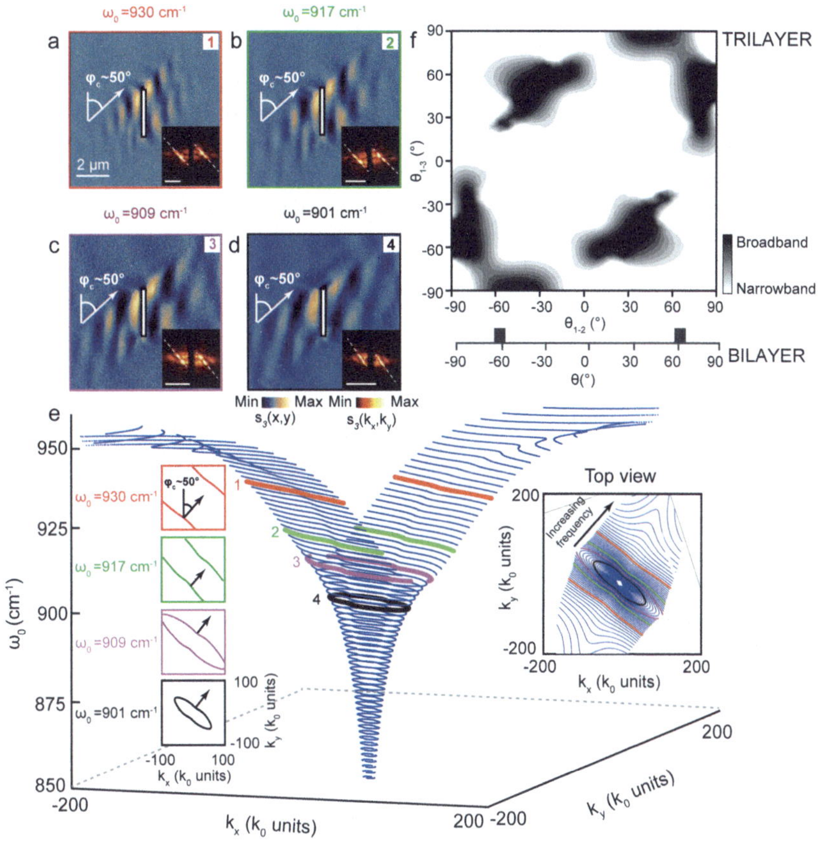

Fig. 8.13 Spectrally robust photonic magic angles and broadband polariton canalization in twisted trilayers. (**a–d**) Near-field amplitude images of a twisted trilayer device ($\theta_{1-2} = 30°$, $\theta_{1-3} \approx 40°$, $d_1 = 250\,\text{nm}$, $d_2 = 240\,\text{nm}$, $d_3 = 180\,\text{nm}$) at $\omega_0 = 930\,\text{cm}^{-1}$ (**a**) $\omega_0 = 917\,\text{cm}^{-1}$ (**b**) $\omega_0 = 909\,\text{cm}^{-1}$ (**c**) and $\omega_0 = 901\,\text{cm}^{-1}$ (**d**). In all these cases, the propagation of PhPs is strongly collimated along a narrow spatial sector. The corresponding experimental IFCs (**insets in a–d**) feature flat branches tilted along the same direction, verifying the emergence of one photonic magic angle within a wide spectral range, giving rise to canalization along the direction of $\varphi_c = 50°$ (white arrows). Scale bar (FFT images), $50k_0$. (**e**) Analytic IFCs of PhPs as a function of ω_0 for the twisted trilayer shown in (**a–d**). The three-dimensional and top view (right inset) show flattened IFCs in a wide frequency range (870–940 cm^{-1}). The left insets show two-dimensional cross sections of IFCs at the four incident frequencies used in (**a–d**), showing polariton canalization along the direction of $\varphi_c = 50°$ (black arrows). (**f**) Comparison of the broadbandness of polariton canalization in twisted trilayers (**top**) and twisted bilayers (**bottom**). The thickness of the α-MoO_3 layers is set to 200 nm. The broadbandness of polariton canalization in a twisted trilayer is calculated in the parameter space (θ_{1-2}, θ_{1-3}), where θ_{1-2} (θ_{1-3}) denotes the twist angle between the [001] crystal direction of the bottom layer and the [001] crystal direction of the middle (top) layer. The broadbandness of polariton canalization in a twisted bilayer is calculated in the parameter space of the twist angle θ between the slabs

quantify the robustness of the broadband canalization of PhPs in twisted trilayers, we calculate the angular spreading in the field-intensity directivity plots for different combinations of twist angles and compare the results obtained with those calculated for twisted bilayers (see Appendix G for the derivation of the analytical framework and Fig. 8.13f for the results). While in twisted bilayers there is only a small domain in the parameter space of twist angles θ giving rise to broadband canalization ($\Delta\omega > 20\,\text{cm}^{-1}$) (Fig. 8.13f, bottom), twisted trilayers (Fig. 8.13f, top) support robust broadband polariton canalization for a much larger area of the parameter space of twist angles (θ_{1-2} and θ_{1-3}).

Conclusions

Overall, the findings presented in this section demonstrate the existence of multiple photonic magic angles in reconfigurable twisted α-MoO_3 trilayers, which enable on-demand polariton canalization along an arbitrary in-plane direction by simply varying the interlayer twist angles. Importantly, the fabrication of a new sample for every twist angle is not needed, but, in turn, the stacks can be reassembled or reconfigured in the same sample. Furthermore, it is found that the photonic magic angles are spectrally robust in twisted trilayers, that is, they arise within a wide frequency range, giving rise to broadband canalization of polaritons, which is crucial for the implementation of twisted vdW heterostructures in optical nanotechnologies. In addition, polariton canalization in twisted trilayers is also robust to variations in layer thickness. Interestingly, we envision that even wider spectral ranges could potentially be covered by incorporating additional twisted layers of α-MoO_3 or other vdW materials supporting polaritons, such as α-V_2O_5 and intercalated α-NaV_2O_5 [33], WTe_2 [34] or graphene, as will be shown in the next chapter. In addition, the use of machine learning [35, 36] or inverse design [37, 38] algorithms to retrieve the critical parameters that generate a given IFC can be a promising approach to exploit the full potential of twisted multilayered structures made of anisotropic slabs. Altogether, these results bring twistoptics to the next level, opening a plethora of new possibilities for controlling and steering the flow of electromagnetic energy at will at the nanoscale.

References

1. Y. Cao et al., Unconventional superconductivity in magic-angle graphene superlattices. Nature **556**, 43–50 (2018). https://doi.org/10.1038/nature26160
2. X. Lu et al., Superconductors, orbital magnets and correlated states in magic-angle bilayer graphene. Nature **574**, 653–657 (2019). https://doi.org/10.1038/s41586-019-1695-0
3. M. Yankowitz et al., Tuning superconductivity in twisted bilayer graphene. Science **363**, 1059–1064 (2019). https://www.science.org/doi/abs/10.1126/science.aav1910

4. Y. Cao et al., Correlated insulator behaviour at half-filling in magic-angle graphene superlattices. Nature **556**, 80–84 (2018). https://doi.org/10.1038/nature26154
5. A.L. Sharpe et al., Emergent ferromagnetism near three-quarters filling in twisted bilayer graphene. Science **365**, 605–608 (2019). https://www.science.org/doi/abs/10.1126/science.aaw3780
6. K. Seo, V.N. Kotov, B. Uchoa, Ferromagnetic Mott state in twisted graphene bilayers at the magic angle. Phys. Rev. Lett. **122**, 246402 (2019). https://link.aps.org/doi/10.1103/PhysRevLett.122.246402
7. R. Bistritzer, A.H. MacDonald, Moiré bands in twisted double-layer graphene. Proc. Natl. Acad. Sci. **108**, 12233–12237 (2011). https://www.pnas.org/doi/abs/10.1073/pnas.1108174108
8. S.S. Sunku et al., Photonic crystals for nano-light in Moiré; graphene superlattices. Science **362**, 1153–1156 (2018). https://www.science.org/doi/abs/10.1126/science.aau5144
9. C. Jin et al., Observation of Moiré excitons in WSe_2/WS_2 heterostructure superlattices. Nature **567**, 76–80 (2019). https://doi.org/10.1038/s41586-019-0976-y
10. K. Tran et al., Evidence for Moiré excitons in van der Waals heterostructures. Nature **567**, 71–75 (2019). https://doi.org/10.1038/s41586-019-0975-z
11. K.L. Seyler et al., Signatures of Moiré-trapped valley excitons in $MoSe_2/WSe_2$ heterobilayers. Nature **567**, 66–70 (2019). https://doi.org/10.1038/s41586-019-0957-1
12. H. Yu, G.-B. Liu, J. Tang, X. Xu, W. Yao, Moiré excitons: From programmable quantum emitter arrays to spin-orbit-coupled artificial lattices. Sci. Adv. **3**, e1701696 (2017). https://www.science.org/doi/abs/10.1126/sciadv.1701696
13. F. Wu, T. Lovorn, A.H. MacDonald, Topological exciton bands in Moiré heterojunctions. Phys. Rev. Lett. **118**, 147401 (2017). https://link.aps.org/doi/10.1103/PhysRevLett.118.147401
14. G.X. Ni et al., Plasmons in graphene Moiré superlattices. Nat. Mater. **14**, 1217–1222 (2015). https://doi.org/10.1038/nmat4425
15. N.C.H. Hesp et al., Observation of interband collective excitations in twisted bilayer graphene. Nat. Phys. **17**, 1162–1168 (2021). https://doi.org/10.1038/s41567-021-01327-8
16. O.V. Kotov, Y.E. Lozovik, Hyperbolic hybrid waves and optical topological transitions in few-layer anisotropic metasurfaces. Phys. Rev. B **100**, 165424 (2019). https://link.aps.org/doi/10.1103/PhysRevB.100.165424
17. G. Hu, A. Krasnok, Y. Mazor, C.-W. Qiu, A. Alù, Moiré hyperbolic metasurfaces. Nano Lett. **20**, 3217–3224 (2020). https://doi.org/10.1021/acs.nanolett.9b05319
18. W. Ma et al., In-plane anisotropic and ultra-low-loss polaritons in a natural van der Waals crystal. Nature **562**, 557 (2018). https://www.nature.com/articles/s41586-018-0618-9
19. Z. Zheng et al., A mid-infrared biaxial hyperbolic van der Waals crystal. Sci. Adv. **5**, eaav8690 (2019). https://advances.sciencemag.org/content/5/5/eaav8690
20. J. Duan et al., Twisted nano-optics: manipulating light at the nanoscale with twisted phonon polaritonic slabs. Nano Lett. **20**, 5323–5329 (2020). https://doi.org/10.1021/acs.nanolett.0c01673
21. G. Álvarez-Pérez, K.V. Voronin, V.S. Volkov, P. Alonso-GonzáLlez, A.Y. Nikitin, Analytical approximations for the dispersion of electromagnetic modes in slabs of biaxial crystals. Phys. Rev. B **100**, 235408 (2019). https://link.aps.org/doi/10.1103/PhysRevB.100.235408
22. D. Correas-Serrano, A. Alù, J.S. Gomez-Diaz, Plasmon canalization and tunneling over anisotropic metasurfaces. Phys. Rev. B **96**, 075436 (2017). https://link.aps.org/doi/10.1103/PhysRevB.96.075436
23. J.S. Gómez-Díaz, M. Tymchenko, A. Alù, Hyperbolic plasmons and topological transitions over uniaxial metasurfaces. Phys. Rev. Lett. **114**, 233901 (2015). https://link.aps.org/doi/10.1103/PhysRevLett.114.233901
24. A.Y. Nikitin et al., Real-space mapping of tailored sheet and edge plasmons in graphene nanoresonators. Nat. Photon. **10**, 239–243 (2016). https://doi.org/10.1038/nphoton.2016.44
25. V. Agranovich, D. Mills, *Surface Polaritons – Electromagnetic Waves at Surfaces and Interfaces* (Elsevier Science, 1982). https://www.elsevier.com/books/surface-polaritons/agranovich/978-0-444-86165-8

26. F.J. Alfaro-Mozaz et al., Deeply subwavelength phonon-polaritonic crystal made of a van der Waals material. Nat. Commun. **10**, 42 (2019). https://doi.org/10.1038/s41467-018-07795-6
27. A. Woessner et al., Highly confined low-loss plasmons in graphene.boron nitride heterostructures. Nat. Mater. **14**, 421–425 (2015). https://doi.org/10.1038/nmat4169
28. G. Álvarez-Pérez et al., Infrared permittivity of the biaxial van der Waals semiconductor α-MoO_3 from near- and far-field correlative studies. Adv. Mater. **32**, 1908176 (2020). https://onlinelibrary.wiley.com/doi/abs/10.1002/adma.201908176
29. H. Herzig Sheinfux, F.H.L. Koppens, The rise of twist-optics. Nano Lett. **20**, 6935–6936 (2020). https://doi.org/10.1021/acs.nanolett.0c03175
30. J. Duan et al., Multiple and spectrally robust photonic magic angles in reconfigurable α-MoO_3 trilayers. Nat. Mater. **22**, 867–872 (2023). https://doi.org/10.1038/s41563-023-01582-5
31. N.C. Passler et al., Hyperbolic shear polaritons in low-symmetry crystals. Nature **602**, 595–600 (2022). https://doi.org/10.1038/s41586-021-04328-y
32. G. Hu et al., Real-space nanoimaging of hyperbolic shear polaritons in a monoclinic crystal. Nat. Nanotechnol. **18**, 64–70 (2023). https://doi.org/10.1038/s41565-022-01264-4
33. J. Taboada-Gutiérrez et al., Broad spectral tuning of ultra-low-loss polaritons in a van der Waals crystal by intercalation. Nat. Mater. **19**, 964–968 (2020). https://doi.org/10.1038/s41563-020-0665-0
34. C. Wang et al., van der Waals thin films of WTe2 for natural hyperbolic plasmonic surfaces. Nat. Commun. **11**, 1158 (2020). https://doi.org/10.1038/s41467-020-15001-9
35. W. Ma et al., Deep learning for the design of photonic structures. Nat. Photon. **15**, 77–90 (2021). https://doi.org/10.1038/s41566-020-0685-y
36. Z.A. Kudyshev, A.V. Kildishev, V.M. Shalaev, A. Boltasseva, Machine learning–assisted global optimization of photonic devices. Nanophotonics **10**, 371–383 (2021). https://doi.org/10.1515/nanoph-2020-0376
37. M. He et al., Deterministic inverse design of Tamm plasmon thermal emitters with multiresonant control. Nat. Mater. **20**, 1663–1669 (2021). https://doi.org/10.1038/s41563-021-01094-0
38. P.R. Wiecha, A. Arbouet, C. Girard, O.L. Muskens, Deep learning in nano-photonics: inverse design and beyond. Photon. Res. **9**, B182–B200 (2021). https://opg.optica.org/prj/abstract.cfm?URI=prj-9-5-B182

Chapter 9
Active Tuning of Hyperbolic Polaritons in van der Waals Materials by Integrating a Gated Graphene Layer

Abstract Although in-plane anisotropic PhPs can be steered and even canalized by twisting individual crystal layers in a van der Waals (vdW) stack, active control of their propagation via external stimuli presents a significant challenge. In this chapter, we report on a technology in which anisotropic PhPs supported by biaxial vdW layers are actively tuned by simply gating a graphene layer. Excitingly, we predict active tuning of optical topological transitions, which enable controlling the canalization of PhPs along different in-plane directions in twisted heterostructures. Apart from their fundamental interest, these findings hold promises for the development of optoelectronic devices, such as sensors or photodetectors, based on PhPs with dynamically controllable properties.

Some of the calculations in this chapter were performed in collaboration with Arturo González-Morán, Nathaniel Capote-Robayna and Kirill Voronin (Donostia International Physics Center, Spain). The results were published in "Active Tuning of Highly Anisotropic Phonon Polaritons in Van der Waals Crystal Slabs by Gated Graphene" by Gonzalo Álvarez-Pérez, Arturo González-Morén et al. ACS Photonics 9 (2), 383–390 (2022).

As shown in the preceding chapters, hyperbolic phonon polaritons (HPhPs) in vdW materials allow confining light at the nanoscale and its propagation only along certain specific directions, thus providing a natural playground to control, route, and direct light at the nanoscale. A series of efforts have been carried out in order to control HPhPs [1] using phase change materials [2] and low-symmetry crystals [3–5]. However, the dynamical control of the propagation properties of in-plane HPhPs, crucial for their potential implementation in optical and optoelectronic applications, such as tunable photodetection or sensing, had remained elusive before the completion of the work presented in this chapter.

As discussed in Chap. 1, graphene supports plasmon polaritons that feature low losses, exceptional lateral confinement, and tunability via modification of the Fermi energy E_F. This makes them a very suitable ingredient in nano-opto-electronic applications requiring active control over the electromagnetic fields. In this chapter [12], we demonstrate that in-plane HPhPs in vdW crystal layers can

G. Álvarez Pérez, *Fundamentals of Nano-Optics in Hyperbolic van der Waals Materials*, Springer Theses, https://doi.org/10.1007/978-3-031-75767-9_9

be effectively manipulated by incorporating a gated graphene layer. Although our general theoretical results are valid for any biaxial layer (with all the main elements of its permittivity tensor being different $\varepsilon_x \neq \varepsilon_y \neq \varepsilon_z$), and hence for the less general case of uniaxial and isotropic layers, we illustrate our idea for molybdenum trioxide (α-MoO_3). Due to coupling between HPhPs in α-MoO_3 layers and graphene plasmon polaritons [6, 7] (GPPs) (see Appendix H.1), hybrid polaritons in graphene/α-MoO_3 heterostructures become actively tunable via electrostatic gating of the graphene layer. Importantly, we show the possibility of active controlling polariton canalization in twisted layers of α-MoO_3. Furthermore, we will show that hybrid polaritons in graphene/α-MoO_3 heterostructures exhibit low ohmic losses, allowing for preserving the long propagation length and lifetime from PhPs in α-MoO_3.

9.1 Active Tuning of Hyperbolic Polaritons in a Single α-MoO_3 Layer by Gated Graphene

Figure 9.1a shows a schematic of the proposed device, in which a gated graphene layer is placed on top of an α-MoO_3 layer of thickness d. The dispersion of polaritons in such graphene/α-MoO_3 heterostructure can be analytically derived by generalizing the dispersion relation of polaritons in biaxial layers surrounded by two semi-infinite media in the high-momenta approximation. This approximation was already introduced in Chap. 3 and the details of its application to a system composed of a biaxial layer and a graphene layer between two semi-infinite isotropic media can be found in Appendix H.2, yielding:

$$q = \frac{\rho}{k_0 d}\left[\arctan\left(\frac{\varepsilon_1 + 2i\alpha_g q}{\varepsilon_z}\rho\right) + \arctan\left(\frac{\varepsilon_3}{\varepsilon_z}\rho\right) + \pi l\right], \quad l = 0, 1, 2..., \tag{9.1}$$

where $\mathbf{q} = \mathbf{k}/k_0$ is the normalized in-plane wavevector (with $k^2 = k_x^2 + k_y^2$), $k_0 = \omega/c$ is the free-space wavevector, with c the speed of light in free space, ε_1 and ε_3 are the dielectric permittivities of the superstrate and substrate, respectively, d is the thickness of the layer, with φ being the angle between the x axis and the in-plane wavevector, and l is the counting number. $\alpha_g = 2\pi\sigma/c$ is the 2D conductivity of graphene, as presented in the introductory section "Plasmon Polaritons in Graphene" in Chap. 1, which was devoted to discuss GPPs and their dispersion. Here σ is the conductivity derived from quantum theory within the random-phase approximation, with contributions from interband and intraband electron–photon scattering processes (see Eq. 1.13 and section "Plasmon Polaritons in Graphene" in Chap. 1 for more details). We will take the value of this conductivity at ambient conditions all throughout the work. Equation (9.1) is an implicit expression, since $\mathbf{q}$ appears both in the left part and in the argument of the arctan, and can be solved numerically (see Appendix H.2). The different polaritonic modes in the

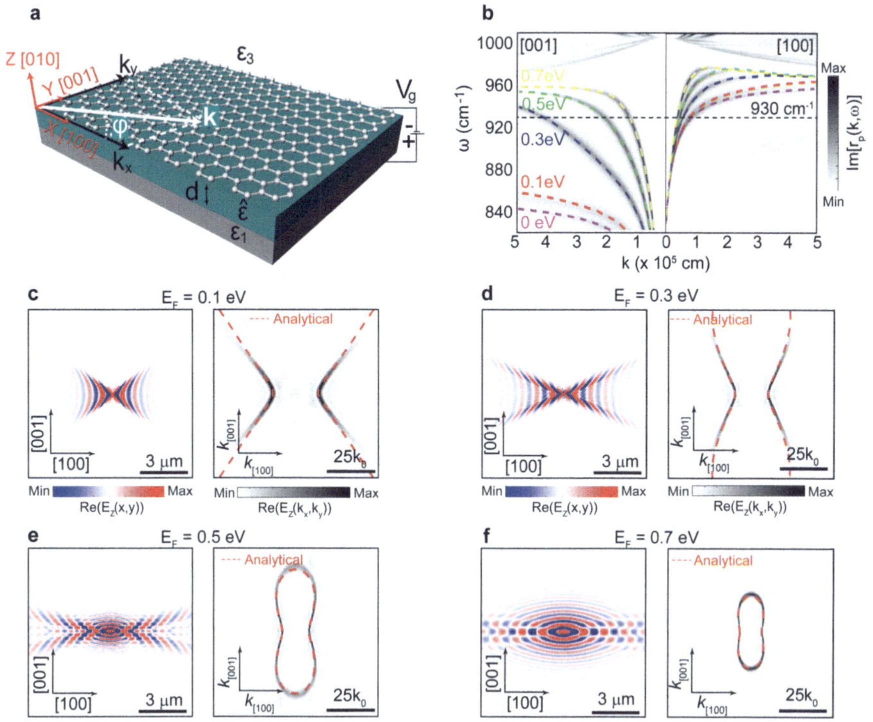

Fig. 9.1 Active tuning of hybrid polaritons in a graphene/α-MoO_3 heterostructure. (**a**) Schematics of an α-MoO_3 biaxial thin layer with a gated graphene sheet placed on top. (**b**) Dispersion of hybrid polaritons in α-MoO_3/graphene for different representative values of $E_F =$ 0.1, 0.3, 0.5, and 0.7 eV. The false color plot displays transfer-matrix calculations showing the imaginary part of Fresnel's reflection coefficient, while the analytical dispersions are depicted as dashed lines. (**c–f**) Simulated electric field spatial distribution (left panels) and corresponding numerical and analytical polaritonic IFCs (colorplots and dashed red lines in the right panels, respectively) of polaritons for $E_F = 0.1$, 0.3, 0.5, and 0.7 eV, showing a topological transition from open hyperbolas (**c**, **d**) to closed curves (**e**, **f**) and canalization (**f**). The incident frequency is fixed to $\omega_0 = 930\,\text{cm}^{-1}$

heterostructure, described by Eq. (9.1), present hybridization between HPhPs in a thin layer of α-MoO_3 and GPPs (see Appendix H.1), with l characterizing the quantization of the electromagnetic field of the mode in the z direction, in such a way that the higher the l, the stronger confinement of the modes and the shorter wavelength and propagation length. An instructive way to represent the dispersion of polaritonic modes is to plot the imaginary part of the corresponding Fresnel reflection coefficient, $\text{Im}[r_p(\omega, \mathbf{k})]$, obtained from transfer-matrix calculations [8], as a function of frequency and wavevector (Fig. 9.1b). In our calculations the thickness of the thin α-MoO_3 layer is set to $d = 120\,\text{nm}$ (thus being in the range of typical thicknesses in recent near-field experiments), while the relaxation time of charge carriers in graphene is set to $\tau_{\text{rel}} = 0.5$ ps. Although this value of τ_{rel} is typical for encapsulated graphene used in optical experiments at ambient conditions [9, 10], we do not consider the encapsulating h-BN layers in our calculations since they do not introduce any significant changes, and particularly, in-plane effects to the PhPs dispersion (see Appendix H.3). In addition, the excitations in our devices lie well outside the Landau damping regime (see Appendix H.4). For clarity, the results for $\mathrm{Im}[r_p(\omega, k)]$ at different Fermi levels of the graphene layer, E_F (or, equivalently, different gate voltages) are superimposed on the same colorplot in Fig. 9.1b. The bright maxima correspond to the dispersion of polaritonic modes in the heterostructure. At the same time, the dashed lines render the lowest-order modes ($l = 0$) calculated according to Eq. (9.1). On the one hand, the perfect matching between the maxima of the colorplot and the curves proves the validity of our simple analytical approximation. On the other hand, we clearly see that the dispersion relation of the mode of the same index l is strongly dependent upon E_F. Namely, the dispersion along both the [100] and [001] crystal directions become steeper as E_F increases from 0 to 0.7 eV, so that the momentum of polaritons decreases with E_F at a fixed frequency. Hence, the tuning of E_F provides an effective method to control the subdiffractional confinement of PhPs in heterostructures made of graphene and a biaxial polaritonic layer, allowing to do so, in addition, in an active way, for instance, by an applied gate voltage.

To study in more detail the polaritonic features as a function of E_F in graphene/α-MoO_3 heterostructures, we calculate the spatial distribution of the vertical component of the electric field, $E_z(x, y)$, created by a vertical electric point dipole at different E_F. For the sake of simplicity, we perform our calculations at a representative frequency $\omega_0 = 930\,\mathrm{cm}^{-1}$ (shown by a horizontal dashed line in Fig. 9.1b) belonging to the hyperbolic RB of α-MoO_3, which spans the spectral region between $\omega_0 = 821.4$ and $963.0\,\mathrm{cm}^{-1}$ (RB2), as shown in Chap. 4 and in Ref. [11]. For a Fermi level $E_F = 0$ eV, i.e., the case without graphene, the near-field distribution features concave wavefronts, as typically observed for HPhPs in α-MoO_3 (see also Appendix H.5). We note that, being rigorous, the inclusion of the undoped graphene layer modifies the electromagnetic (polaritonic) response, as it changes the environment of the biaxial layer. However, this contribution is negligible for the thickness of the biaxial layer considered, of several tenths of nanometers, since the average effective thickness of a graphene layer is below 1 nm. For $E_F = 0.1$ eV, PhPs also propagate within a hyperbolic sector centered along the [100] direction, which, however, spans now a smaller angle (Fig. 9.1c, left panel). By increasing E_F to 0.3 and 0.5 eV, such propagating sector becomes much narrower (left panels in Fig. 9.1d and e, respectively), indicating a further reduced angle of the hyperbolic sector, while, on the other hand, the PhPs wavelength increases. Interestingly, in the case of $E_F = 0.5$ eV, additional elliptical wavefronts appear in the neighborhood of the origin. Excitingly, when reaching $E_F = 0.7$ eV, the propagating hyperbolic sector virtually converts into a line (Fig. 9.1f, left panel), so that PhPs appear to propagate in their majority only along one specific direction (canalization regime). Again, elliptical wavefronts are also present in the electric field distribution. These results unambiguously demonstrate that the E_z distribution of polaritons in a graphene/α-MoO_3 heterostructure exhibits E_F-

dependent wavefronts, thus corroborating that such hybrid polaritons inherit the extremely directional nature of HPhPs in α-MoO_3 while gaining a unique virtue of GPPs: tunability via E_F.

In order to further characterize the propagation properties of hybrid PhPs in a graphene/α-MoO_3 heterostructure, we analyze the polaritonic IFCs in momentum space. To that end we perform the Fourier-transform (FT) of the near-field $E_z(x, y)$, represented by colorplots (Fig. 9.1c–f, right panels). The FTs for $E_F = 0$ eV (see Appendix H.5) and 0.1 eV (Fig. 9.1c, right panel) show open hyperbolas, as expected from the concave shape of the wavefronts in the field distributions, and in excellent agreement with our analytical calculations (dashed red curves). For $E_F = 0.3$ eV (Fig. 9.1d, right panel), the opening angle of the hyperbola (given by the asymptote) increases, consistently with the field distribution, still showing concave wavefronts (Fig. 9.1d, left panel). Conversely, for $E_F = 0.5$ eV (Fig. 9.1e, right panel), the FT presents a closed curve, revealing both a hyperbolic-like shape IFC along the [100] direction and an elliptic-like shape along the [001] direction. Finally, for $E_F = 0.7$ eV, the FT reveals a closed IFC that covers a smaller area (Fig. 9.1f, right panel), consistently with the smaller momenta of GPPs for larger E_F. Note that the IFC is not a perfectly closed ellipse owing to the circular wavefronts around the origin, yet it exhibits a strong flattening along the [100] direction, explaining the canalization of PhPs along this direction. We note that all the numerical IFCs (colorplots) in Fig. 9.1c–f show an excellent agreement with the analytical curves given by Eq. (9.1) (red dashed curves), thus reconfirming its versatility. Importantly, apart from the strong dependence of the IFCs upon E_F, their evolution unveils a topological transition. Such transition can be characterized by a topological quantity: the genus of the IFC, which describes the number of holes in the curve. The genus of the IFCs in Fig. 9.1c,d (left panels) is equal to 0, while for the IFCs in Fig. 9.1e,f (left panels) is equal to 1. The transition frequency ω_T is highly dependent upon E_F, i.e., for a given ω_T we can find a certain E_F at which canalization of polaritons takes place. In particular, at $\omega_0 = 930\,\text{cm}^{-1}$, the transition takes place at $E_F = 0.32$ eV, and it is well captured by our analytical equation (see Appendix H.5). From a practical perspective, these results show that the topology of the polaritonic IFCs in a graphene/α-MoO_3 heterostructure can be actively controlled by electrical means. Note that, prior to this work, published at the beginning of 2022 [12], such a dramatic modification of the topology of the IFCs for in-plane anisotropic PhPs had only been obtained via coupling with polar substrates, as shown in Chap. 7 and in Refs. [13, 14] or by fabricating twisted bilayers of the polaritonic material, as shown in Chap. 8 and Refs. [15–18], i.e., by approaches precluding any direct active control. After the completion of the work presented here, another theoretical study also proposed the use of integrated graphene to modulate the propagation of HPhPs in α-MoO_3 [19] and several groups experimentally demonstrated the tuning of HPhPs by integrating a doped graphene layer [20–22]. Recently, gate-tunable negative refraction of mid-IR polaritons [23] has also been demonstrated in graphene/α-MoO_3 heterostructures.

Although doping graphene up to the Fermi level of 0.7 eV to achieve canalization in a single thin layer is, in principle, feasible in some practical devices (particularly,

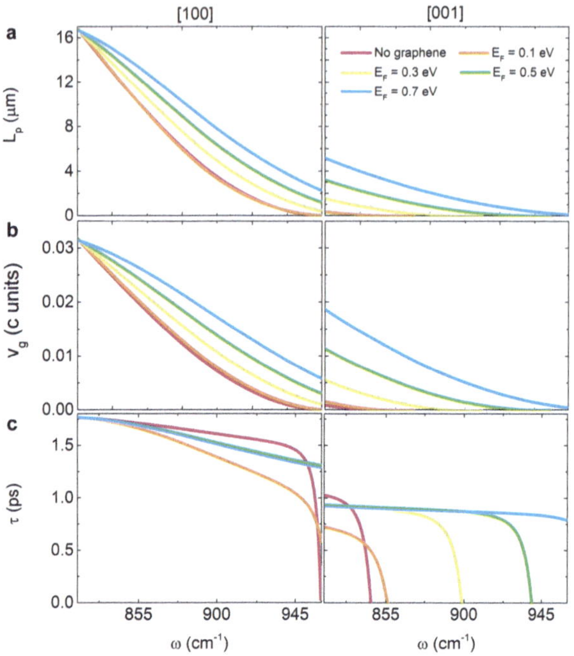

Fig. 9.2 Propagation properties of hybrid polaritons in a graphene/α-MoO$_3$ heterostructure. (**a**) Analytical propagation length L_p, (**b**) group velocity v_g, and (**c**) lifetime τ as a function of frequency for different Fermi energies $E_F = 0.1, 0.3, 0.5$, and 0.7 eV, together with the calculation for the case without graphene. The left (respectively, right) panel shows the calculations for polaritons propagating along the [100] (respectively, [001]) direction

in intercalated [24], chemically doped [22], or top-gated [25, 26] graphene), we also note that, given the dependence of the polaritonic dispersion given by Eq. (9.1) on the thickness of the biaxial thin layer, it is possible to achieve canalization at the same or higher frequencies for smaller, more experimentally accessible Fermi energies (0.4 and 0.5 eV) by decreasing the thickness of the α-MoO$_3$ thin layer (see Appendix H.5).

To gain further insights into the properties of hybrid polaritons in a graphene/α-MoO$_3$ heterostructure, we quantitatively characterize some of their crucial propagation parameters: propagation length, group velocity, and lifetime as a function of the incident frequency ω and Fermi energy E_F. Figure 9.2a shows the propagation length, $L_p = 1/\mathrm{Im}(k)$, as a function of frequency, ω. L_p monotonically decreases with ω for polaritons propagating along both the [100] and [001] crystal directions (left and right panels of Fig. 9.2a, respectively), which can be explained by the higher confinement of polaritons with increasing ω. In contrast, L_p increases with E_F, with this tendency being more dramatic for the [001] crystal direction, along which L_p increases by a factor of 10 as E_F increases from 0.1 to 0.7 eV (as can be seen, e.g., at $\omega_0 = 820\,\mathrm{cm}^{-1}$). This is due to a loss of confinement (longer polaritonic wavelength), consistently with the results shown in Fig. 9.1b. Similarly, the group velocity, $v_g = \partial\omega/\partial k$, increases with E_F (Fig. 9.2b) while decreasing with ω, showing the same trend along both in-plane crystal directions. Having calculated both L_p and v_g, the polariton lifetime, τ, can be easily analyzed following the relation $\tau = L_p/v_g$. In contrast to L_p and v_g, the dependence of

τ upon ω is rather flat (Fig. 9.2c), reaching appreciable values of $\approx$2 ps along the [100] crystal direction and of $\approx$1 ps along the [001] direction. Importantly, although τ shows a steep drop along the [001] direction, which blueshifts with increasing E_F, the maximum value of τ is virtually independent upon E_F (remaining high up to the dropping frequency even for large values of E_F) in the frequency range analyzed. In fact, the spectral range where τ takes an appreciable value along the [001] direction ($\approx$1 ps, i.e., as long as when considering the thin layer without graphene) increases with E_F. Hence, the polaritonic propagation properties in heterostructures made of graphene and α-MoO_3, such as propagation length and lifetime, are not undermined by the active tuning capability, but instead can be substantially improved. As a result, the tunable propagating hybrid polaritons in graphene/α-MoO_3 heterostructures preserve the low-loss character of PhPs in α-MoO_3, which is crucial for applications requiring long-lived light-matter interactions, such as mid-infrared (bio)sensing or strong coupling between polaritons and molecules.

9.2 Active Tuning of Hyperbolic Polaritons in a Twisted α-MoO_3 Bilayer by Gated Graphene

Going one step further, we also study active manipulation of polaritons by considering vdW stacks formed by two mutually twisted α-MoO_3 thin layers covered by a graphene layer. As shown in Chap. 8 and in Refs. [15–18], PhPs in two twisted α-MoO_3 thin layers display canalization due to the emergence of a topological transition in the IFC. However, in such structures, the twist angle at which the canalization of PhPs takes place (as well as the in-plane canalization angle itself) is predefined by the thicknesses of the thin layers and the incident frequency. Specifically, for a given thickness and incident frequency, canalization of PhPs in twisted bilayers can only be achieved for one specific twist angle and one specific in-plane direction of propagation, thus severely limiting the tunability of these directional polaritons. Here, we introduce an additional degree of freedom for active tuning of canalized polaritons in twisted vdW bilayers by adding a gated graphene layer. In particular, we consider a vdW heterostructure formed by two 50 nm-thick layers of α-MoO_3 (the choice of this representative thickness does not affect the generality of our results) twisted by an angle θ with respect to each other and a graphene layer placed on top (schematics in Fig. 9.3a). In order to illustrate the propagation of polaritons in such stack, we calculate the spatial distributions of the electric field generated by a point dipole source for different E_F (Fig. 9.3c–f) at $\omega = 930\,\text{cm}^{-1}$, as well as the corresponding FTs (analogously to the case of a single α-MoO_3 thin layer shown in Fig. 9.1).

For the case $E_F = 0$ (i.e., no graphene) and by monitoring the spatial electric field distributions at different twist angles θ, we find that only one critical twist angle, $\theta_c = 75°$, yields a canalization regime in which the excited polaritons propagate in a rather narrow angular sector (see Appendix H.5). The corresponding

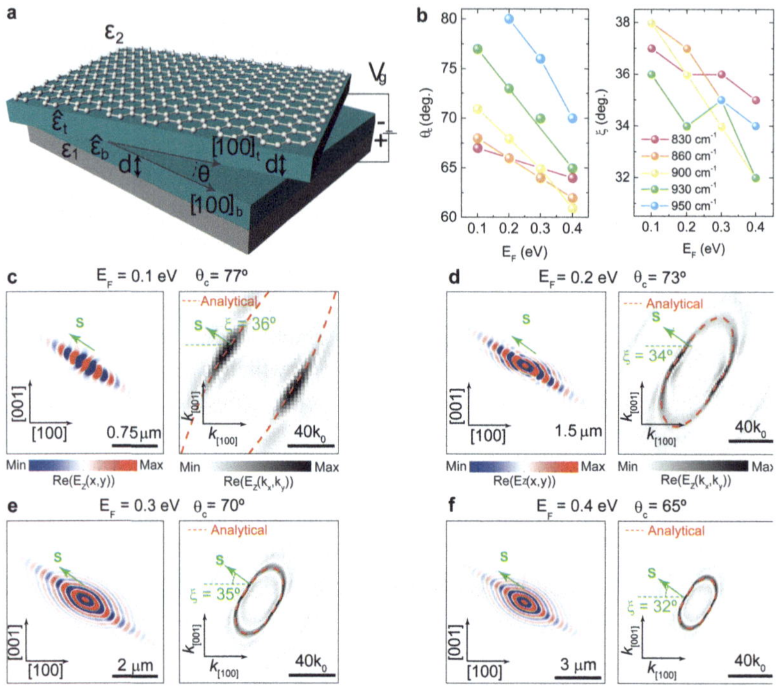

Fig. 9.3 **Active tuning of hybrid polaritons in twisted vdW stacks.** (**a**) Schematics of two α-MoO$_3$ thin layers superimposed with their in-plane crystalline directions rotated by a critical twist angle θ_c and with a gated graphene layer placed on top. (**b**) Dependence of θ_c, and the in-plane angle at which canalization takes place, ξ, as a function of E_F for different ω. (**c–f**) Simulated electric field spatial distribution (left panels) and corresponding numerical and analytical polaritonic IFCs (color plots and dashed blue lines in the right panels, respectively) of PhPs propagating in twisted α-MoO$_3$ stacks with a gated graphene layer placed on top for $E_F = 0.1, 0.2,$ 0.3, and 0.4 eV, and $\theta_c = 77°, 73°, 70°$, and 65°. The incident frequency is fixed to $\omega = 930\,\text{cm}^{-1}$

direction along which the canalized polaritons propagate forms an angle $\xi = 41°$ with the x axis. Interestingly, by adding a slightly doped graphene layer ($E_F =$ 0.1 eV), the critical angle decreases to $\theta_c = 77°$, as observed in Fig. 9.3c (left panel). The corresponding FT (colorplot in Fig. 9.3c, right panel) shows a highly flat IFC, matching well the corresponding analytical result (red dashed lines). Due to the flatness of the IFC, the Poynting vectors (oriented perpendicularly to the IFC and denoted as **S**) of the plane waves with different wavevectors are parallel, forming an angle $\xi = 33°$ with the x axis (Fig. 9.3c, right panel), which reconfirms that the polaritons propagate in the canalization regime. By increasing the doping of the graphene layer ($E_F = 0.2$ eV), the critical twist angle decreases to $\theta_c = 73°$ and the canalized polaritons propagate within a narrower angular sector (Fig. 9.3d, left panel), corresponding to $\xi = 35°$ (Fig. 9.3d, right panel). Interestingly, elliptical wavefronts appear close to the origin. The emergence of these wavefronts becomes clear after analyzing the corresponding IFC (colorplot and red dashed lines in

Fig. 9.3c, right panel), which although still giving rise to parallel Poynting vectors along its flat part, as expected for a canalized regime, also shows an elongated ellipse-like closed curve. As we further increase E_F up to 0.3 eV (Fig. 9.3e), the critical twist angle at which canalization takes place further decreases to $\theta_c = 70°$. For these parameters, the near field pattern reveals canalized polaritons with longer wavelengths and featuring elliptical wavefronts in the neighborhood of the origin (Fig. 9.3e, left panel).

Accordingly, the corresponding FT reveals a smaller closed IFC, which is flattened at an angle $\xi = 35°$ (Fig. 9.3e, right panel). Dashed red lines show the analytical IFCs, which are derived in Appendix H.6 for the case of two twisted layers and one graphene layer on top. By increasing E_F up to 0.4 eV (Fig. 9.1f, left panel), the canalization regime requires decreasing the critical twist angle to $\theta_c = 65°$. This is in accordance with the numerical and analytical IFCs, which are flattened at an angle $\xi = 37°$ (Fig. 9.3f, right panel). Analogous calculations are shown in Appendix H.5 for $E_F = 0.5$ eV. These results clearly indicate that for any given Fermi energy E_F, it is possible to find a critical twist angle θ_c at which canalization emerges. Conversely, for a predetermined twist angle, θ, we can actively achieve canalization of polaritons by properly matching E_F, simply varying it between 0.1 and 0.5 eV for a rather wide range of twist angles. The results shown in Fig. 9.3c–f for an incident frequency $\omega_0 = 930\,\text{cm}^{-1}$ are collected in Fig. 9.3b (left panel), along with analogous calculations for other incident frequencies ω within the Reststrahlen band RB2 of α-MoO_3, spanning the range between $\omega =$ 821.4 and $963.0\,\text{cm}^{-1}$. These results clearly reveal that canalization of polaritons in our actively tunable stack can be attained for a rather wide range of twist angles, between 60° and 80°, by simply varying E_F between experimentally attainable values of 0 and 0.5 eV. In addition, canalization can occur at different in-plane angles ξ. In fact, the results shown in the right panel of Fig. 9.3b reveal that the in-plane direction along which the hybrid polaritons are canalized can be actively manipulated in an angular sector of more than 12° by simply varying ω and E_F. Note that while all these calculations are performed for fixed layer thicknesses of 50 nm, the twist and canalization angles θ_c and ξ depend dramatically on such thicknesses. Therefore, these angles could be further manipulated for other combinations of parameters (see Appendix H.5). In addition, we note that polaritons behave the same way no matter whether we consider the graphene layer on top of the twisted heterostructure or sandwiched between the two twisted layers of α-MoO_3 (see Appendix H.5). Altogether, these findings open the door for the manipulation of highly directional (canalized) PhPs in a robust and dynamic way for a wide range of frequencies, angles, and Fermi energies.

9.3 Conclusions

In conclusion, we report a pathway through which hyperbolic PhPs supported by biaxial vdW layers can be tuned actively and in situ by integrating a gated graphene layer. We have unveiled tunable topological transitions of the polaritonic IFCs, leading to canalization of polaritons, which are highly dependent on the Fermi level of charge carriers in graphene. Excitingly, the propagating properties of the resulting hybrid polaritons, such as propagation length or lifetime, key for applications requiring long-lived light-matter interactions such as strong coupling or mid-infrared (bio)sensing, are not undermined by the ohmic losses in graphene. In addition, our findings demonstrate that low-loss canalization of polaritons can be also actively tuned in twisted vdW heterostructures, where the in-plane canalization direction can be controlled in a certain range of angles, opening the door to steer infrared light along specific directions on demand. Apart from their fundamental relevance, our findings provide a basis to design near-field optical experiments and hold promises for fundamental studies, for instance chiral polaritons [27], and optoelectronic devices, such as sensors and photodetectors, based on PhPs with dynamically controllable properties.

References

1. Q. Zhang et al., Interface nano-optics with van der Waals polaritons. Nature **597**, 187–195 (2021). https://doi.org/10.1038/s41586-021-03581-5
2. T.G. Folland et al., Reconfigurable infrared hyperbolic metasurfaces using phase change materials. Nat. Commun. **9**, 4371 (2018). https://doi.org/10.1038/s41467-018-06858-y
3. W. Ma et al., Ghost hyperbolic surface polaritons in bulk anisotropic crystals. Nature **596**, 362–366 (2021). https://doi.org/10.1038/s41586-021-03755-1
4. N.C. Passler et al., Hyperbolic shear polaritons in low-symmetry crystals. Nature **602**, 595–600 (2022). https://doi.org/10.1038/s41586-021-04328-y
5. G. Hu et al., Real-space nanoimaging of hyperbolic shear polaritons in a monoclinic crystal. Nat. Nanotechnol. **18**, 64–70 (2023). https://doi.org/10.1038/s41565-022-01264-4
6. S. Dai et al., Graphene on hexagonal boron nitride as a tunable hyperbolic metamaterial. Nat. Nanotechnol. **10**, 682–686 (2015). https://doi.org/10.1038/nnano.2015.131
7. A. Woessner et al., Highly confined low-loss plasmons in graphene–boron nitride heterostructures. Nat. Mater. **14**, 421–425 (2015). https://doi.org/10.1038/nmat4169
8. N.C. Passler, A. Paarmann, Generalized 4×4 matrix formalism for light propagation in anisotropic stratified media: study of surface phonon polaritons in polar dielectric heterostructures. J. Opt. Soc. Am. B **34**, 2128–2139 (2017). https://opg.optica.org/josab/abstract.cfm?URI=josab-34-10-2128
9. G.X. Ni et al., Ultrafast optical switching of infrared plasmon polaritons in high-mobility graphene. Nat. Photon. **10**, 244–247 (2016). https://doi.org/10.1038/nphoton.2016.45
10. G.X. Ni et al., Fundamental limits to graphene plasmonics. Nature **557**, 530–533 (2018). https://doi.org/10.1038/s41586-018-0136-9
11. G. Álvarez-Pérez et al., Infrared permittivity of the biaxial van der Waals semiconductor α-MoO_3 from near- and far-field correlative studies. Adv. Mater. **32**, 1908176 (2020). https://onlinelibrary.wiley.com/doi/abs/10.1002/adma.201908176
12. G. Álvarez-Pérez et al., Active tuning of highly anisotropic phonon polaritons in van der Waals crystal slabs by gated graphene. ACS Photon. **9**, 383–390 (2022). https://doi.org/10.1021/acsphotonics.1c01549
13. J. Duan et al., Enabling propagation of anisotropic polaritons along forbidden directions via a topological transition. Sci. Adv. **7**, eabf2690 (2021). https://www.science.org/doi/abs/10.1126/sciadv.abf2690

14. Q. Zhang et al., Hybridized hyperbolic surface phonon polaritons at α-MoO_3 and polar dielectric interfaces. Nano Lett. **21**, 3112–3119 (2021). https://doi.org/10.1021/acs.nanolett.1c00281
15. G. Hu, A. Krasnok, Y. Mazor, C.-W. Qiu, A. Alù, Moiré hyperbolic metasurfaces. Nano Lett. **20**, 3217–3224 (2020). https://doi.org/10.1021/acs.nanolett.9b05319
16. J. Duan et al., Twisted nano-optics: manipulating light at the nanoscale with twisted phonon polaritonic slabs. Nano Lett. **20**, 5323–5329 (2020). https://doi.org/10.1021/acs.nanolett.0c01673
17. Z. Zheng et al., Phonon polaritons in twisted double-layers of hyperbolic van der Waals crystals. Nano Lett. **20**, 5301–5308 (2020). https://doi.org/10.1021/acs.nanolett.0c01627
18. M. Chen et al., Configurable phonon polaritons in twisted α-MoO_3. Nat. Mater. **19**, 1307–1311 (2020). https://doi.org/10.1038/s41563-020-0732-6
19. A. Bapat, S. Dixit, Y. Gupta, T. Low, A. Kumar, Gate tunable light–matter interaction in natural biaxial hyperbolic van der Waals heterostructures. Nanophotonics **11**, 2329–2340 (2022). https://doi.org/10.1515/nanoph-2022-0034
20. F.L. Ruta et al., Surface plasmons induce topological transition in graphene/α-MoO_3 heterostructures. Nat. Commun. **13**, 3719 (2022). https://doi.org/10.1038/s41467-022-31477-z
21. Y. Zeng et al., Tailoring topological transitions of anisotropic polaritons by interface engineering in biaxial crystals. Nano Lett. **22**, 4260–4268 (2022). https://doi.org/10.1021/acs.nanolett.2c00399
22. H. Hu et al., Doping-driven topological polaritons in graphene/α-MoO_3 heterostructures. Nat. Nanotechnol. **17**, 940–946 (2022). https://doi.org/10.1038/s41565-022-01185-2
23. H. Hu et al., Gate-tunable negative refraction of mid-infrared polaritons. Science **379**, 558–561 (2023). https://www.science.org/doi/abs/10.1126/science.adf1251
24. F.J. Bezares et al., Intrinsic plasmon–phonon interactions in highly doped graphene: a near-field imaging study. Nano Lett. **17**, 5908–5913 (2017). https://doi.org/10.1021/acs.nanolett.7b01603
25. A. Das et al., Monitoring dopants by Raman scattering in an electrochemically top-gated graphene transistor. Nat. Nanotechnol. **3**, 210–215 (2008). https://doi.org/10.1038/nnano.2008.67
26. N. Kakenov et al., Observation of gate-tunable coherent perfect absorption of terahertz radiation in graphene. ACS Photon. **3**, 1531–1535 (2016). https://doi.org/10.1021/acsphotonics.6b00240
27. X. Lin et al., Chiral plasmons with twisted atomic bilayers. Phys. Rev. Lett. **125**, 077401 (2020). https://link.aps.org/doi/10.1103/PhysRevLett.125.077401

Chapter 10
Twistable Polaritonics with In-Operando Rotatable van der Waals Bilayers

Abstract As shown in previous chapters, in structures assembled from two-dimensional hyperbolic materials, like α-MoO_3, the characteristics of propagating polaritons vary strongly with the rotation angle between the crystal lattices. These variations give rise to phenomena such as topological transitions and polariton canalization. However, usually the twist angle between layers is fixed after assembly. While the reconfiguration of layers has been successfully demonstrated in twisted α-MoO_3 trilayers in Chap. 8, this methodology does not allow dynamically adjusting the twist angle during operation, and can also result in sample degradation. In this chapter, we develop several strategies aimed at achieving mechanically in-operando rotatable bilayers. Firstly, we unsuccessfully try to exert pressure with atomic force microscope (AFM) tip on the top slab of a bilayer to induce rotation and vary the twist angle. Then, we employ a methodology wherein synthesized microribbons of α-MoO_3 are transferred onto larger exfoliated α-MoO_3 flakes, and, again, we push the microribbon with the AFM tip to vary the twist angle. Through this approach, we observe variations with twist angle in properties such as the polariton wavelength within the same sample, which would be difficult to tune in static rotated structures. However, a clear elucidation of these effects is hindered due to the small microribbons' dimensions, which function as microresonators, thereby introducing complexity to the result interpretation. Consequently, as a third avenue of exploration, we assemble a twisted bilayer from two exfoliated α-MoO_3 flakes, and we shape the upper α-MoO_3 layer via focused ion beam (FIB) to facilitate pushing with the AFM tip. Nonetheless, complications stemming from friction and dangling bonds between the α-MoO_3 slabs come into play into this approach, which remains a work in progress. The chapter culminates with a discussion on the conclusions and prospects of this research project.

The experiments discussed in this chapter were carried out in Prof. Basov's laboratory during a Fulbright stay at Columbia University (USA) in 2022, in collaboration with Dr. Maëlle Kapfer, Francesco L. Ruta and Samuel L. Moore. Sample fabrication was performed in collaboration with Ana Isabel Fernández Tresguerres-Mata. Focused Ion Beam was carried out by Dr. Lucía Herrer and Prof. José María de Teresa (Instituto de Nanociencia y Materiales de Aragón, Spain). The

G. Álvarez Pérez, *Fundamentals of Nano-Optics in Hyperbolic van der Waals Materials*, Springer Theses, https://doi.org/10.1007/978-3-031-75767-9_10

synthesized α-MoO_3 microribbons were provided by Shang-Jie Yu and Jonathan A. Fan (Stanford University, USA).

The weak van der Waals (vdW) forces between atomic planes within two-dimensional (2D) materials open the possibility of constructing devices with arbitrary rotational configurations. The possibility to design and fabricate devices wherein the electronic properties are modulated through manipulation of the relative twist angle between the layers has given rise to the domain of *twistronics*, as elaborated upon in Chap. 8. This emerging field investigates how the angular misalignment (twist) between vertically stacked 2D layers influences the properties of the stack [1]. This effect is unique to the realm of 2D materials and has showcased its potential to profoundly alter the material's band structure and other characteristics. In certain 2D configurations, this impact is so profound that the corresponding twist angles are termed *magic angles*, exhibiting effects such as unconventional superconductivity [2], which arises as a consequence of angle-dependent interlayer interactions.

A parallel stride in the realm of nanophotonics was explored in Chap. 8, demonstrating that the dispersion characteristics of hyperbolic phonon polaritons (PhPs) can be dramatically manipulated by altering the angular orientation between two flakes of α-MoO_3 and superimposing them [3–7]. A transition from hyperbolic to elliptic dispersion, marked by a flattened dispersion precisely at the transition point, results in the canalization of PhPs. The resultant lifetimes of PhPs in twisted α-MoO_3 are comparable to those of a single α-MoO_3 flake, representing among the longest PhP lifetimes observed up to date. Just as the electronic version of twistronics has given rise to a flurry of research on superconductivity and exotic electronic states, the emerging field of *twist-optics* promises to stimulate a plethora of optical studies on manipulating the electromagnetic energy at the nanoscale and to enable the design of novel optical components.

The advent of twistronics and twist-optics has catalyzed the development of diverse techniques aimed at fabricating layered heterostructures with precisely controlled rotational alignments between the layers. These techniques include optical alignment of crystal edges [8–11], rotational positioning during assembly [12], and self-alignment via thermal annealing [13, 14]. However, all these approaches necessitate a priori understanding of the crystallographic orientation of each layer prior to assembly, which can be hard to identify at naked eye in some materials. Crucially, once assembled, the twist angle between the layers cannot be further adjusted. In Chap. 8, we described a methodology to reconfigure twisted trilayers. However, this approach does not allow for a dynamical control of the twist angle during operation. A pioneering experimental methodology within the realm of twistronics achieved real-time manipulation of the orientation between layers within a hBN/graphene/hBN vdW heterostructure [15]. In that work, relative easiness to mechanically rotate and translate hBN over graphene, attributed to their minimal friction, enables dynamic tunability of the optical, mechanical, and electronic characteristics of the system, achieved through pushing with an AFM tip.

In this chapter, we aim to enable analogous capabilities in the field of twist-optics. To do so, we design and fabricate different samples that are intended to be pushed with the tip of an atomic force microscope (AFM) to dynamically vary the twist angle, thereby enabling in-operando rotation for twistable polaritonics.

10.1 AFM-Pushing of an Exfoliated α-MoO_3 Bilayer

We start with the naivest approach: we fabricate a bilayer structure of twisted α-MoO_3 by assembling two exfoliated flakes on a silica (SiO_2) substrate using the conventional dry-transfer technique (see Sect. 2.1 and Ref. [16]), with both thicknesses around 120 nm (Fig. 10.1a). In an attempt to induce rotational manipulation, we laterally push the upper layer using a stiff silicon (Si) AFM tip, akin to the methodology outlined in Ref. [15]. Specifically, we employ the Nanoman software, a product of Bruker, which facilitates transition from tapping mode to contact mode without retracting the AFM tip, while simultaneously disabling the AFM feedback mechanism. The pushing process starts by capturing an image of the device using tapping mode. Then, within the Nanoman software, we define a vector in the initial tapping image. The AFM tip subsequently traces this vector in contact mode with the feedback disabled. The force exerted by the tip can be calculated by assuming a linear model where the deflection voltage is related to the force on the cantilever as $F_{\text{tip}} = V_{\text{tip}}k/S$ [17], where V_{tip} is the lateral deflection voltage of the AFM tip, k is the tip spring constant, and S is the AFM sensitivity. In our case, we used tips with $k \approx 250\,\text{N/m}$, with a tip deflection setpoint of up to 4 V [15] while S was usually between 2.5 and 3 mV/m, yielding $F_{\text{tip}} \approx 4\ \mu\text{N}$. To enable rotation, we push on a corner of the flake, and to assess any displacement of the top α-MoO_3 slab, a

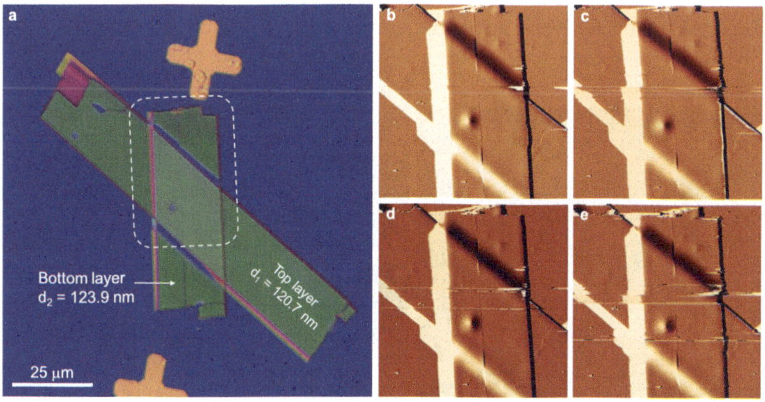

Fig. 10.1 AFM pushing of synthesized α-MoO_3 microribbons on exfoliated α-MoO_3 flakes. (**a**) Optical image of an exfoliated α-MoO_3 bilayer, with thicknesses around 120 nm. (**b–e**) AFM topography images showing iterative attempts to rotate the top layer

secondary scan in tapping mode is performed. The outcomes of iterative attempts following this procedure are illustrated in Fig. 10.1b–e, revealing no discernible rotation or movement of the upper layer. We also observe horizontal artifacts, likely attributed to the AFM tip experiencing damage during the pushing process.

Following the same procedure, we tried to rotate individual slabs of α-MoO_3 on a SiO_2 substrate. Since higher temperatures tend to mitigate frictional forces, we performed analogous experiments using a heating stage between 75 °C and 100 °C. Both attempts, not shown here for space reasons, did not yield any appreciable movement. Altogether, these unfavorable outcomes serve to underscore the presence of substantial frictional interactions between the layers of α-MoO_3, as well as between these layers and the underlying substrate. Such frictional forces may be attributed to the existence of dangling bonds between the α-MoO_3 layers and between the α-MoO_3 bottom layer and the SiO_2 substrate which increase friction in the system. Pre-treating the substrate with dodecanol to remove such dangling bonds does not yield any positive effects either. It thus becomes apparent that what is preventing rotation is an excessive friction between the α-MoO_3 layers. Consequently, a different arrangement within the α-MoO_3 bilayer itself is required to facilitate in-situ rotational manipulation successfully.

10.2 S-SNOM in-Operando Rotation of Synthesized α-MoO_3 Microribbons on Top of Exfoliated α-MoO_3 Flakes

In Ref. [18], the authors report bottom-up-synthesized α-MoO_3 structures grown by flame vapor deposition (FVD) as phonon polaritonic systems that feature tailorable morphologies and crystal qualities similar to those found in bulk single crystals. In an attempt to enable in-operando rotatable bilayer structures, we transferred FVD-grown α-MoO_3 microribbons on top of larger exfoliated α-MoO_3 flakes (Fig. 10.2a). After assembly, such microribbons can be pushed with the s-SNOM

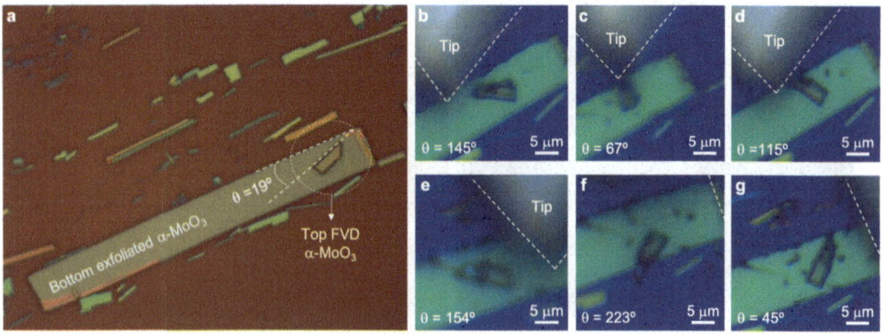

Fig. 10.2 In-operando rotation of synthesized α-MoO_3 microribbons on exfoliated α-MoO_3 flakes. (**a**) Optical image of a synthesized α-MoO_3 microribbon on top of an exfoliated α-MoO_3 flake, twisted at an angle $\theta = 19°$. (**b–e**) Optical images showing in-operando rotation of the top layer with the s-SNOM tip

tip. Here, we no longer use Bruker's Nanoman software, but instead a Neaspec system, where the region of interest is visualized through an optical camera, less precise than the Nanoman mode. By deactivating the feedback mechanism, the AFM operates in contact mode, enabling the system to exert pressure on the uppermost flake. To facilitate rotation, pressure is strategically applied to a corner of the flake. It should be emphasized that the s-SNOM tip utilized in this context is notably softer in comparison to the one utilized in the preceding section, attributed to its specific characteristics geared towards enhancing the near-field signal. In particular, the force constant of s-SNOM tips is below $k = 100$ N/m, while those utilized in conventional AFMs, as the ones described in the previous section, which reach up to $k = 250$ N/m.

Figure 10.2b–g shows the iterative rotational manipulation of the upper flake through tip-induced pressure, demonstrating the feasibility of in-operando rotational manipulation over a wide range of angles from $\theta = 19°$ up to $\theta = 223°$. Despite the lower force constant of the s-SNOM tip, rotational manipulation can be achieved, presumably due to the lower friction between the synthesized and exfoliated α-MoO_3 flakes, in comparison with the exfoliated bilayers shown in the previous section. In addition, the introduction of dirt as a consequence of dragging the tip around the sample is also appreciable.

To explore the effects of the in-operando rotation on the properties of PhPs, we perform near field nanoimaging of the microribbons for different twist angles using s-SNOM (see Sect. 2.4). Figure 10.3a–d (top panels) shows the near-field amplitude $s_4(x, y)$ for four different twist angles, namely $\theta = 67°$, 145°, 154° and 223°, showing an appreciable difference in the fringe spacing as a function of θ. This difference is further corroborated by extracting line profiles along the

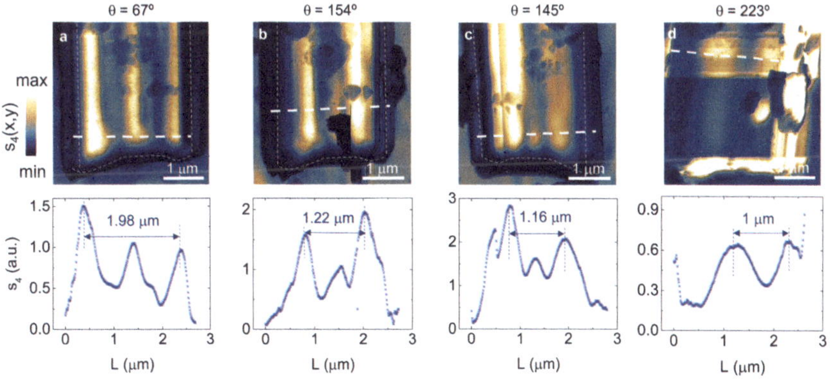

Fig. 10.3 Near-field nanoimaging of in-operando rotatable synthesized α-MoO_3 microribbons on exfoliated α-MoO_3 flakes. (**Top panels**) Near-field amplitude $s_4(x, y)$ of a synthesized α-MoO_3 microribbon on top of an exfoliated α-MoO_3 flake, in-operando twisted at different angles $\theta = 67°$, 145°, 154° and 223°. (**Bottom panels**) Line profiles $s_4(L)$ extracted along the white dashed lines depicted in the top panels, validating the different fringe spacing as a function of the twist angle θ

directions marked by the white dashed lines, which are shown in the lower panels. At $\theta = 67°$, the bottom panel in Fig. 10.3a shows three maxima, corresponding to the three bright fringes in the top panel. The two outermost maxima are separated by a distance $L = 1.98$ μm. This distance is substantially reduced down to $L = 1.22$ μm at a twist angle $\theta = 145°$ (Fig. 10.3b), while the near-field signal of the central maximum decreases. We can clearly notice the introduction of dirt as a consequence of dragging the tip. Figure 10.3c shows a slight decrease in the distance between the peaks at $\theta = 154°$, with $L = 1.16$ μm. The double peak on the left can be likely attributed to an artifact due to damages in the AFM tip. Finally, at $\theta = 223°$, the two maxima are separated by a distance $L = 1$ μm, while the central maximum disappears. Nevertheless, the introduction of dirt, together with the damage in the tip, already have dramatic effects, thereby posing substantial difficulties in the interpretability of this image.

However, it is clear that the different values of θ, a direct consequence of the in-operando rotation, result in the misalignment between the IFCs of both slabs in the bilayer. Similarly to what was shown in Chap. 8, this implies that the electromagnetic coupling between both layers changes as a function of the twist angle, and thus gives rise to different values of the polaritonic wavelength along the direction that is perpendicular to the microribbon's edges, which is the one probed by s-SNOM in polariton interferometry. This variation in wavelength, which can go up to a factor of 5, accounts for the dissimilar fringe spacing exhibited depending on the twist angle.

Nonetheless, due to the small dimensions of the microribbon, it operates as a cavity, thereby selecting a specific resonant wavelength for each twist angle θ. This inherent cavity effect introduces intricacies into the quantitative analysis of the images presented in Fig. 10.3a–d, thereby impeding the unambiguous determination of the polaritonic wavelength and consequently complicating the validation of the observed phenomenon. Hence, it becomes imperative to utilize bilayer structures that possess a balance between facile rotatability and dimensions that preclude the introduction of extraneous effects not central to the phenomenon of in-operando rotation, such as the aforementioned cavity-induced influences.

10.3 AFM-Pushing of FIB-Patterned α-MoO_3 Rotators on Top of Exfoliated α-MoO_3 Flakes

In order to attain a more clear and unequivocal demonstration of the polaritonic wavelength modulation through the process of in-operando rotation, we designed disk-like bilayer α-MoO_3 structures characterized by four extended segments, referred to as *arms*, in order to facilitate movement and thus rotation.

To achieve this configuration, we initiated the fabrication process by assembling a twisted α-MoO_3 bilayer utilizing mechanical exfoliation and the dry-transfer methodology, akin to the approach followed to stack the sample shown in Fig. 10.1a.

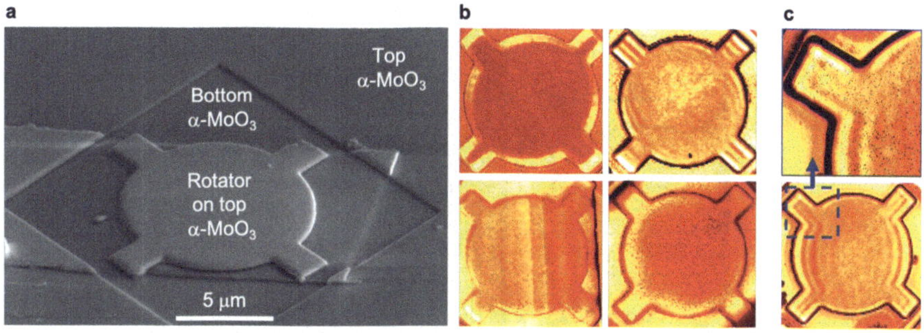

Fig. 10.4 FIB-patterned α-MoO_3 rotators on top of exfoliated α-MoO_3 flakes. (**a**) SEM image of a FIB-patterned rotator fabricated on an α-MoO_3 twisted bilayer. (**b**) Near-field optical amplitude of several FIB-patterned rotators after annealing the sample at 400 °C for four hours. (**c**) Near-field optical amplitude of a FIB-patterned rotator, the properties of which were well recovered after annealing the sample at 400 °C for four hours. The top panel shows a zoom of the dashed area below

Subsequently, by focused ion beam (FIB), we selectively etched away specific regions of the upper layer, thereby retaining only the intended disk-shaped morphology. The configuration of this engineered structure, which we called a *rotator*, having a radius of approximately 10 μm, is depicted in the scanning electron microscope (SEM) image presented in Fig. 10.4a. The FIB fabrication was performed in collaboration with Dr. Lucía Herrer and Prof. José María de Teresa (Instituto de Nanociencia y Materiales de Aragón, Spain).

Before delving into our efforts to move this rotator structure, we characterize its near-field optical response using s-SNOM. In Fig. 10.4b we plot the near-field amplitude $s_4(x, y)$. This representation brings to light that the exceptional polaritonic attributes typically exhibited by the α-MoO_3 bilayers are frequently compromised due to the aggressiveness of the FIB fabrication process. Notably, subjecting the samples to an annealing process at 400 °C for a duration of four hours yields occasional, albeit inconsistent, restoration of their high-quality polaritonic signal. Figure 10.4c showcases a particular sample manifesting good optical quality subsequent to such treatment. Furthermore, this particular disk-like configuration, characterized by its circular morphology, offers a clear advantage with respect to the approaches intended in the previous sections. As reported in prior works, the circular geometry allows to measure the polaritonic wavelength along all in-plane directions, thereby directly affirming the hyperbolic characteristics [19]. As such, this rotator geometry would allow to directly map the IFC for the different twist angles as a function of the in-operando rotation.

Figure 10.5 shows our efforts aimed at inducing movement on one of such rotator structures. The utilized sample, the near-field signal of which is shown in Fig. 10.5a, has a radius of about 5 μm. Figure 10.5b–e offers a sequence of AFM images documenting several attempts to induce rotation through the procedure outlined in Sect. 10.1, employing Bruker's Nanoman mode. The result reveals an

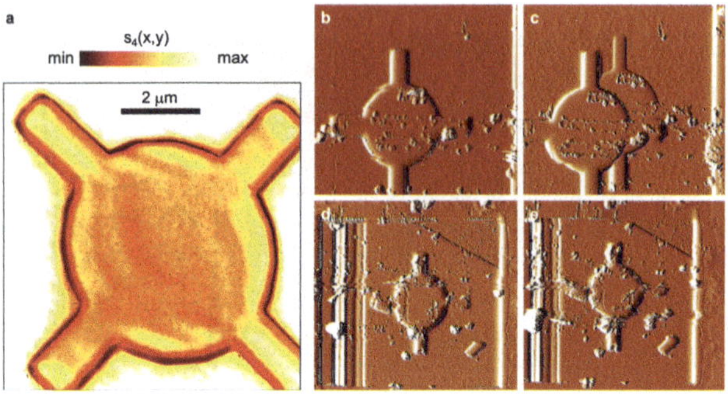

Fig. 10.5 AFM pushing of FIB-patterned α-MoO_3 rotators on top of exfoliated α-MoO_3 flakes. (**a**) Near-field optical amplitude of several FIB-patterned rotators after annealing the sample at 400 °C for four hours. (**b**–**e**) AFM topography images showing iterative attempts to rotate the top layer

absence of appreciable rotation. Notably, the motion of the AFM tip aggregates dirt in the vicinity of the rotator, thus introducing an undesired element to the correct acquisition of high-quality optical images. As our efforts proceeded, the sample broke, and also the AFM tip, evident through the duplication of the structure (Fig. 10.5c). All this can be ascribed to the notable frictional forces between the two layers of α-MoO_3. Such high frictional interaction prompts the need to explore strategies aimed at mitigating this frictional effect between the constituent layers, an avenue for future work.

10.4 Conclusions and Outlook

In this chapter, we have expounded upon several approaches aimed at achieving mechanical in-operando rotation of vdW twisted layers for polariton manipulation. The three strategies described here have been confronted with challenges, mostly stemming from the high inter-layer friction. While the primary objective was successfully attained through the utilization of synthesized α-MoO_3 microribbons on exfoliated α-MoO_3 flakes, the small dimensions of these microribbons hindered the unequivocal determination of the polaritonic wavelengths. Consequently, this added a layer of complexity to the process of validation. Hence, to provide a clear demonstration of the effect, further investigations are required. Currently, our focus resides in the fabrication of analogous structures to those expounded upon in Sect. 10.3, this is, the design and fabrication of a rotator on a twisted α-MoO_3 bilayer. However, a novel element will be introduced in the form of an intermediate layer composed of a lubricating vdW material, such as graphene, hBN, or WSe_2. The purpose of this interlayer is to alleviate the frictional forces between the constituent layers of α-MoO_3. A second approach that we are currently investigating involves the fabrication of micro-manipulators, few-layer graphite structures shaped into a gear-like geometry. Using an AFM tip to slide the manipulator allows to apply an in-plane load onto the layer that is intended to be moved or rotated and has been successfully applied to bend graphene on top of hBN [20].

All in all, the successful realization of in-operando rotation, if ultimately achieved, would mark a pioneering advancement, enabling the dynamic modulation of electromagnetic coupling between layers through mechanical manipulation. Consequently, the potential applications of such an achievement are anticipated to have a strong impact within the realm of twist-optics.

References

1. S. Carr et al., Twistronics: Manipulating the electronic properties of two-dimensional layered structures through their twist angle. Phys. Rev. B **95**, 075420 (2017). https://link.aps.org/doi/10.1103/PhysRevB.95.075420
2. Y. Cao et al., Unconventional superconductivity in magic-angle graphene superlattices. Nature **556**, 43–50 (2018). https://doi.org/10.1038/nature26160
3. G. Hu, A. Krasnok, Y. Mazor, C.-W. Qiu, A. Alù, Moiré hyperbolic metasurfaces. Nano Lett. **20**, 3217–3224 (2020). https://doi.org/10.1021/acs.nanolett.9b05319
4. J. Duan et al., Twisted nano-optics: manipulating light at the nanoscale with twisted phonon polaritonic slabs. Nano Lett. **20**, 5323–5329 (2020). https://doi.org/10.1021/acs.nanolett.0c01673
5. Z. Zheng et al., Phonon polaritons in twisted double-layers of hyperbolic van der Waals crystals. Nano Lett. **20**, 5301–5308 (2020). https://doi.org/10.1021/acs.nanolett.0c01627
6. M. Chen et al., Configurable phonon polaritons in twisted α-MoO_3. Nat. Mater. **19**, 1307–1311 (2020). https://doi.org/10.1038/s41563-020-0732-6
7. J. Duan et al., Multiple and spectrally robust photonic magic angles in reconfigurable α-MoO_3 trilayers. Nat. Mater. **22**, 867–872 (2023). https://doi.org/10.1038/s41563-023-01582-5
8. M. Yankowitz et al., Emergence of superlattice Dirac points in graphene on hexagonal boron nitride. Nat. Phys. **8**, 382–386 (2012). https://doi.org/10.1038/nphys2272
9. B. Hunt et al., Massive Dirac fermions and Hofstadter butterfly in a van der Waals heterostructure. Science **340**, 1427–1430 (2013). https://www.science.org/doi/abs/10.1126/science.1237240
10. L.A. Ponomarenko et al., Cloning of Dirac fermions in graphene superlattices. Nature **497**, 594–597 (2013). https://doi.org/10.1038/nature12187
11. C.R. Dean et al., Hofstadter's butterfly and the fractal quantum Hall effect in Moiré superlattices. Nature **497**, 598–602 (2013). https://doi.org/10.1038/nature12186
12. K. Kim et al., van der Waals heterostructures with high accuracy rotational alignment. Nano Lett. **16**, 1989–1995 (2016). https://doi.org/10.1021/acs.nanolett.5b05263
13. D. Wang et al., Thermally induced graphene rotation on hexagonal boron nitride. Phys. Rev. Lett. **116**, 126101 (2016). https://link.aps.org/doi/10.1103/PhysRevLett.116.126101
14. C.R. Woods et al., Macroscopic self-reorientation of interacting two-dimensional crystals. Nat. Commun. **7**, 10800 (2016). https://doi.org/10.1038/ncomms10800
15. R. Ribeiro-Palau et al., Twistable electronics with dynamically rotatable heterostructures. Science **361**, 690–693 (2018). https://www.science.org/doi/abs/10.1126/science.aat6981
16. A. Castellanos-Gomez et al., Deterministic transfer of two-dimensional materials by all-dry viscoelastic stamping. 2D Mater. **1**, 011002 (2014). https://doi.org/10.1088/2053-1583/1/1/011002

17. A.Z. Barabas et al., Mechanically reconfigurable van der Waals devices via low-friction gold sliding. Sci. Adv. **9**, eadf9558 (2023). https://www.science.org/doi/abs/10.1126/sciadv.adf9558
18. S.-J. Yu et al., Ultrahigh-quality infrared polaritonic resonators based on bottom-up-synthesized van der Waals nanoribbons. ACS Nano **16**, 3027–3035 (2022). https://doi.org/10.1021/acsnano.1c10489
19. W. Ma et al., In-plane anisotropic and ultra-low-loss polaritons in a natural van der Waals crystal. Nature **562**, 557 (2018). https://www.nature.com/articles/s41586-018-0618-9
20. M. Kapfer et al., Programming twist angle and strain profiles in 2D materials. Science **381**, 677–681 (2023). https://www.science.org/doi/abs/10.1126/science.ade9995

Chapter 11
Conclusions and Outlook

The work that comprises this thesis undertakes an extensive exploration into the fundamental principles of nano-optics in anisotropic 2D, or quasi-2D, materials. Initiated in 2018, this research was spurred by the discovery of α-MoO_3 as the first natural material supporting in-plane anisotropic phonon polaritons. An extensive revision of the field, as well as a discussion of the experimental and modelling techniques used throughout this thesis is carried out in the introductory Part I. The exploration of α-MoO_3, which upon the discovery of in-plane hyperbolic polaritons became one of the prototypical mid-infrared 2D materials, constitutes a central block of this thesis. Part II therefore ranges from the comprehensive analytical framework of polaritons in thin slabs of α-MoO_3 and other biaxial crystals, in Chap. 3, to the precise experimental characterization of this material's infrared optical response, combining far- and near-field methods, in Chap. 4. Part II therefore establishes the building blocks for the study of hyperbolic-related nano-optical phenomena using α-MoO_3, which not only are exploited in the subsequent chapters of the thesis, but also have been widely used in the scientific community working on nano-optics in 2D materials.

Building upon these findings, the following chapters cover a wide spectrum of optical studies. Part III delves into the investigation of two inherently fundamental and intriguing optical phenomena: negative reflection and anomalous refraction of nanoscale-confined polaritons, in Chaps. 5 and 6, respectively. Achieving the first direct visualization of these effects at the nanoscale extends the influence of our work beyond the realm of mid-infrared nanophotonics and endows our work with generality. These revelations not only expand our physical understanding of reflection and refraction, arguably two of the most basic optical phenomena, but also bare practical applications for controlling light at the nanoscale. Our results unveil the possibility to manipulate and tune both the polaritonic wavelength and direction of propagation upon negative reflection and anomalous refraction in a broad range. Based on these properties, we design the first in-plane hyperbolic nanoresonator, and design, fabricate and visualize the first planar polaritonic hyperlens, with a deeply

G. Álvarez Pérez, *Fundamentals of Nano-Optics in Hyperbolic van der Waals Materials*, Springer Theses, https://doi.org/10.1007/978-3-031-75767-9_11

sub-diffractional focus size 50 times smaller than the wavelength if the incident radiation.

Part IV unveils a toolkit of strategies to control and manipulate hyperbolic phonon polaritons. Notably, some of these advances have given rise to a novel research domain, nowadays widely referred to as twist-optics, extensively discussed in Chap. 8. The visualization of canalization of nanoscale-confined polaritons, originated from a topological transition of the isofrequency curve, provides an exciting route for the control of the flow of energy at the nanoscale. In addition, we have shown how to control and manipulate anisotropic phonon polaritons both passively via the coupling with other polaritonic materials (Chap. 7) and dynamically, through electric (Chap. 9) as well as mechanical (Chap. 10) means, which constitutes an imperative stride towards their practical integration in optical nanotechnologies.

Indeed, further exploring the emergent field of twist-optics, as well as controlling hyperbolic polaritons through external stimuli, present exciting technological frontiers in polaritonics and nano-optics. Recent breakthroughs, like the implementation of dynamically tunable canalization [1] and negative refraction [2] in α-MoO_3, hold promise for manipulating hyperbolic phonon polaritons, and thus the flow of energy at the nanoscale. On the other hand, the in-operando mechanical tunability of polaritonic properties, such as the wavelength, confinement or directions of propagation, is still a challenge. In this regard, a more unequivocal exploration of twistable bilayers than the one introduced in Chap. 10 would introduce a mechanical tuning element alongside the existing electrical control. Additional tuning mechanisms could be realized through response to controlled strain fields [3–5], the application of magnetic fields and interaction with magnetic excitations [6, 7], as well as the regulation of carriers via photoinjection [8], charge-transfer mechanisms [9, 10] and more. If combined with the optical phenomena and devices previously described—such as hyperbolic nanoresonators, hyperlenses, and polariton canalization—these capabilities could pave the way for manipulating light-matter interactions at the nanoscale in an unprecedented way.

Another intriguing avenue for future research lies in exploring systems with even lower crystal symmetry [11, 12]. This reduction in symmetry could enhance the directionality of optical phonons, and thus the anisotropy in the optical response, potentially unlocking even more exotic optical phenomena. Studies on off-cut uniaxial crystals, like calcite, have unveiled the existence of ghost polaritons [13] with complex-valued wavevectors, and leaky polaritons [14] featuring lenticular isofrequency curves. Furthermore, investigations into monoclinic crystals in the oxides bGO [15] and $CdWO_4$ [16] have revealed shear polaritons with even lower symmetry due to one of the optical axes being frequency-dependent. Further reductions in crystal symmetry, as seen in triclinic crystals like ReS_2 [17], or hybrids of the aforementioned low-symmetry crystals [18], offer great opportunities for exploring light-matter interactions and developing associated technologies like cavities, lenses, polarizers, waveguides, photodetectors, and more [19].

In addition, the development of stable and reliable laser sources in the far-IR regime has enabled the visualization of hyperbolic phonon polaritons and polariton canalization at unprecedentedly low frequencies. In fact, the pursuit of visualizing polaritons at progressively lower energies stands as a current focus in nanophotonics, due to the associated fundamental and technological relevance. Phonon polaritons have been observed down to 6.5 THz in single slabs of α-MoO_3 [20] and GeS [21], as well as in twisted bilayers of α-MoO_3 [22]. In addition, acoustic plasmon polaritons have been visualized in monoclinic Ag_2Te platelets down to 2.5 THz [23].

Not only expanding, but also controlling the frequency range in which these excitations occur constitutes another avenue in nanophotonics. Such control is essential for the advancement of nano-optical devices, such as sensors and photodetectors. In this regard, the successful intercalation of hydrogen and alkali atoms has allowed a broad spectral tuning of *Reststrahlen* bands in α-MoO_3 [24] and α-V_2O_5 [25], respectively, and thus the frequency ranges where phonon polaritonic excitations take place in these materials.

Likewise, the exploration of strong coupling between polaritons and molecular vibrations in the mid-IR and THz regimes could be helpful in the development of polariton chemistry [26, 27]. Vibrational strong coupling has been successfully demonstrated in hBN micro-ribbons and flakes, both in the far- [28] and near-field [29]. Furthermore, to achieve a more direct detection of vibrational strong coupling, the method of remote probing has been introduced, circumventing the direct near-field interaction between the tip and the molecules [30]. Investigating the characteristics of vibrational strong coupling effects in materials with lower symmetry than hBN, such as α-MoO_3, thus opens exciting prospects, both in the understanding of strong coupling when the underlying polaritons are very directional or hyperbolic and in the development of sensors and detectors that exploit the advantages of hyperbolicity.

In this context, the development of nanoresonators that enhance confinement and harness the high density of optical states characteristic of hyperbolic polaritons emerges as a promising pathway. The experimental realization of the proof-of-concept in-plane hyperbolic nanoresonator introduced in Chap. 5 sets the stage for creating a structure with unparalleled quality and Purcell factors. This could be helpful in exploring new regimes of light-matter interaction, such as ultrastrong [31, 32] or even deep strong [33] coupling, where the fundamental properties of light and matter undergo drastic transformations. The extreme confinement of polaritons could catalyze investigations into non-linear, non-local and quantum effects, or even studies of phonon-photon interactions away from the Γ point, which could benefit from the integration of polaritonics in picocavities [34] or STM systems. Likewise, the fabrication of nanostructures, including ribbons and resonators [35–37], antennas [38], and polaritonic crystals [39, 40], from highly anisotropic vdW materials, represents a highly promising avenue of investigation for manipulating, directing, and managing light at the nanoscale.

Furthermore, due to their hybrid light-matter nature, polaritons can serve as probes for specific physical properties of condensed matter systems, such as insulator—metal phase transitions [41, 42], ferroelectricity [43], or magnetic effects [7, 44]. Similarly, they can serve to unveil novel phenomena in condensed matter physics and optics, that go beyond the realm of polaritonics. This was accomplished in this thesis through the visualization of negative reflection and canalization, both being general optical phenomena which hold not only for polaritons. Demonstrations of Fizeau drag of graphene plasmons [45], as well as the emergence of topologically protected states [46] in hBN, also exemplify the prospects of exploring interesting physical phenomena with the use of polaritons in 2D materials. Remarkably, some of these effects can be experimentally observable thanks to the parallel development of sophisticated setups, such as cryo-SNOM or, more recently, magneto-cryo-SNOM [6] systems.

Similarly, advancement in ultrafast laser technology have enabled the generation of pulses at subpicosecond timescales, or even shorter durations, such as attosecond pulses [47], an achievement that was recently awarded with a Nobel Prize as it enables to resolve electronic dynamics. In combination with s-SNOM, ultrafast lasers have allowed investigating the dynamics of hyperbolic nanolight pulse propagation [48], as well as the controlled switching of hyperbolic dispersion on and off and on demand [49] and the observation of polaritonic vortices [50]. Investigating hyperbolic polariton dynamics through ultrafast nanoimaging represents an exciting avenue for future research, offering the potential to experimentally access and understand the interactions between photons and matter within anisotropic polaritonic systems at incredibly short time scales.

Furthermore, strongly confined polaritons display unique thermal emission properties and can carry thermal energy in unconventional manners, including super-Planckian, narrowband, coherent, and even non-reciprocal thermal emission [51]. In the near field and within the nanometer scale, thermally excited polaritons can carry thermal power density exceeding the blackbody limit, opening up novel avenues for manipulating heat flow. In particular, hyperbolic phonon polaritons, spanning the entire infrared spectrum, serve as highly efficient carriers of thermal radiation. By harnessing their active tunability and their strong anisotropy, highly compact thermal emitters with reduced structural complexity could be developed [51]. These emitters hold potential for diverse applications in renewable energy and storage, coherent thermal sources, remote cooling, thermal regulation, and camouflage.

In summary, the realm of polaritons in 2D materials, and particularly the emergence of hyperbolic phonon polaritons in recent years, present an exciting frontier at the intersection of condensed matter physics, optics, nanotechnology, and material science. Investigating both the properties of hyperbolic phonon polaritons and their potential as a versatile tool for uncovering new physical phenomena offers intriguing research prospects. Moreover, the development of precise techniques for guiding, controlling, and steering polariton propagation is key for the realization of all-optical computing, which could overcome the fundamental limits of modern semiconductor-based electronics—such as interconnect delays and large heat generation. In this regard, polaritonic devices and circuits offer a promising platform, since they possess intrinsically higher information-carrying capacity, produce low heat loads, and can reach sub-nanometric sizes. In this direction, customized applications and devices, such as all-optical logic gates [52] and routers [53], have already been demonstrated. Within this context, ongoing efforts to enhance structural design flexibility, facilitated by the user-friendly nature and adaptable functionalities of vdW and 2D materials, amplify the potential for advancements. The symbiotic relationship between scientific exploration—experimentally, analytically, and numerically—and continued technological development promises the practical realization of these progressions. As the field of polaritonics in 2D materials continues to advance, the integration of fundamental studies, optical manipulation, nanofabrication methodologies, and innovative setups is poised to unlock new frontiers in nanotechnology, condensed matter physics, and optics. In this landscape, hyperbolic phonon polaritons are taking center stage, as evidenced by the large amount of outstanding studies reported in the last few years. The field of mid-infrared nano-optics in hyperbolic materials is thus set to deepen our comprehension of light and its interactions with matter, and thesis contributes its grain of sand to the exploration of its essential principles.

References

1. H. Hu et al., Doping-driven topological polaritons in graphene/α-MoO_3 heterostructures. Nat. Nanotechnol. **17**, 940–946 (2022). https://doi.org/10.1038/s41565-022-01185-2
2. H. Hu et al., Gate-tunable negative refraction of mid-infrared polaritons. Science **379**, 558–561 (2023). https://www.science.org/doi/abs/10.1126/science.adf1251
3. R. Frisenda et al., Biaxial strain tuning of the optical properties of single-layer transition metal dichalcogenides. npj 2D Mater. Appl. **1**, 10 (2017). https://doi.org/10.1038/s41699-017-0013-7
4. J. Martín-Sánchez et al., Semicond. Sci. Technol. **33**, 013001 (2017). https://dx.doi.org/10.1088/1361-6641/aa9b53
5. E. Blundo, E. Cappelluti, M. Felici, G. Pettinari, A. Polimeni, Strain-tuning of the electronic, optical, and vibrational properties of two-dimensional crystals. Appl. Phys. Rev. **8**, 021318 (2021). https://doi.org/10.1063/5.0037852
6. M. Dapolito et al., Infrared nano-imaging of Dirac magnetoexcitons in graphene. Nat. Nanotechnol. (2023). https://doi.org/10.1038/s41565-023-01488-y
7. F. Dirnberger et al., Magneto-optics in a van der Waals magnet tuned by self-hybridized polaritons. Nature **620**, 533–537 (2023). https://doi.org/10.1038/s41586-023-06275-2
8. A.D. Dunkelberger et al., Active tuning of surface phonon polariton resonances via carrier photoinjection. Nat. Photon. **12**, 50–56 (2018). https://doi.org/10.1038/s41566-017-0069-0
9. D.J. Rizzo et al., Charge-transfer plasmon polaritons at graphene/α-$RuCl_3$ interfaces. Nano Lett. **20**, 8438–8445 (2020). https://doi.org/10.1021/acs.nanolett.0c03466
10. B.S.Y. Kim et al., Ambipolar charge-transfer graphene plasmonic cavities. Nat. Mater. **22**, 838–843 (2023). https://doi.org/10.1038/s41563-023-01520-5
11. M. He et al., Anisotropy and modal hybridization in infrared nanophotonics using low-symmetry materials. ACS Photon. **9**, 1078–1095 (2022). https://doi.org/10.1021/acsphotonics.1c01486
12. E. Galiffi et al., Extreme light confinement and control in low-symmetry phonon-polaritonic crystals. Nat. Rev. Mater. (2023). https://doi.org/10.1038/s41578-023-00620-7
13. W. Ma et al., Ghost hyperbolic surface polaritons in bulk anisotropic crystals. Nature **596**, 362–366 (2021). https://doi.org/10.1038/s41586-021-03755-1

14. X. Ni et al., Observation of directional leaky polaritons at anisotropic crystal interfaces. Nat. Commun. **14**, 2845 (2023). https://doi.org/10.1038/s41467-023-38326-7
15. N.C. Passler et al., Hyperbolic shear polaritons in low-symmetry crystals. Nature **602**, 595–600 (2022). https://doi.org/10.1038/s41586-021-04328-y
16. G. Hu et al., Real-space nanoimaging of hyperbolic shear polaritons in a monoclinic crystal. Nat. Nanotechnol. **18**, 64–70 (2023). https://doi.org/10.1038/s41565-022-01264-4
17. F. Mooshammer et al., In-plane anisotropy in biaxial ReS_2 crystals probed by nano-optical imaging of waveguide modes. ACS Photon. **9**, 443–451 (2022). https://doi.org/10.1021/acsphotonics.1c01841
18. A.J. Sternbach et al., Negative refraction in hyperbolic hetero-bicrystals. Science **379**, 555–557 (2023). https://www.science.org/doi/abs/10.1126/science.adf1065
19. S. Castilla et al., Plasmonic antenna coupling to hyperbolic phonon-polaritons for sensitive and fast mid-infrared photodetection with graphene. Nat. Commun. **11**, 4872 (2020). https://doi.org/10.1038/s41467-020-18544-z
20. T.V.A.G. De Oliveira et al., Nanoscale-confined terahertz polaritons in a van der Waals crystal. Adv. Mater. **33**, 2005777 (2021). https://onlinelibrary.wiley.com/doi/abs/10.1002/adma.202005777
21. T. Nörenberg et al., Germanium monosulfide as a natural platform for highly anisotropic THz polaritons. ACS Nano **16**, 20174–20185 (2022). https://doi.org/10.1021/acsnano.2c05376
22. M. Obst et al., Terahertz twistoptics–engineering canalized phonon polaritons. ACS Nano **17**, 19313–19322 (2023). https://doi.org/10.1021/acsnano.3c06477
23. S. Chen et al., Real-space observation of ultraconfined in-plane anisotropic acoustic terahertz plasmon polaritons. Nat. Mater. **22**, 860–866 (2023). https://doi.org/10.1038/s41563-023-01547-8
24. Y. Wu et al., Chemical switching of low-loss phonon polaritons in α-MoO_3 by hydrogen intercalation. Nat. Commun. **11**, 2646 (2020). https://doi.org/10.1038/s41467-020-16459-3
25. J. Taboada-Gutiérrez et al., Broad spectral tuning of ultra-low-loss polaritons in a van der Waals crystal by intercalation. Nat. Mater. **19**, 964–968 (2020). https://doi.org/10.1038/s41563-020-0665-0
26. R.F. Ribeiro, L.A. Martínez-Martínez, M. Du, J. Campos-Gonzalez-Angulo, J. Yuen-Zhou, Polariton chemistry: controlling molecular dynamics with optical cavities. Chem. Sci. **9**, 6325–6339 (2018). https://doi.org/10.1039/C8SC01043A
27. D. Sidler, C. Schäfer, M. Ruggenthaler, A. Rubio, Polaritonic chemistry: collective strong coupling implies strong local modification of chemical properties. J. Phys. Chem. Lett. **12**, 508–516 (2021). https://doi.org/10.1021/acs.jpclett.0c03436
28. M. Autore et al., Boron nitride nanoresonators for phonon-enhanced molecular vibrational spectroscopy at the strong coupling limit. Light Sci. Appl. **7**, 17172–17172 (2018). https://doi.org/10.1038/lsa.2017.172
29. A. Bylinkin et al., Real-space observation of vibrational strong coupling between propagating phonon polaritons and organic molecules. Nat. Photon. **15**, 197–202 (2021). https://doi.org/10.1038/s41566-020-00725-3
30. I. Dolado et al., Remote near-field spectroscopy of vibrational strong coupling between organic molecules and phononic nanoresonators. Nat. Commun. **13**, 6850 (2022). https://doi.org/10.1038/s41467-022-34393-4
31. M. Barra-Burillo et al., Microcavity phonon polaritons from the weak to the ultrastrong phonon–photon coupling regime. Nat. Commun. **12**, 6206 (2021). https://doi.org/10.1038/s41467-021-26060-x
32. D. Yoo et al., Ultrastrong plasmon.phonon coupling via epsilon-near-zero nanocavities. Nat. Photon. **15**, 125–130 (2021). https://doi.org/10.1038/s41566-020-00731-5
33. N.S. Mueller et al., Deep strong light–matter coupling in plasmonic nanoparticle crystals. Nature **583**, 780–784 (2020). https://doi.org/10.1038/s41586-020-2508-1
34. J.J. Baumberg, Picocavities: a primer. Nano Lett. **22**, 5859—5865 (2022). https://doi.org/10.1021/acs.nanolett.2c01695

35. J.D. Caldwell et al., Sub-diffractional volume-confined polaritons in the natural hyperbolic material hexagonal boron nitride. Nat. Commun. **5**, 5221 (2014). https://doi.org/10.1038/ncomms6221
36. I. Dolado et al., Nanoscale guiding of infrared light with hyperbolic volume and surface polaritons in van der Waals material ribbons. Adv. Mater. **32**, 1906530 (2020). https://onlinelibrary.wiley.com/doi/abs/10.1002/adma.201906530
37. O.G. Matveeva et al., Twist-tunable polaritonic nanoresonators in a van der Waals crystal. npj 2D Mater. Appl. **7**, 31 (2023). https://doi.org/10.1038/s41699-023-00387-z
38. P. Pons-Valencia et al., Launching of hyperbolic phonon-polaritons in h-BN slabs by resonant metal plasmonic antennas. Nat. Commun. **10**, 3242 (2019). https://doi.org/10.1038/s41467-019-11143-7
39. F.J. Alfaro-Mozaz et al., Deeply subwavelength phonon-polaritonic crystal made of a van der Waals material. Nat. Commun. **10**, 42 (2019). https://doi.org/10.1038/s41467-018-07795-6
40. N. Capote-Robayna, O.G. Matveeva, V.S. Volkov, P. Alonso-González, A.Y. Nikitin, Twisted polaritonic crystals in thin van der Waals slabs. Laser Photon. Rev. **16**, 2200428 (2022). https://onlinelibrary.wiley.com/doi/abs/10.1002/lpor.202200428
41. A.S. McLeod et al., Nanotextured phase coexistence in the correlated insulator V_2O_3. Nat. Phys. **13**, 80–86 (2017). https://doi.org/10.1038/nphys3882
42. K.W. Post et al., Coexisting first- and second-order electronic phase transitions in a correlated oxide. Nat. Phys. **14** (2018). https://doi.org/10.1038/s41567-018-0201-1
43. S. Zhang et al., Visualizing moiré ferroelectricity via plasmons and nano-photocurrent in graphene/twisted-WSe2 structures. Nat. Commun. **14**, 6200 (2023). https://doi.org/10.1038/s41467-023-41773-x
44. A.S. McLeod et al., Multi-messenger nanoprobes of hidden magnetism in a strained manganite. Nat. Mater. **19**, 397–404 (2020). https://doi.org/10.1038/s41563-019-0533-y
45. Y. Dong et al., Fizeau drag in graphene plasmonics. Nature **594**, 513–516 (2021). https://doi.org/10.1038/s41586-021-03640-x
46. S. Guddala et al., Topological phonon-polariton funneling in midinfrared metasurfaces. Science **374**, 225–227 (2021). https://www.science.org/doi/abs/10.1126/science.abj5488
47. P.M. Paul et al., Observation of a train of attosecond pulses from high harmonic generation. Science **292**, 1689–1692 (2001). https://www.science.org/doi/abs/10.1126/science.1059413
48. X. Zhang et al., Ultrafast anisotropic dynamics of hyperbolic nanolight pulse propagation. Sci. Adv. **9**, eadi4407 (2023). https://www.science.org/doi/abs/10.1126/sciadv.adi4407
49. A.J. Sternbach et al., Programmable hyperbolic polaritons in van der Waals semiconductors. Science **371**, 617–620 (2021). https://www.science.org/doi/abs/10.1126/science.abe9163
50. Y. Kurman et al., Dynamics of optical vortices in van der Waals materials. Optica **10**, 612–618 (2023). https://opg.optica.org/optica/abstract.cfm?URI=optica-10-5-612
51. G.T. Papadakis, A. Davoyan, P. Yeh, H.A. Atwater, Mimicking surface polaritons for unpolarized light with high-permittivity materials. Phys. Rev. Mater. **3**, 015202 (2019). https://link.aps.org/doi/10.1103/PhysRevMaterials.3.015202
52. H. Wei, Z. Wang, X. Tian, M. Käll, H. Xu, Cascaded logic gates in nanophotonic plasmon networks. Nat. Commun. **2**, 387 (2011). https://doi.org/10.1038/ncomms1388
53. F. Marsault et al., Realization of an all optical exciton-polariton router. Appl. Phys. Lett. **107**, 201115 (2015). https://doi.org/10.1063/1.4936158

Appendix A
Infrared Dielectric Functions of AgCl and BaF_2

The FTIR spectra shown in this appendix and their modelling to extract the infrared dielectric functions of AgCl and BaF_2 were performed in collaboration with Dr. Thomas G. Folland and Prof. Joshua D. Caldwell (Vanderbilt University, USA), and published in "Infrared Permittivity of the Biaxial van der Waals Semiconductor α-MoO_3 from Near- and Far-Field Correlative Studies" by Gonzalo Álvarez-Pérez et al., in Advanced Materials 32(29), 1908176 (2020) [1].

A.1 Infrared Dielectric Function of AgCl

The infrared dielectric function of AgCl was extracted using analogous methods to those outlined in Sect. 4.1 for the extraction of the permittivity of α-MoO_3. In particular, a 2-mm-thick AgCl window is mounted in the bench of a Fourier transform infrared (FTIR) spectrometer. In this case, unpolarized reflection and transmission data were taken using a room temperature pyroelectric DLaTGs detector into the far infrared (Fig. A.1). These data were then fit using JA Wollams VASE software, using a phenomenological model with multiple Gaussian oscillators of the functional form

G. Álvarez Pérez, *Fundamentals of Nano-Optics in Hyperbolic van der Waals Materials*, Springer Theses, https://doi.org/10.1007/978-3-031-75767-9

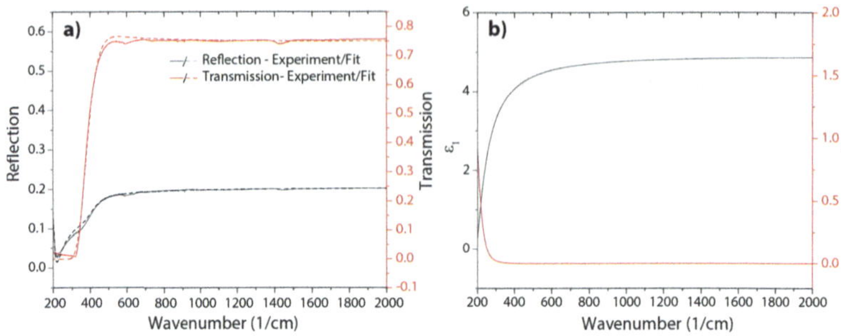

Fig. A.1 IR optical response of AgCl. (**a**) FTIR reflection and transmission spectra of an AgCl substrate. (**b**) Fitted infrared dielectric function of AgCl

Table A.1 Fitting parameters for our AgCl dielectric function model

$\varepsilon_\infty = 4.89$	A_n	E_n	Br_n
Oscillator 1	1.14	12.3	266
Oscillator 2	11.5	94.7	30.1
Oscillator 3	69.0	13.3	153

$$\begin{aligned} \varepsilon(E) &= \varepsilon_\infty + \sum_{i=1}^{3} \varepsilon_1^n + i\varepsilon_2^n, \\ \varepsilon_1^n(E) &= \frac{2}{\pi} P \int_0^\infty \frac{\xi \varepsilon_2^n(\xi)}{\xi^2 - E^2} d\xi, \\ \varepsilon_2^n(E) &= A_n e^{-\left(\frac{E-E_n}{\sigma_n}\right)^2} - A_n e^{-\left(\frac{E+E_n}{\sigma_n}\right)^2}, \\ \sigma &= \frac{Br_n}{2\sqrt{\ln(2)}} \end{aligned} \tag{A.1}$$

where A_n is the amplitude, Br_n is the full width at half maximum (FWHM) of the gaussian, E is the frequency, E_n is the center frequency and P is the Cauchy principal value. This model was deliberately over-parameterized, and whilst it provides a good dielectric function fit, it does not provide physical information (such as phonon energies) about AgCl itself. The tabulated values for the fit are shown in Table A.1. The dielectric function associated with this material is shown in Fig. A.1, showing minimal dispersion in the region of interest for our experiments on α-MoO_3 samples.

A.2 Dielectric Function of BaF_2

The infrared dielectric function of BaF_2 was extracted using analogous methods as the ones described above for extracting the dielectric function of AgCl: using a

Table A.2 Fitting parameters for our BaF_2 dielectric function model

$\varepsilon_\infty = 2.68$	A_n	E_n	Br_n
Oscillator 1	1551.5	189.46	9.77
Oscillator 2	2389.9	17.69	388.91
Oscillator 3	0.1110	34.31	560.57

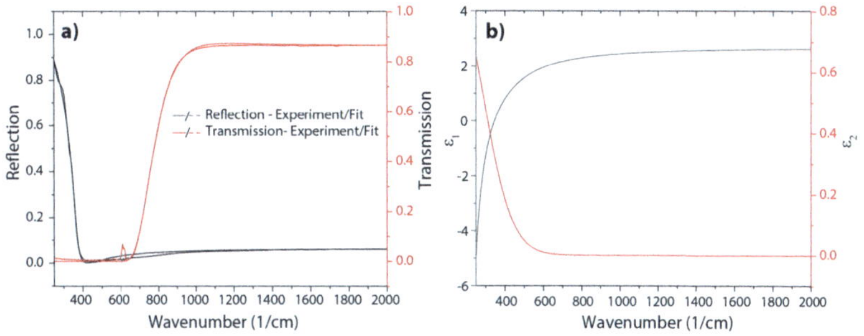

Fig. A.2 **IR optical response of BaF_2.** (**a**) FTIR reflection and transmission spectra of an BaF_2 substrate. (**b**) Fitted infrared dielectric function of BaF_2

phenomenological model with multiple Gaussian oscillators of the functional form in Eq. (A.1). As that, this model for BaF_2 has been deliberately over-parameterized, and whilst it provides a good dielectric function fit, it does not provide physical information about BaF_2 itself. The tabulated values are provided in Table A.2, while the infrared optical response and the fitted IR dielectric function of this material are shown in Fig. A.2, showing minimal dispersion in the region of interest for our experiments on α-MoO_3 samples.

Appendix B
Ab Initio Infrared Dielectric Function of α-MoO_3

In quantum mechanics, the behavior of electrons is typically described using wave functions, which provide information about the probability of finding an electron at a particular location. However, dealing with wave functions for large systems with many electrons becomes computationally challenging. A scheme that reduces the calculations of ground-state properties of systems of interacting particles exactly to the solution of single-particle Hartree-type equations has obvious advantages [2]. Density functional theory (DFT) simplifies the problem by focusing on the electron density, which describes the distribution of electrons throughout a system. The basic idea is that the electron density contains all the necessary information to determine the system's properties, such as total energy, electron density distribution, and forces between atoms.

DFT is based on the Hohenberg-Kohn theorems [3], which state that the ground-state energy of a system is uniquely determined by its electron density, and vice versa. This means that if we know the electron density, we can calculate the system's total energy, and if we know the total energy, we can determine the electron density.

In practice, DFT involves solving the Kohn-Sham equations [4], which are a set of equations that describe the behavior of non-interacting electrons in an effective potential. These equations are similar to the Schrödinger equation but with an additional exchange-correlation term that accounts for the effects of electron-electron interactions.

To calculate the exchange-correlation term, various approximations are used, such as the local density approximation (LDA) or the generalized gradient approximation (GGA) [5]. These approximations make DFT computationally feasible for practical applications.

The dielectric function of α-MoO_3 was calculated fully ab-initio by Prof. Ion Errea (University of the Basque Country, Spain) and published in "Infrared Permittivity of the Biaxial van der Waals Semiconductor α-MoO_3 from Near- and Far-Field Correlative Studies" by Gonzalo Álvarez-Pérez et al., in Advanced Materials 32 (29), 1908176 (2020) [1]. Use is made of DFT within the Perdew,

G. Álvarez Pérez, *Fundamentals of Nano-Optics in Hyperbolic van der Waals Materials*, Springer Theses, https://doi.org/10.1007/978-3-031-75767-9

Burke, and Ernzerhof parametrization of the exchange-correlation functional [5]. Ultra-soft pseudopotentials are employed in the calculations including 6 valence electrons for O and 14 valence electrons for Mo. The dielectric tensor, which becomes diagonal in an orthorhombic crystal, is calculated within perturbation theory [6]. In this framework, the dielectric tensor for Cartesian directions α and β is given by

$$\varepsilon_{\alpha\beta}(\omega) = \varepsilon^{\infty}_{\alpha\beta} + 4\pi\chi_{\alpha\beta}(\omega), \tag{B.1}$$

where $\varepsilon^{\infty}_{\alpha\beta}$ is the electronic contribution to the dielectric tensor (the high-frequency limit) and $\chi_{\alpha\beta}(\omega)$ is the susceptibility. The latter is expressed in atomic units as

$$\chi_{\alpha\beta}(\omega) = \frac{1}{\Omega}\sum_i \frac{2\omega_i M_i^{\alpha} M_i^{\beta}}{\omega^2 - \omega_i^2 - 2\omega_i \Pi_i(\omega_i)}, \tag{B.2}$$

where Ω is the unit cell volume, the phonon frequency of the i-th mode (the sum over i is restricted to phonon modes at the Γ point), $\Pi_i(\omega_i)$ the i-th phonon mode's self-energy due to phonon-phonon interaction taken at the phonon frequency, and M_i^{α} is the contribution of the i-th mode to the dipole moment:

$$M_i^{\alpha} = \sum_{\beta s} \frac{Z_s^{\alpha\beta} e_{is}^{\beta}}{\sqrt{2M_s\omega_i}} \tag{B.3}$$

In Eq. (B.3), $Z_s^{\alpha\beta}$ is the effective-charge tensor for atom s, M_s is mass of the atom s, and e_{is}^{β} the polarization vector of mode i for atom s along the Cartesian direction β.

Phonon frequencies, polarization vectors, effective charges, and the high-frequency limit of the dielectric function are calculated making use of density functional perturbation theory (DFPT) [7] as implemented in Quantum Espresso [8, 9] An 80 Ry energy cutoff was used for the plane-wave basis and a 800 Ry cutoff for the density. Electronic integrations were performed on an $8 \times 2 \times 8$ grid. The phonon self-energy was calculated just including the so-called bubble contribution [10, 11]:

$$\begin{aligned}\Pi_i(\omega) = &-\frac{1}{2N_q}\sum_{qjk} |\Phi_{ijk}(0, q, -q)|^2 \\ &\times \left\{ \frac{2(\omega_{jq} + \omega_{k-q})\left[1 + n_B(\omega_{jq}) + +n_B(\omega_{k-q})\right]}{(\omega_{jq} + \omega_{k-q})^2 - (\omega + i\delta)^2} \right. \\ &\left. + \frac{2(\omega_{jq} - \omega_{k-q})\left[n_B(\omega_{k-q}) + +n_B(\omega_{jq})\right]}{(\omega_{k-q} + \omega_{jq})^2 - (\omega - i\delta)^2} \right\}.\end{aligned} \tag{B.4}$$

In Eq. (B.4) represents the j-th mode at the **q** point of the Brillouin zone, N_q the number of **q** points in the sum, $n_B(\omega)$ is the Bose-Einstein occupation factor, δ is a small number ($10\,\mathrm{cm}^{-1}$ in our case), and $\Phi_{ijk}(0, q, -q)$ are the anharmonic third-order force constants transformed to the phonon mode basis [11]. The third-order anharmonic force- constants are calculated by finite-differences calculating atomic forces on displaced supercells created by the ShengBTE code [12, 13]. The third-order force constants are calculated in a $2 \times 1 \times 2$ supercell including interaction terms up to 5 nearest-neighbors. The self-energy is then calculated at 300 K by including in the sum over the **q** points a $16 \times 4 \times 16$ phonon grid. The phonon frequencies and third-order force constants in this grid are obtained by Fourier interpolation. The dynamical matrices are calculated originally in a $6 \times 2 \times 6$ point grid. The ab-initio damping factors of the phonon mode reported in the main text correspond to the imaginary part of the phonon self-energy taken at the phonon frequency:

$$\gamma_i = -\mathrm{Im}\,[\Pi_i(\omega_i)] \tag{B.5}$$

Note: the derivations shown in this appendix were carried out by Prof. Ion Errea (University of the Basque Country, Spain).

Appendix C
Analytical Shape of a Total Hyperbolic Retroreflector and a Planar Hyperlens for Polaritons

C.1 Analytical Shape of a Total Hyperbolic Retroreflector

In this appendix, we derive the analytical shape of a mirror capable of reflecting all the incident hyperbolic waves back to their source, i.e., a hyperbolic retroreflector. These derivations were performed in collaboration with Kirill Voronin (Moscow Institute of Physics and Technology, Russia).

We start by considering a plane wave propagating inside a biaxial half-space (that, without loss of generality, we will assume infinite along the z direction). We will assume that the dielectric permittivity tensor of the medium has the following diagonal form:

$$\hat{\varepsilon} = \begin{pmatrix} \varepsilon_x & 0 & 0 \\ 0 & \varepsilon_y & 0 \\ 0 & 0 & \varepsilon_z \end{pmatrix}.$$

We will also consider that the plane wave, polarized in the xy plane and with wavevector $\mathbf{k}$ and Poynting vector $\mathbf{S}$, is incident on the surface $f(x, y) = 0$ (see Fig. C.1). The dispersion of an electromagnetic wave propagating in the medium is given by Fresnel's equation, which gives rise to the following IFC:

$$F \equiv \frac{k_x^2}{\varepsilon_y} + \frac{k_y^2}{\varepsilon_x} - k_0^2 = 0, \tag{C.1}$$

where $k_0 = \omega/c$, where ω and c are the frequency of the wave and the speed of light, respectively. In the lossless case, i.e., when the dielectric permittivities are completely real, the Poynting vector is parallel to the group velocity. Therefore, $\mathbf{S}$ is given by $\mathbf{S} \approx \nabla F(k_x, k_y)$, or, more explicitly:

G. Álvarez Pérez, *Fundamentals of Nano-Optics in Hyperbolic van der Waals Materials*, Springer Theses, https://doi.org/10.1007/978-3-031-75767-9

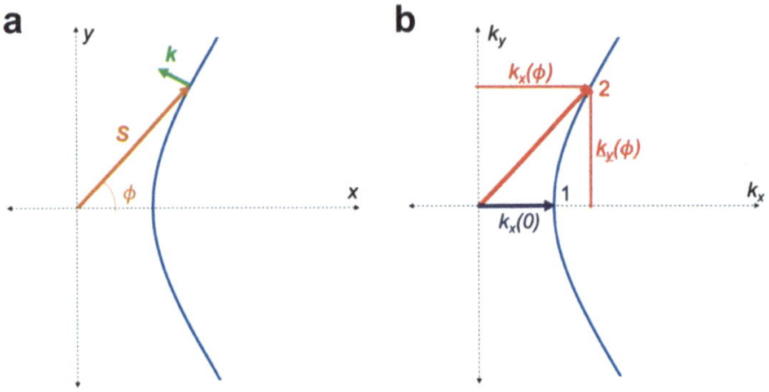

Fig. C.1 Optical paths of HPhPs back-reflecting on a hyperbolic retroreflector. Schematics of a retroreflector in (**a**) real space and (**b**) momentum space. **S** and **k** illustrate the directions of the Poynting and wave vectors, respectively. In (**b**), the waves back-reflecting at the points 1 and 2 follow different optical paths

$$S_x \approx \frac{k_x}{\varepsilon_y} \qquad S_y \approx \frac{k_y}{\varepsilon_x}.$$

When considering the optical losses of the medium, the direction of the Poynting vector will be slightly tilted with respect to the one predicted by the previous equation, but we will neglect this effect here for the sake for simplicity, i.e. we will restrict ourselves to the lossless case.

To fulfill momentum conservation at the boundary, any wave whose wave vector is perpendicular to the boundary surface will be reflected along the same direction but in the opposite direction. Therefore, the shape of the total reflector must be perpendicular to the IFC given by Eq. (C.1).

$$\frac{x^2}{\varepsilon_x} - \frac{y^2}{\varepsilon_y} = C. \tag{C.2}$$

In order to determine C, we consider the phase of the wave. For the electric field to be maximized at the location of the source, the wave must propagate to the reflector and reflect back with constructive interference at the source. Let us consider now two different optical paths to the point (0, 0): one along the x axis and the other starting from an arbitrary point of the lens surface (labeled as "2" in Fig. C.1). The phase of the waves propagating along these paths are the following:

$$\begin{cases} \varphi_1 = 2k_x(0)d, \\ \varphi_2 = 2[k_y(\varphi)y + k_x(\varphi)x)], \end{cases} \tag{C.3}$$

where d is the distance from the origin to point 1; $k_x(\varphi)$ and $k_y(\varphi)$ are the wave vector components in the direction marked by the angle φ. On the other hand, the angle φ can be calculated geometrically as $\tan\varphi = \varepsilon_x k_x / \varepsilon_y k_y$. These values are determined by the IFC of the medium in which the wave propagates, given by Eq. (C.1). At $\varphi = 0$, $k_y = 0$ and hence $k_x(0) = \sqrt{\varepsilon_y}k_0$. Substituting these two expressions into Eq. (C.3) we get:

$$\begin{cases} \varphi_1 = 2\sqrt{\varepsilon_y}k_0 d \\ \varphi_2 = 2\frac{\sqrt{\varepsilon_x}}{1+\frac{\varepsilon_x}{\varepsilon_y}\frac{x^2}{y^2}}k_0\left(y + \frac{\varepsilon_x}{\varepsilon_y}\frac{x^2}{y^2}\right) = 2\sqrt{\varepsilon_x y^2 + \varepsilon_y x^2 k_0)} \end{cases}$$

For these waves to focus on the point (0, 0), and, consequently, for the curve $f(x, y) = 0$ to be a total retroreflector, both phases must be equal, i.e. $\varphi_1 = \varphi_2$. Under this condition, the phases of all the waves coming through the lens to the point (0, 0) will be equal, as the point "2" is arbitrary. The shape of the lens in real space is thus defined by

$$\frac{x^2}{\varepsilon_x} - \frac{y^2}{\varepsilon_y} = -\left(\frac{d}{\sqrt{\varepsilon_x}}\right)^2.$$

Note that the parameter d/ε_x corresponds to the distance from the origin to the vertex of the hyperbola. For the interference to be constructive, the condition $\varphi_1 = \varphi_2 = 2\sqrt{\varepsilon_y}k_0 d = 2\pi m$, with m an integer, must be fulfilled:

$$d = \frac{\pi m}{\sqrt{\varepsilon_y}k_0}.$$

In all the simulations performed in the work we have set $m = 1$, i.e., the first interference order. With the help of the derived hyperbola, we can perform full-wave simulations, as shown in Fig. 5.9.

We note that this derivation assumes that the shape of the IFC is given by Eq. (5.1), which does not consider the finite thickness of the slab. A more realistic, accurate derivation, should consider such finite thickness, e.g. as described by the dispersion of polaritons in biaxial slabs given in Ref. [14].

C.2 Analytical Shape of a Planar Hyperlens for Polaritons

In this section, organized in a tutorial style, we perform the analytical calculation of the optimal shape of a lens in hyperbolic media, also in collaboration with Kirill Voronin (Moscow Institute of Physics and Technology, Russia). In the first part, we consider a two-dimensional (2D) model, in which the lens is a curved boundary separating two different biaxial half-spaces. This case is introduced to illustrate our approach to engineering a hyperbolic lens. In the second part, we take a more realistic and rigorous approach, in which confined polaritons propagate in a thin film of a biaxial crystal, namely α-MoO_3.

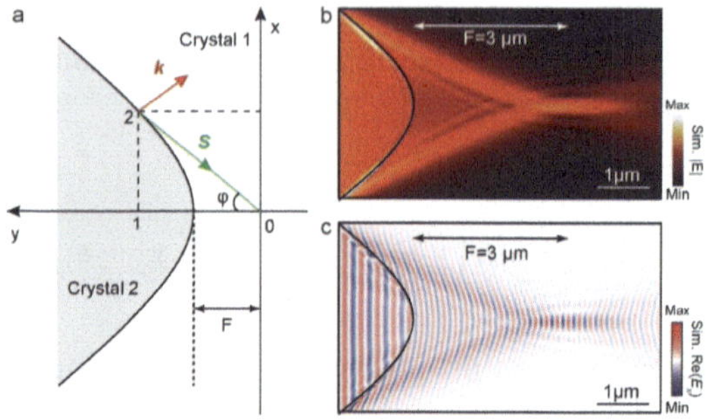

Fig. C.2 Analytic derivation of the lens shape between two different biaxial half-spaces. (**a**) Schematics of the lens between two different biaxial half-spaces: crystal 1 (white region) and crystal 2 (grey region). (**b–c**), Distribution of the absolute value and the real part of the x-component of the electric field respectively, based on full-wave numerical simulations

Two Dimensional Case

In this subsection, we consider a system composed of two different biaxial half-spaces (without loss of generality, we will assume that they are infinite along the z direction) and separated by a surface whose shape is given by the equation $f(x, y) = 0$ (see Fig. C.2). We will assume that the dielectric permittivity tensors of both media have the following diagonal form:

$$\hat{\varepsilon_i} = \begin{pmatrix} \varepsilon_{ix} & 0 & 0 \\ 0 & \varepsilon_{iy} & 0 \\ 0 & 0 & \varepsilon_{iz} \end{pmatrix},$$

where $i = 1, 2$ stands for the number of the medium. It means that we align the optics axes of the first crystal parallel to the corresponding optics axes of the second crystal and one of these axis parallel to the z direction. Also, x and y coordinate axes are chosen along the main crystal axes. A plane wave, polarized in the xy plane and with wavevector $\mathbf{k}$ and Poynting vector $\mathbf{S}$, is incident on the surface $f(x, y) = 0$. To mimic our experiment, we assume that this wave satisfies the conditions: $\mathbf{S} \parallel \mathbf{k} \parallel y$ axis. Therefore, as in the previous section, the dispersion of an electromagnetic wave propagating in the first medium is given by Fresnel's equation [15]:

$$F(k_{1x}, k_{1y}) \equiv \frac{k_{1x}^2}{\varepsilon_y} + \frac{k_{1y}^2}{\varepsilon_x} - k_0^2 = 0, \tag{C.4}$$

In the lossless case, when the dielectric permittivities are purely real, the Poynting vector is parallel to the group velocity. Therefore, $\mathbf{S}$ is given by $\mathbf{S}_1 \approx \nabla F(k_{1x}, k_{1y})$, or, more explicitly:

$$S_{1x} \approx \frac{k_{1x}}{\varepsilon_{1y}} \qquad S_{1y} \approx \frac{k_{1y}}{\varepsilon_{1x}}. \tag{C.5}$$

When taking into account the optical losses of the medium, the direction of the Poynting vector will be slightly tilted with respect to the one predicted by Eq. (C.5), but we will neglect this effect here for the sake for simplicity, i.e. we will restrict ourselves to the lossless case. Let us consider now two different optical paths to the point (0,0): one along the y axis and the other starting from an arbitrary point of the lens surface (labeled as "2" in Fig. C.2a). The phase of the waves propagating along these paths are the following:

$$\begin{cases} \Phi_{10} = -k_{2y}(0)(y-F) - k_{1y}(0)F = \sqrt{\varepsilon_{2x}}(y-F) + \sqrt{\varepsilon_{1x}}F \\ \Phi_{20} = -k_{1y}(\varphi)y - k_{1x}(\varphi)x = -k_{1y}(\varphi)\left(y + \frac{x^2\varepsilon_{1y}}{y\varepsilon_{1x}}\right) \end{cases}$$

where φ can be geometrically calculated as:

$$\tan\varphi = \frac{x}{y} = \frac{S_x}{S_y} = \frac{\varepsilon_{1x}k_{1x}}{\varepsilon_{1y}k_{1y}}. \tag{C.6}$$

Therefore, we can immediately calculate $k_{1y}(\varphi)$ using Eq. (C.6) and Fresnel's equation, given by Eq. (C.4), according to:

$$k_{1y}(\varphi) = -\sqrt{\frac{\varepsilon_{1x}}{1 + \frac{\varepsilon_{1y}x^2}{\varepsilon_{1x}x^2}}}$$

For these waves to focus at the point (0, 0), and, consequently, for the curve $f(x, y) = 0$ to be a hyperbolic refractive lens with focal distance F, both phases must be equal, i.e. $\Phi_{10} = \Phi_{20}$. Under this condition, the phases of all the waves coming through the lens to the point (0, 0) will be equal, as the point "2" is arbitrary. The shape of the lens is thus defined by:

$$\sqrt{\varepsilon_{1x}}F + \sqrt{\varepsilon_{2x}}(y-F) = \sqrt{\frac{\varepsilon_{1x}y}{\varepsilon_{1x}y^2 + \varepsilon_{1y}x^2}}\left(y + \frac{\varepsilon_{1y}x^2}{\varepsilon_{1x}y}\right),$$

After some simplification, we obtain that the lens should be have a hyperbolic shape:

$$\left(y - \frac{\sqrt{\varepsilon_{2x}}}{\sqrt{\varepsilon_{1x}} + \sqrt{\varepsilon_{2x}}}F\right)^2 + \frac{\varepsilon_{1y}}{\varepsilon_{1x} - \varepsilon_{2x}}x^2 = \frac{\varepsilon_{1x}}{(\sqrt{\varepsilon_{1x}} + \sqrt{\varepsilon_{2x}})^2}F^2,$$

To verify this result, we perform a full-wave numerical simulation of this system using the finite- element method COMSOL Multiphysics. We chose $F = 3\,\mu\text{m}$, $\varepsilon_{1x} = 20+0.3i$, $\varepsilon_{1y} = -20+0.3i$, $\varepsilon_{2x} = 10+0.3i$, $\varepsilon_{2y} = -10+0.3i$. Figure C.2b,c

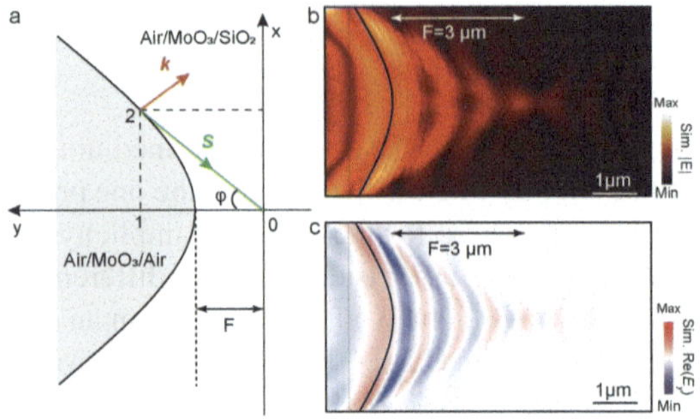

Fig. C.3 **Analytic derivation of the lens shape between two hyperbolic media.** (**a**) Schematics of the lens between two hyperbolic media: air/α-MoO_3/SiO_2 (white region) and air/α-MoO_3/air (grey region). (**b–c**) Distribution of the absolute value and the real part of the y-component of the electric field respectively, based on full-wave numerical simulations

show waves focusing at the (0, 0) point, in excellent agreement with our theoretical prediction.

Thin Film Case

However, the goal of this work is not the lensing of free-space light, but the lensing of highly confined polaritons propagating in finite-thickness-slab of biaxial crystal. Therefore, let us consider now a homogeneous biaxial slab with thickness d embedded between two different semi-infinite isotropic media. In other words, we consider a system which is similar to the one shown in Fig C.2a, but with the same biaxial film and superstrate in both regions, and with different substrates (Fig. C.3a). In this setting, we approximate the slab by a thin-film with vanishing thickness $d \to 0$. This limit has been extensively utilized during the course of this thesis. Moreover, we consider waves with high momentum, $k \gg k_0$, which has also been consistently used, and is justified here as we are studying highly confined polaritons. Under such approximations, the dispersion of polaritons propagating in the film is given by:

$$F\left(q_x, q_y\right) = iq\left(\alpha_x \cos^2\varphi + \alpha_y \sin^2\varphi\right) + \varepsilon = 0, \tag{C.7}$$

where $\mathbf{q} = \mathbf{k}/k_0$ is the normalized wavevector, and $q = \sqrt{q_x^2 + q_y^2}$. Furthermore, $\alpha = \frac{2\pi}{c}\sigma = \frac{k_0 d\varepsilon}{2i}$ is the normalized effective 2D conductivity of the film, and $\varepsilon_{1,2}$ is an arithmetical mean of the substrate and superstrate dielectric permittivities. Equation (C.7) can also be written as:

$$k_0 d\left(\varepsilon_x q_x^2+\varepsilon_y q_y^2\right)+2\varepsilon q=0. \tag{C.8}$$

In analogy to the previous section, we derive the direction of the Poynting vector as $S_1 \sim \nabla F\left(k_{1x}, k_{1y}\right)$:

$$S_{1x} \sim q_{1x}\left(\varepsilon_x+\frac{\varepsilon_1}{q k_0 d}\right); \quad S_{1y} \sim q_{1y}\left(\varepsilon_y+\frac{\varepsilon_1}{q k_0 d}\right). \tag{C.9}$$

Finally, we impose that the polaritonic phase along two different optical paths must be equal, i.e. $\Phi_{10}=\Phi_{20}$, yielding:

$$\begin{cases}\Phi_{10}=-k_{2y}(0)(y-F)-k_{1y}(0)F=-\frac{2\varepsilon_2}{k_0 d\varepsilon_y}(y-F)-\frac{2\varepsilon_1}{k_0 d\varepsilon_y}F; \\ \Phi_{20}=-k_{1y}(\varphi)y-k_{1x}(\varphi)x.\end{cases} \tag{C.10}$$

Combining Eqs. (C.8), (C.9), and (C.10) with the relation $\frac{x}{y}=\frac{s_x}{s_y}$, we obtain a system of equations which defines the shape of the lens:

$$\begin{cases}-\frac{2\varepsilon_1}{\varepsilon_y}F-\frac{2\varepsilon_2}{\varepsilon_y}(\mathrm{y}-F)=\frac{\kappa\left(\varepsilon\left(x^2+y^2\right)+\kappa\left(\varepsilon_y x^2+\varepsilon_x y^2\right)\right)}{\sqrt{\left(\varepsilon_1+\varepsilon_y\kappa\right)^2x^2+\left(\varepsilon+\varepsilon_x\kappa\right)^2y^2}}; \\ \kappa=2\varepsilon_1\frac{\left(\varepsilon_x\kappa+\varepsilon\right)^2y^2+\left(\varepsilon_y\kappa+\varepsilon\right)^2x^2}{\varepsilon_y\left(\varepsilon_x\kappa+\varepsilon\right)^2y^2+\varepsilon_x\left(\varepsilon_y\kappa+\varepsilon\right)^2x^2},\end{cases} \tag{C.11}$$

with $\kappa=qk_0d$. To plot the shape of the lens, one should solve the system (C.11) numerically. To verify our analytical definition of the shape of the lens, we perform full-wave numerical simulations. As a representative example, we consider a 100-nm-thick α-MoO_3 film5 on top of SiO_2 (white region in Fig. C.3) and air (grey region in Fig. C.3), both covered by air. The simulation was performed at $\lambda_0=11.11\,\mu\text{m}$ and $F=3\,\mu\text{m}$. Figure C.3b–c shows a good agreement with the theoretically predicted focal point.

C.3 Optimization of a Lens for Polaritons in Hyperbolic Media

Focusing of polaritons is obtained upon the interference of the outgoing refracted polaritons as they pass through the refractive lens (prism). The wavevectors ($\mathbf{k}_{out}$) and direction ($\mathbf{S}_{out}$) of the refracted polaritons depend on the angle of the boundaries defining the lens (θ), or equivalently, on the angle of incidence, as shown in Fig. 6.4. The size of the focus spot is intimately related to the wavelength of the refracted polaritons ($\lambda_{out}=2\pi|\mathbf{k}_{out}|$): smaller sizes are obtained for smaller λ_{out}. Moreover, $|\mathbf{k}_{out}|$ increases as the angle that the wavevector forms with the x axis, α, increases up to the asymptote of the hyperbolic IFC. Therefore, the polaritonic wavelength

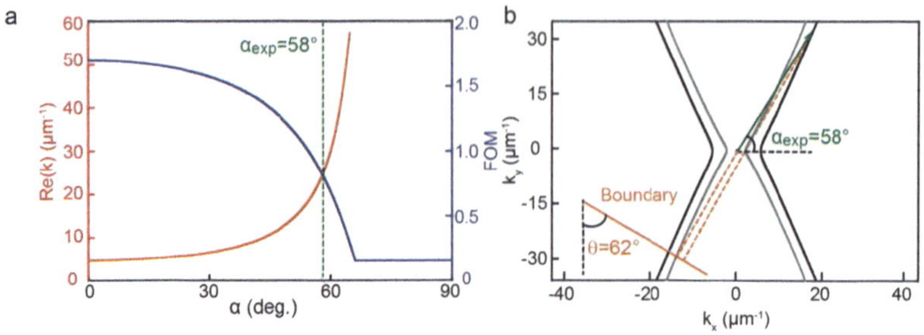

Fig. C.4 Optimization of a lens for polaritons in hyperbolic media. (**a**) Calculated modulus of the wavevector (**k**) and the figure of merit (FOM) of polaritons in a 170-nm-thick α-MoO_3 flake as a function of the in-plane angle (α) between **k** and the direction of $k_y = 0$ in the IFCs. (**b**) Corresponding analytic IFC of polaritons in the same α-MoO_3 flake as in (**a**) on top of SiO_2 (black curve) and air (grey curve).When the angle of the boundary (orange line) is chosen $\theta = 62°$, we obtain an angle $\alpha_{exp} = 58°$ based on momentum conservation at the boundary

(and, therefore, the size of the focus) is smaller as we approach the angle of the asymptote of the hyperbolic IFC.

On the other hand, since polaritons decay exponentially as they propagate, the intensity at a distant focus can be rather weak. Propagation losses are more remarkable for polaritons with larger wavevectors (smaller wavelengths).

There is thus a compromise between momentum (directly related to the focus size) and propagation loss (directly related to the intensity at the focus spot). The propagation loss of polaritons can be evaluated by the value of their figure of merit (FOM), which is defined by L_p/λ_p, L_p and λ_p being the polaritonic propagation length and wavelength, respectively. It should be noted that for a given design of the lens where its boundaries are perpendicular to the asymptote of the hyperbolic IFC, the refracted polaritons possess infinitely large momentum, thus setting a lower limit to the size of the focus spot, however, with relatively high losses as mentioned above.

In the following, we optimize the design of the lens (angle of their boundaries, θ) for enhanced focusing of polaritons. To do that, we analytically calculate the modulus of the polaritonic wavevector and FOM as a function of the angle (α) between the wavevector (**k**) of the refracted polaritons and the direction of the x axis, i.e. the axis $k_y = 0$, as shown in Fig. C.4a. The point at $\alpha = 58°$ gives us a good compromise between small optical losses and small wavelength as it is close to the asymptote of the hyperbolic IFC. While this crossing of course depends on the relative scales of the axes, it yields an optimal value for the design of the lens, since it sets a good trade-off both the focus size and the optical losses. The experimental results presented in Fig. 6.6 of this thesis correspond to a lens with a boundary tilted an angle $\theta = 62°$, nearly perpendicular to the asymptote of the hyperbolic IFC, which yields an angle $\alpha_{exp} = 58°$ based on momentum conservation at the boundary (Fig. C.4b). Remarkably, by using the optimized design, a focus with a FWHM of $\lambda_p/6$, λ_p being the polaritonic wavelength along [100] crystal direction of α-MoO_3, is obtained.

Appendix D
Rotation of the Propagation Direction of In-Plane Hyperbolic Polaritons in a Slab Embedded Between Two Semi-infinite Media

D.1 Theoretical Derivation

In this appendix we derive the conditions under which propagation of hyperbolic polaritons is possible along either of the two in-plane axes of a biaxial slab surrounded between two semiinfinite media. These conditions, and the approximations upon which they are derived, will determine the requirements for the dielectric permittivities of the different media in the three-layer system, and therefore the frequency ranges in which the propagation can occur along each in-plane axis. To derive such requirements, we consider the dispersion relation of polaritons propagating in a biaxial slab embedded between two semi-infinite media, which was derived in Chap. 3 [14] and is reproduced here for completeness:

$$q = \frac{\rho}{k_0 d}\left[\arctan\left(\frac{\varepsilon_1 \rho}{\varepsilon_z}\right) + \arctan\left(\frac{\varepsilon_3 \rho}{\varepsilon_z}\right) + \pi l\right], \quad l = 0, 1, 2... \tag{D.1}$$

where $q = k_t/k_0$ is the normalized in-plane momentum ($k_t^2 = k_x^2 + k_y^2$), ε_1 and ε_3 are the permittivities of the superstrate and substrate, respectively, d is the thickness of the α-MoO_3 flake, l is the mode index, and $k_0 = \omega/c$ is the wavevector in free space and

$$\rho = i\sqrt{\frac{\varepsilon_z q^2}{\varepsilon_x q_x^2 + \varepsilon_y q_y^2}} = i\sqrt{\frac{\varepsilon_z}{\varepsilon_x \cos^2\alpha + \varepsilon_y \sin^2\alpha}},$$

with α being the angle between the x axis and the in-plane component vector.

Since $\arctan(x) + \arctan(y) = \arctan((x + y)/(1 - xy))$, Eq. (D.1) can be conveniently rewritten as:

G. Álvarez Pérez, *Fundamentals of Nano-Optics in Hyperbolic van der Waals Materials*, Springer Theses, https://doi.org/10.1007/978-3-031-75767-9

$$\tan\left(\frac{kd}{\rho}\right) = \frac{\frac{\rho(\varepsilon_1+\varepsilon_3)}{\varepsilon_z}}{1-\frac{\rho^2\varepsilon_1\varepsilon_3}{\varepsilon_z}} \tag{D.2}$$

Equations (D.1) and (D.2) are applicable when the polaritonic momenta are large ($k \gg k_0$, k_0 being the wavevector of light in free space). However, even under the approximation of large momenta, the analysis of the propagation directions and the shape of the isofrequency curves is not trivial for the general case of arbitrary complex dielectric permittivities. Nevertheless, the analysis can be significantly simplified in a certain parametric range, studied in the present work. Therefore, let us first assume that the permittivities in the biaxial slab are purely real. This assumption is reasonable under small losses (see the next section), i.e. when $\mathrm{Re}(\varepsilon_j) \gg \mathrm{Im}(\varepsilon_j)$ for $j \in \{1, 3, x, y, z\}$. Second, we assume that the dielectric permittivities fulfill $\varepsilon_x < 0$, $\varepsilon_y > 0$, $\varepsilon_z > 0$, with $\varepsilon_1 > 0$, $\varepsilon_3 < 0$, in analogy with biaxial materials in which in-plane hyperbolic polaritons have been reported, such as α-MoO_3 [1] and α-V_2O_5 [16]. Finally, we consider only the lowest-order mode, $l = 0$, since higher-order modes possess a much larger wavevector and, therefore, the mode $l = 0$ is excited easily and prevails.

Requirements for Propagation of Polaritons Along the x Axis in the Slab

Let us study propagation of polaritons along the axis, according to Eq. (D.1). In this case we set $\alpha = 0$ and therefore $\rho = i\sqrt{\varepsilon_z/\varepsilon_x}$. We can thus simplify Eq. (D.2) as follows:

$$\tan\left(\frac{kd}{i\sqrt{\varepsilon_z/\varepsilon_x}}\right) = \frac{\frac{i\sqrt{\frac{\varepsilon_z}{\varepsilon_x}}(\varepsilon_1+\varepsilon_3)}{\varepsilon_z}}{1+\frac{\varepsilon_1\varepsilon_3}{\varepsilon_x\varepsilon_z}}.$$

Since we have assumed that $\varepsilon_x < 0$, then $i\sqrt{\varepsilon_z/\varepsilon_x} = \sqrt{\varepsilon_z/|\varepsilon_x|} > 0$ and:

$$\tan\left(\frac{kd}{\sqrt{\varepsilon_z/|\varepsilon_x|}}\right) = \frac{\frac{\sqrt{\frac{\varepsilon_z}{|\varepsilon_x|}}(\varepsilon_1+\varepsilon_3)}{\varepsilon_z}}{1+\frac{\varepsilon_1\varepsilon_3}{\varepsilon_x\varepsilon_z}}. \tag{D.3}$$

Here $kd/\sqrt{\varepsilon_z/|\varepsilon_x|} > 0$, but also $kd/\sqrt{\varepsilon_z/|\varepsilon_x|} < \pi/2$, as we consider the mode $l = 0$; consequently, the left hand side of Eq. (S3) is positive. Therefore, for Eq. (D.3) to have a solution, the right-hand side must be also positive. Under the conditions $\varepsilon_x < 0$, $\varepsilon_z > 0$, $\varepsilon_1 > 0$, and $\varepsilon_3 < 0$, the denominator of the fraction in the right-hand side of Eq. (D.2) is positive (i.e. $1 + \varepsilon_1\varepsilon_3/\varepsilon_x\varepsilon_z > 0$). Thus, for Eq. (D.3) to have a solution, the numerator in the right-hand side of Eq. (D.3) must be also positive. Hence:

$$\frac{\sqrt{\frac{\varepsilon_z}{|\varepsilon_x|}}(\varepsilon_1+\varepsilon_3)}{\varepsilon_z} > 0,$$

which, as $|\varepsilon_x|, \varepsilon_z > 0$, yielding:

$$\varepsilon_1+\varepsilon_3 > 0.$$

Requirements for Propagation of Polaritons Along the y Axis in the Slab

Analogously, let us study propagation of polaritons along the axis. In this case we set $\alpha = \pi/2$ and therefore $\rho = i\sqrt{\varepsilon_z/\varepsilon_y}$. We can thus simplify Eq. (D.2) as:

$$\tan\left(\frac{kd}{i\sqrt{\varepsilon_z/\varepsilon_y}}\right) = \frac{\frac{i\sqrt{\frac{\varepsilon_z}{\varepsilon_y}}(\varepsilon_1+\varepsilon_3)}{\varepsilon_z}}{1+\frac{\varepsilon_1\varepsilon_3}{\varepsilon_y\varepsilon_z}}. \tag{D.4}$$

Since the argument of the tangent is purely imaginary, as $\sqrt{\varepsilon_z/\varepsilon_y}$ is real and positive, and bearing in mind that $\tan(i\xi) = i\cdot\tanh(\xi)$ (with ξ also real and positive), Eq. (D.4) can be rewritten as:

$$i\cdot\tanh\left(\frac{kd}{-\sqrt{\varepsilon_z/\varepsilon_y}}\right) = \frac{\frac{i\sqrt{\frac{\varepsilon_z}{\varepsilon_y}}(\varepsilon_1+\varepsilon_3)}{\varepsilon_z}}{1+\frac{\varepsilon_1\varepsilon_3}{\varepsilon_y\varepsilon_z}}.$$

Hence:

$$\tanh\left(\frac{kd}{-\sqrt{\varepsilon_z/\varepsilon_y}}\right) = \frac{\frac{\sqrt{\frac{\varepsilon_z}{\varepsilon_y}}(\varepsilon_1+\varepsilon_3)}{\varepsilon_z}}{1+\frac{\varepsilon_1\varepsilon_3}{\varepsilon_y\varepsilon_z}},$$

and, taking into account that tanh is an odd function, i.e. $\tanh(-u) = \tanh(u)$, we get:

$$\tanh\left(\frac{kd}{\sqrt{\varepsilon_z/\varepsilon_y}}\right) = -\frac{\frac{\sqrt{\frac{\varepsilon_z}{\varepsilon_y}}(\varepsilon_1+\varepsilon_3)}{\varepsilon_z}}{1+\frac{\varepsilon_1\varepsilon_3}{\varepsilon_y\varepsilon_z}}. \tag{D.5}$$

Recall that, as we have assumed that $\varepsilon_y, \varepsilon_z > 0$, then $kd/\sqrt{\varepsilon_z/\varepsilon_y} > 0$ and the left-hand side of Eq. (D.5) is positive. Therefore, for Eq. (D.5) to have a solution, the right-hand side must be also positive. If $\varepsilon_y\varepsilon_z > |\varepsilon_1\varepsilon_3|$, which might be reasonable as the superstrate is usually a low-permittivity medium, as air (see in the next section, see Fig. D.2), then the denominator in the right-hand side of Eq. (D.5) is positive. Hence, for Eq. (D.5) to have a solution the numerator must be negative:

$$\frac{\sqrt{\frac{\varepsilon_z}{\varepsilon_y}}(\varepsilon_1+\varepsilon_3)}{\varepsilon_z} < 0,$$

which, as $\varepsilon_y, \varepsilon_z > 0$, translates into:

$$\varepsilon_1 + \varepsilon_3 < 0.$$

Our analysis shows that anisotropic polaritons can propagate along the x axis in the range in which $\varepsilon_1 + \varepsilon_3 > 0$, while they can propagate along the y axis in the range $\varepsilon_1 + \varepsilon_3 < 0$ (provided that the assumptions $\varepsilon_x < 0$, $\varepsilon_y > 0$, $\varepsilon_z > 0$, $\varepsilon_1 > 0$, $\varepsilon_3 < 0$, $\mathrm{Re}(\varepsilon_j) \gg \mathrm{Im}(\varepsilon_j)$ for $j \in \{1, 3, x, y, z\}$, and $\varepsilon_y\varepsilon_z > |\varepsilon_1\varepsilon_3|$ fulfilled by the different media). As such, a 90°-reorientation of the propagation direction of hyperbolic polaritons (or equivalently of the polaritonic isofrequency curve) from propagation along the axis to propagation along the axis is possible in a polaritonic heterostructures in a neighborhood of the condition $\varepsilon_1 + \varepsilon_3 = 0$.

D.2 Verification for the Particular Configuration α-MoO_3/4H-SiC

In this section we analyze the conditions under which the propagation of hyperbolic polaritons is possible along either of the two in-plane axes of an α-MoO_3 biaxial slab on top of a 4H-SiC substrate, based on the derivation carried out in the previous section. These conditions will determine the frequency range in which the propagation can occur along each in-plane axis. These conditions are the following:

1. $\varepsilon_x < 0$, $\varepsilon_y > 0$, $\varepsilon_z > 0$, for the biaxial slab, fulfilled by α-MoO_3 in the frequency range 821.4–956.7 cm^{-1} (see Fig. D.1). The permittivity of α-MoO_3 is taken from Ref. [1].
2. $\varepsilon_1 > 0$ for the superstrate, fulfilled by air as $\varepsilon_{air} = 1$.
3. $\varepsilon_3 < 0$ for the substrate, fulfilled by 4H-SiC in the frequency range 782–965 cm^{-1} (see Fig. D.1). The permittivity of 4H-SiC is taken from Ref. [17].
4. $\mathrm{Re}(\varepsilon_j) \gg \mathrm{Im}(\varepsilon_j)$ for $j \in \{1, 3, x, y, z\}$, fulfilled in the frequency range 910–955 cm^{-1} (see Fig. D.1).
5. $\varepsilon_y\varepsilon_z > |\varepsilon_1\varepsilon_3|$, fulfilled in the frequency range 885–956.7 cm^{-1} (see Fig. D.2).

A rotation of the propagation direction of in-plane hyperbolic polaritons can therefore occur in α-MoO_3/4H-SiC heterostructures in a neighborhood of the condition $\varepsilon_1 + \varepsilon_3 = 0$, which occurs at 943 cm^{-1} for the heterostructure system considered in this work.

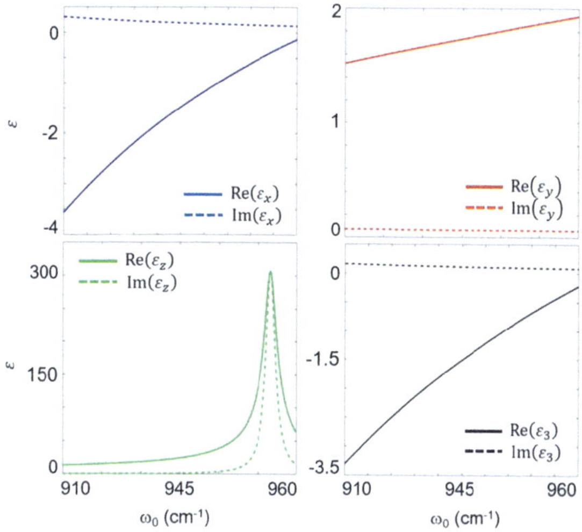

Fig. D.1 Comparison of the real and imaginary parts of the dielectric permittivity tensor components of α-MoO_3 along the three crystallographic axes (x axis in blue, y axis in red, z axis in green), and 4H-SiC (in black) in the frequency range 910–960 cm^{-1}

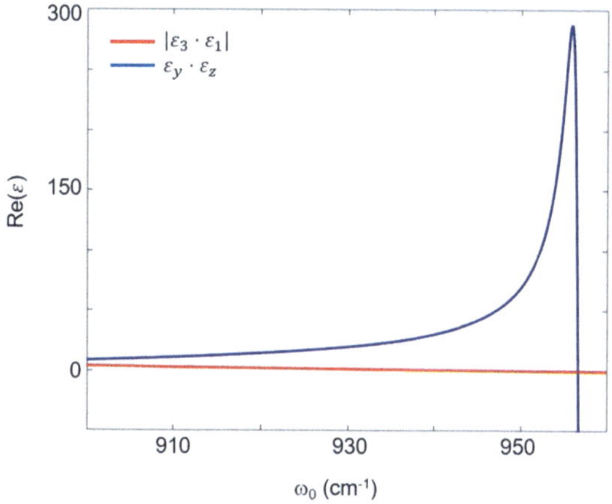

Fig. D.2 Verification of the condition $\varepsilon_y \varepsilon_z > |\varepsilon_1 \varepsilon_3|$ in a polaritonic heterostructure α-MoO_3/4H-SiC in the frequency range 900–960 cm^{-1}

Appendix E
Dyadic Green's Function for a Two-Dimensional Conductivity Biaxial Layer

In this appendix, we present the analytical calculation of the polaritonic IFC in an α-MoO_3 biaxial slab on top of 4H-SiC, where PhPs are excited by a point dipole. These calculations were carried out in collaboration with Kirill Voronin (Moscow Institute of Technology) and published in "Enabling propagation of anisotropic polaritons along forbidden directions via a topological transition" by Jiahua Duan, Gonzalo Álvarez-Pérez et al. Science Advances 7 (14), eabf2690 (2021) [18].

We make several assumptions to do the calculations. First, a thin-film approximation is used, such that the biaxial slab is treated as a 2D-conductivity layer. This approximation has been widely utilized throughout this thesis. Second, we assume that the exciting electric field is parallel to the (out-of-plane) z axis and, therefore, we will consider only the z component of the electric field in all our derivations. Due to space limitations, we will not present the full derivation here. In turn, we direct interested readers to the Supplementary Materials (Section S6) of Ref. [18] for more details.

To calculate the distribution of the electric field created by any source, in general, we need to solve a non-homogenous linear differential equation (non-homogeneous wave equation) of the form $\hat{\mathcal{L}} f = f_0$, where $\hat{\mathcal{L}}$ is a linear differential operator, f_0 is the "initial condition" (source), and f is the variable to be calculated (in our case, the electric field). To solve this non-homogeneous equation, we can make use of the Green's function formalism (analogously to the Huygens principle in the case of propagating waves). To that end, we calculate the Green's function, $G(\mathbf{r}-\mathbf{r}')$, for the linear differential operator and get an electric field distribution as the convolution of this function with the initial field: $f(r) = \iiint d^3\mathbf{r}' G(\mathbf{r}-\mathbf{r}') f_0(\mathbf{r}')$. The physical meaning of the Green's function is the field created at the point with radius-vector $\mathbf{r}$ by the point source placed at the point $\mathbf{r}'$. The Green's function is also related to the local density of electromagnetic states (LDOS). Applying Fourier transformation to the equation for $f(\mathbf{r})$, we obtain $f(\mathbf{k}) = G(\mathbf{k}) f_0(\mathbf{k})$, that means that $G(\mathbf{k})$ describes an efficiency of excitation of the mode with momentum $\mathbf{k}$, and is also related to the density of electromagnetic states (DOS). As we are dealing with a vector field, we

G. Álvarez Pérez, *Fundamentals of Nano-Optics in Hyperbolic van der Waals Materials*, Springer Theses, https://doi.org/10.1007/978-3-031-75767-9

consider the generalization of the Green's function, the dyadic Green's function (DGF), $\hat{G}(\mathbf{r}-\mathbf{r}')$, constructed for the vectorial wave equation. Using the DGF, the overall electric field of polaritons emitted by a source and propagating along the film can be calculated at any point near the film according to the Lippmann-Schwinger integral equation [19]:

$$E(\mathbf{r}) = k_0^2 \int d^3\mathbf{r}' \hat{G}(\mathbf{r}-\mathbf{r}') E_0(\mathbf{r}')$$

where $E(\mathbf{r})$ is the field created by the source at the point $\mathbf{r}$ and $E_0(\mathbf{r}')$ is the initial field (at the surface of the source. As we consider only the z component of the electric field, we need only the zz component of the DGF. In case of the point dipole, oriented along the z axis, and, according to the definition of the DGF, the z component of the field is given by $E_z(\mathbf{r}) \approx \hat{G}_{zz}(\mathbf{r}-\mathbf{r}')$, where $\mathbf{r}'$ is the position of the dipole. Since s-polarized plane waves in dielectric media have zero projection on the z axis, we consider only the contribution of p polarized waves to the DGF, which we denote as $\hat{G}_{pp}$. As such, $\hat{G}_{pp}$ simply provides the p-polarized component of the electromagnetic wave excited by a p-polarized point source.

Expressing the DGF corresponding to the field *transmitted* through the layer in Dirac notation [20], and writing the electric fields in the media surrounding the layer, as well in the layer itself (all explicitly derived in Chap. 3) and performing some algebra, we get:

$$G_{zz}(\mathbf{r},\mathbf{r}') \approx \frac{-i}{8\pi^2 k_0\sqrt{\varepsilon_1\varepsilon_2}} \int d^2k_{\|} G_{zz}(k_x,k_y) e^{i(k_{1z}z'-k_{2z}z)} e^{ik_{\|}\Delta_{\|}}, \tag{E.1}$$

where

$$G_{zz}(k_x,k_y) = \frac{k_{\|}^2}{k_{1z}} T_{pp},$$

and the indices $1, 2$ denoted the superstrate and substrate, respectively.

The Fresnel's coefficients T_{pp} can be straightforwardly obtained from Maxwell's equations. We can apply the boundary conditions of Maxwell's equations to match the in-plane components of the magnetic and electric fields at the interface between the superstrate and substrate. In addition, assuming that the point source and the *probing* point (at which we wish to know the value of the electromagnetic fields) are located in the same semi-infinite half-space, the *reflected* DGF must be used instead of the transmitted DGF. However, the electromagnetic fields of propagating polaritons (the eigenmodes supported by our 2D layer) are mainly determined by the poles of Fresnel's coefficients (and thus the poles of the DGF). Consequently, we can safely approximate:

$$G_{Rzz}(\mathbf{r},\mathbf{r}') \approx \frac{-i}{8\pi^2 k_0\sqrt{\varepsilon_1\varepsilon_2}} \int d^2k_{\|} \frac{k_{\|^2}}{k_{1z}} T_{pp} \frac{k_{2z}}{k_{1z}} e^{i(k_{1z}z'+k_{2z}z)} e^{ik_{\|}\Delta_{\|}}. \tag{E.2}$$

Equation (E.2) is equivalent to $G_{zz}(\mathbf{r} - \mathbf{r}')$ given by Eq. (E.1) up to the sign, phase factor $e^{2i(k_{1z}z'}$ (≈ 1 for $z' = 0$) and a factor k_{2z}/k_{1z} (which in our case is close to unity). Consequently, we will use Eq. (E.2) as the DGF for the 2D conductivity layer. This function is related to the DOS of propagating hyperbolic polaritons in the layer. The absolute value of $G_{zz}(k_x, k_y)$ is plotted in Fig. 7.7 (Chap. 7) for the illuminating wavelengths $\omega_0 = 948, 943, 941$, and $937\,\mathrm{cm}^{-1}$.

Appendix F
Derivation of the Dispersion Relation of Polaritons in Twisted Anisotropic Conductivity Bilayers

Here we derive the dispersion relation of polaritons propagating in a structure made of two stacked anisotropic conducting layers of thickness d which are rotated a relative angle θ and separated a distance $2h$, with dielectric permittivity tensors $\hat{\varepsilon}_t$ (top layer) and $\hat{\varepsilon}_b$ (bottom layer), respectively, as depicted in Fig. F.1. These calculations were carried out together with Nathaniel Capote-Robayna (Donostia International Physics Center, Spain) and published in "Twisted Nano-Optics: Manipulating Light at the Nanoscale with Twisted Phonon Polaritonic Slabs" by Jiahua Duan et al. Nano Letters 2020, 20, 5323–5329 (2020) [21].

The coordinate systems in which $\hat{\varepsilon}_b$ and $\hat{\varepsilon}_t$ are diagonal are denoted as $\{X, Y, Z\}$ and $\{X', Y', Z'\}$, respectively. Each of the layers is modeled as a two-dimensional conductivity sheet of zero thickness, as schematized in Fig. F.2. This avoids the calculation of the fields inside the slab [22], and has been proven to be valid for graphene [23], transition layer polaritons [24], single layers of biaxial crystals [14, 25], polaritonic crystals of hexagonal boron nitride [26] and Moiré hyperbolic metasurfaces [27] with a layer thickness that is much smaller than the polariton wavelength ($d \gg \lambda_p$). In the model, the effective conductivity for each uniaxial layer is given by:

$$\hat{\sigma}_i = \frac{cd}{2i\lambda_0}\hat{\varepsilon}_i, \quad i \in \{t, b\}, \tag{F.1}$$

where c is the speed of light in vacuum, λ_0 its wavelength and $\hat{\varepsilon}$ the in-plane uniaxial permittivity of the layer (both $\hat{\varepsilon}_i$ and $\hat{\sigma}_i$ are 2×2 matrices). Note that $\hat{\sigma}_i$ scale linearly with d, thus considering the effect of the small slab thickness.

The conductivity tensor of the bottom layer $\hat{\sigma}_b$ is diagonal in the system of coordinates $\{X, Y, Z\}$:

$$\hat{\sigma}_b = \begin{pmatrix} \sigma_{bx} & 0 \\ 0 & \sigma_{by} \end{pmatrix}. \tag{F.2}$$

G. Álvarez Pérez, *Fundamentals of Nano-Optics in Hyperbolic van der Waals Materials*, Springer Theses, https://doi.org/10.1007/978-3-031-75767-9

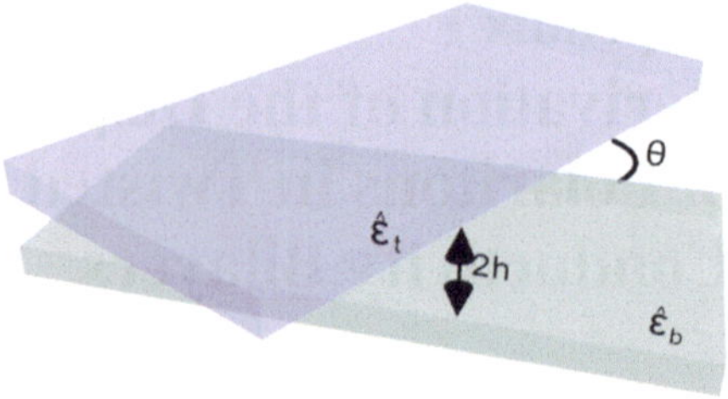

Fig. F.1 Twisted layers separated a distance $2h$ and twisted an angle θ, with dielectric permittivity tensors $\hat{\varepsilon}_t$ and $\hat{\varepsilon}_b$

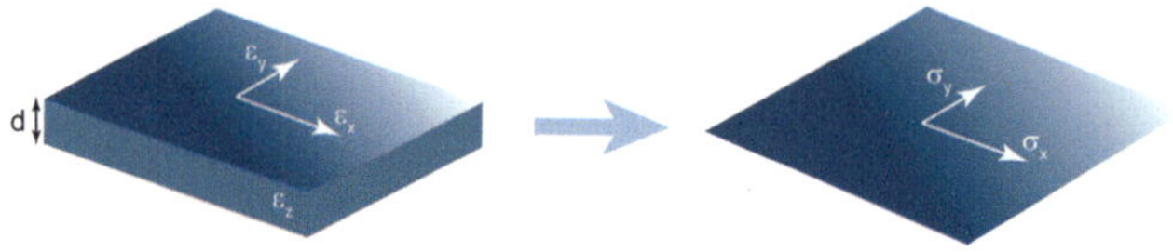

Fig. F.2 The thickness of each layer is set to zero in the model

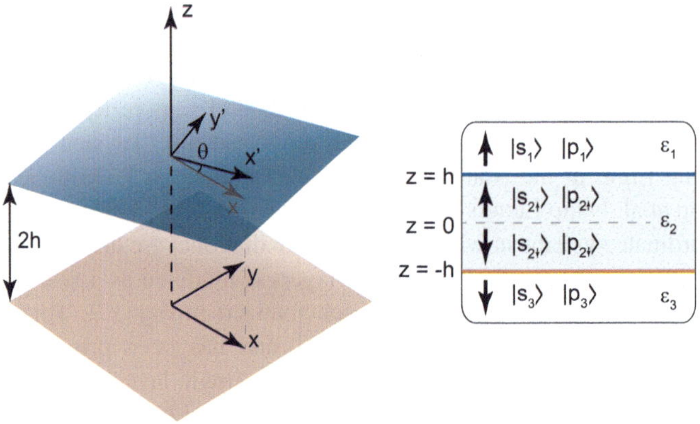

Fig. F.3 (**a**) Systems of coordinates chosen to model twisted layers separated a distance $2h$ and twisted an angle θ, with dielectric permittivity tensors $\hat{\varepsilon}_b$ and $\hat{\varepsilon}_t$. (**b**) Field representation inside each medium: superstrate (brown), space between the twisted anisotropic sheets (blue) and substrate (red), with respective isotropic permittivities ε_j (j = 1,2,3). The two-dimensional sheets are located at $z = h$ and $z = -h$, respectively

In this system, the conductivity of the top layer $\hat{\sigma}_t$ is given by (see Fig. F.3a):

$$\hat{\sigma}_t = \hat{R}^*(\theta)\hat{\sigma}_t'\hat{R}(\theta) =$$

$$\begin{pmatrix} \cos(\theta) & \sin(\theta) \\ -\sin(\theta) & \cos(\theta) \end{pmatrix} \begin{pmatrix} \sigma_{tx} & 0 \\ 0 & \sigma_{ty} \end{pmatrix} \begin{pmatrix} \cos(\theta) & -\sin(\theta) \\ \sin(\theta) & \cos(\theta) \end{pmatrix} = \tag{F.3}$$

$$\begin{pmatrix} \sigma_{tx}\cos^2(\theta) + \sigma_{ty}\sin^2(\theta) & \sin(\theta)\cos(\theta)(\sigma_{ty} - \sigma_{tx}) \\ \sin(\theta)\cos(\theta)(\sigma_{ty} - \sigma_{tx}) & \sigma_{tx}\sin^2(\theta) + \sigma_{ty}\cos^2(\theta) \end{pmatrix}.$$

We represent the fields inside the three media in Dirac notation, as shown in Fig. F.3b:

$$|s(1,3)\rangle = \frac{1}{k_t}\begin{pmatrix}-k_y\\ k_x\\ 0\end{pmatrix} e^{i(xk_x+yk_y)}, \qquad |p(1,3)\rangle = \frac{1}{k_t}\begin{pmatrix}k_x\\ k_y\\ \frac{k_t^2}{\mp k_{1,3z}}\end{pmatrix} e^{i(xk_x+yk_y)},$$

$$|s(2\uparrow,\downarrow)\rangle = \frac{1}{k_t}\begin{pmatrix}-k_y\\ k_x\\ 0\end{pmatrix} e^{i(xk_x+yk_y)}, \qquad |p(2\uparrow,\downarrow)\rangle = \frac{1}{k_t}\begin{pmatrix}k_x\\ k_y\\ \frac{k_t^2}{\mp k_{2z}}\end{pmatrix} e^{i(xk_x+yk_y)}, \tag{F.4}$$

where k_x and k_y are the in-plane wavevector components, with $k_t^2 = k_x^2 + k_y^2$ and $k_{iz} = \sqrt{\epsilon_i k_0^2 - k_t^2}$, $i = \{1, 2, 3\}$ (with $k_0 = \frac{\omega}{c}$) is the out-of-plane wavevector component. s and p denote the s- and p-polarizations, $1, 2, 3$ stand for to the different media and the arrows distinguish the propagation direction inside them. The $-(+)$ sign in the definition of the p-polarization basis vectors should be taken for 1 (3) or $\uparrow$ ($\downarrow$) in the left side, respectively. We also introduce the subvectors of the vectors in the Eq. (F.4) to represent the in-plane components of the fields:

$$|s\rangle = \frac{1}{k_t}\begin{pmatrix}-k_y\\ k_x\end{pmatrix} e^{i(xk_x+yk_y)}, \qquad |p\rangle = \frac{1}{k_t}\begin{pmatrix}k_x\\ k_y\end{pmatrix} e^{i(xk_x+yk_y)}. \tag{F.5}$$

The electric fields can be compactly written for each media as:

$$\begin{aligned}
\mathbf{E}^{(1)} &= (A_{1s}|s(1)\rangle + A_{1p}|p(1)\rangle)e^{izk_{1z}} \quad z \geq h\\
\mathbf{E}^{(2)} &= (A_{2s\uparrow}|s(2\uparrow)\rangle + A_{2p\uparrow}|p(2\uparrow)\rangle)e^{izk_{2z}}\\
&\quad + (A_{2s\downarrow}|s(2\downarrow)\rangle + A_{2p\downarrow}|p(2\downarrow)\rangle)e^{-izk_{2z}} \quad h > z > -h,\\
\mathbf{E}^{(3)} &= (A_{3s}|s(3)\rangle + A_{3p}|p(3)\rangle)e^{-izk_{3z}} \quad z \leq -h,
\end{aligned} \tag{F.6}$$

where A_β are unknown amplitudes. By applying the boundary conditions at $z = h$ and $z = -h$—which imply the continuity of the in-plane components of both electric and magnetic fields at each interface—, we get the following system of equations:

$$E_{1t}(z = h) = E_{2t}(z = h), \qquad E_{2t}(z = -h) = E_{3t}(z = -h), \tag{F.7}$$

$$\mathbf{e}_z \times (\mathbf{H}_1 - \mathbf{H}_2)_t = 2\hat{\alpha}_t \mathbf{E}_{1t}, \qquad \mathbf{e}_z \times (\mathbf{H}_2 - \mathbf{H}_3) = 2\hat{\alpha}_b \mathbf{E}_{3t}, \tag{F.8}$$

where E_{it} and H_{it} ($j = 1, 2, 3$) are the in-plane electric and magnetic fields, respectively and $\mathbf{e}_z$ is the basis vector along the z direction. Using the field representation (F.6), we can rewrite the boundary conditions for the electric field at $z = h$ and $z = -h$ in Eqs. (F.7) in a more explicit way:

$$A_{1s}e^{ik_{1z}h} = A_{2s\uparrow}e^{ik_{2z}h} + A_{2s\downarrow}e^{-ik_{2z}h} \tag{F.9}$$

$$A_{1p}e^{ik_{1z}h} = A_{2p\uparrow}e^{ik_{2z}h} + A_{2p\downarrow}e^{-ik_{2z}h} \tag{F.10}$$

$$A_{3s}e^{ik_{3z}h} = A_{2s\uparrow}e^{-ik_{2z}h} + A_{2s\downarrow}e^{ik_{2z}h} \tag{F.11}$$

$$A_{3p}e^{ik_{3z}h} = A_{2p\uparrow}e^{-ik_{2z}h} + A_{2p\downarrow}e^{ik_{2z}h} \tag{F.12}$$

Since the magnetic field can be expressed as $\mathbf{H} = \mathbf{q} \times \mathbf{E}$ (being $\mathbf{q} = \frac{k_z}{k_0}$), the in-plane magnetic field can be written as $\mathbf{H}_t = \mathbf{e}_z \times \mathbf{q} \times E$. In order to simplify, let us notice that:

$$(\mathbf{e}_z \times (\pm \mathbf{q}_{iz} \times |s(\beta)\rangle))_t = \mp q_{iz}|s\rangle, \qquad (\mathbf{e}_z \times (\pm \mathbf{q}_{iz} \times |p(\beta)\rangle))_t = \mp \frac{\varepsilon_i}{q_{iz}}|p\rangle, \tag{F.13}$$

where the sign of the vector $\mathbf{q}$ gives the direction of the vector $|s, p(\beta)\rangle$, for $\beta = \{1, 2\uparrow, 2\downarrow, 3\}$ (positive sign stands for the positive direction along the Z axis). Introducing these expressions into (F.8), and taking into account that:

$$\hat{\alpha}_i = \frac{2\pi}{c}\hat{\sigma}_i \quad i \in \{t, b\}, \tag{F.14}$$

we obtain more explicit versions of Eqs. (F.9)–(F.12):

$$\begin{aligned} &-A_{1s}q_{1z}e^{ik_{1z}h}|s\rangle - A_{1p}\frac{\varepsilon_1}{q_{1z}}e^{ik_{1z}h}|p\rangle + A_{2s\uparrow}q_{2z}e^{ik_{2z}h}|s\rangle + A_{2p\uparrow}\frac{\varepsilon_2}{q_{2z}}e^{ik_{2z}h}|p\rangle \\ &-A_{2s\downarrow}q_{2z}e^{-ik_{2z}h}|s\rangle - A_{2p\downarrow}\frac{\varepsilon_2}{q_{2z}}e^{-ik_{2z}h}|p\rangle = 2\hat{\alpha}_t\left(A_{1s}|s\rangle + A_{1p}|p\rangle\right)e^{ik_{1z}h} \end{aligned} \tag{F.15}$$

$$\begin{aligned} &-A_{2s\uparrow}q_{2z}e^{-ik_{2z}h}|s\rangle - A_{2p\uparrow}\frac{\varepsilon_2}{q_{2z}}e^{-ik_{2z}h}|p\rangle + A_{2s\downarrow}q_{2z}e^{ik_{2z}h}|s\rangle + A_{2p\downarrow}\frac{\varepsilon_2}{q_{2z}}e^{ik_{2z}h}|p\rangle \\ &-A_{3s}q_{3z}e^{ik_{3z}h}|s\rangle - A_{3p}\frac{\varepsilon_3}{q_{3z}}e^{ik_{3z}h}|p\rangle = 2\hat{\alpha}_b\left(A_{3s}|s\rangle + A_{3p}|p\rangle\right)e^{ik_{3z}h} \end{aligned} \tag{F.16}$$

Multiplying Eqs. (F.15) and (F.16) by $\langle s|$ and $\langle p|$, we get:

$$\begin{aligned} &-A_{1s}q_{1z}e^{ik_{1z}h} + A_{2s\uparrow}q_{2z}e^{ik_{2z}h} - A_{2s\downarrow}q_{2z}e^{-ik_{2z}h} \\ &= 2e^{ik_{1z}h}\left(A_{1s}\langle s|\hat{\alpha}_t|s\rangle + A_{1p}\langle s|\hat{\alpha}_t|p\rangle\right) \end{aligned} \tag{F.17}$$

$$\begin{aligned} &-A_{1p}\frac{\varepsilon_1}{q_{1z}}e^{ik_{1z}h} + A_{2p\uparrow}\frac{\varepsilon_2}{q_{2z}}e^{ik_{2z}h} - A_{2p\downarrow}\frac{\varepsilon_2}{q_{2z}}e^{-ik_{2z}h} \\ &= 2e^{ik_{1z}h}\left(A_{1s}\langle p|\hat{\alpha}_t|s\rangle + A_{1p}\langle p|\hat{\alpha}_t|p\rangle\right) \end{aligned} \tag{F.18}$$

$$
\begin{aligned}
&- A_{2s\uparrow} q_{2z} e^{-ik_{2z}h} + A_{2s\downarrow} q_{2z} e^{ik_{2z}h} - A_{3s} q_{3z} e^{ik_{3z}h} \\
&= 2e^{ik_{3z}h}\left(A_{3s}\langle s|\hat{\alpha}_b|s\rangle + A_{3p}\langle s|\hat{\alpha}_b|p\rangle\right)
\end{aligned}
\tag{F.19}
$$

$$
\begin{aligned}
&- A_{2p\uparrow}\frac{\varepsilon_2}{q_{2z}} e^{-ik_{2z}h} + A_{2p\downarrow}\frac{\varepsilon_2}{q_{2z}} e^{ik_{2z}h} - A_{3p}\frac{\varepsilon_3}{q_{3z}} e^{ik_{3z}h} \\
&= 2e^{ik_{3z}h}\left(A_{3s}\langle p|\hat{\alpha}_b|s\rangle + A_{3p}\langle p|\hat{\alpha}_b|p\rangle\right)
\end{aligned}
\tag{F.20}
$$

Substituting Eqs. (F.9)–(F.12) into Eqs. (F.17)–(F.20), we obtain:

$$
\begin{aligned}
&-q_{1z}\left(A_{2s\uparrow}e^{ik_{2z}h} + A_{2s\downarrow}e^{-ik_{2z}h}\right) + A_{2s\uparrow}q_{2z}e^{ik_{2z}h} - A_{2s\downarrow}q_{2z}e^{-ik_{2z}h} \\
&= 2\langle s|\hat{\alpha}_t|s\rangle\left(A_{2s\uparrow}e^{ik_{2z}h} + A_{2s\downarrow}e^{-ik_{2z}h}\right) + 2\langle s|\hat{\alpha}_t|p\rangle\left(A_{2p\uparrow}e^{ik_{2z}h} + A_{2p\downarrow}e^{-ik_{2z}h}\right)
\end{aligned}
\tag{F.21}
$$

$$
\begin{aligned}
&-\frac{\varepsilon_1}{q_{1z}}\left(A_{2p\uparrow}e^{ik_{2z}h} + A_{2p\downarrow}e^{-ik_{2z}h}\right) + A_{2p\uparrow}\frac{\varepsilon_2}{q_{2z}}e^{ik_{2z}h} - A_{2p\downarrow}\frac{\varepsilon_2}{q_{2z}}e^{-ik_{2z}h} \\
&= 2\langle p|\hat{\alpha}_t|s\rangle\left(A_{2s\uparrow}e^{ik_{2z}h} + A_{2s\downarrow}e^{-ik_{2z}h}\right) + 2\langle p|\hat{\alpha}_t|p\rangle\left(A_{2p\uparrow}e^{ik_{2z}h} + A_{2p\downarrow}e^{-ik_{2z}h}\right)
\end{aligned}
\tag{F.22}
$$

$$
\begin{aligned}
&-A_{2s\uparrow}q_{2z}e^{-ik_{2z}h} + A_{2s\downarrow}q_{2z}e^{ik_{2z}h} - q_{3z}\left(A_{2s\uparrow}e^{-ik_{2z}h} + A_{2s\downarrow}e^{ik_{2z}h}\right) \\
&= 2\langle s|\hat{\alpha}_b|s\rangle\left(A_{2s\uparrow}e^{-ik_{2z}h} + A_{2s\downarrow}e^{ik_{2z}h}\right) + 2\langle s|\hat{\alpha}_b|p\rangle\left(A_{2p\uparrow}e^{-ik_{2z}h} + A_{2p\downarrow}e^{ik_{2z}h}\right)
\end{aligned}
\tag{F.23}
$$

$$
\begin{aligned}
&-A_{2p\uparrow}\frac{\varepsilon_2}{q_{2z}}e^{-ik_{2z}h} + A_{2p\downarrow}\frac{\varepsilon_2}{q_{2z}}e^{ik_{2z}h} - \frac{\varepsilon_3}{q_{3z}}\left(A_{2p\uparrow}e^{-ik_{2z}h} + A_{2p\downarrow}e^{ik_{2z}h}\right) \\
&= 2\langle p|\hat{\alpha}_b|s\rangle\left(A_{2s\uparrow}e^{-ik_{2z}h} + A_{2s\downarrow}e^{ik_{2z}h}\right) + 2\langle p|\hat{\alpha}_b|p\rangle\left(A_{2p\uparrow}e^{-ik_{2z}h} + A_{2p\downarrow}e^{ik_{2z}h}\right)
\end{aligned}
\tag{F.24}
$$

After some algebra we arrive to:

$$
\begin{aligned}
&A_{2s\uparrow}e^{ik_{2z}h}\left[2\langle s|\hat{\alpha}_t|s\rangle + q_{1z} - q_{2z}\right] + A_{2p\uparrow}e^{ik_{2z}h}\left[2\langle s|\hat{\alpha}_t|p\rangle\right] \\
&+A_{2s\downarrow}e^{-ik_{2z}h}\left[2\langle s|\hat{\alpha}_t|s\rangle + q_{1z} + q_{2z}\right] + A_{2p\downarrow}e^{-ik_{2z}h}\left[2\langle s|\hat{\alpha}_t|p\rangle\right] = 0
\end{aligned}
\tag{F.25}
$$

$$
\begin{aligned}
&A_{2s\uparrow}e^{ik_{2z}h}\left[2\langle p|\hat{\alpha}_t|s\rangle\right] + A_{2p\uparrow}e^{ik_{2z}h}\left[2\langle p|\hat{\alpha}_t|p\rangle + \frac{\varepsilon_1}{q_{1z}} - \frac{\varepsilon_2}{q_{2z}}\right] \\
&+A_{2s\downarrow}e^{-ik_{2z}h}\left[2\langle p|\hat{\alpha}_t|s\rangle\right] + A_{2p\downarrow}e^{-ik_{2z}h}\left[2\langle p|\hat{\alpha}_t|p\rangle + \frac{\varepsilon_1}{q_{1z}} + \frac{\varepsilon_2}{q_{2z}}\right] = 0
\end{aligned}
\tag{F.26}
$$

$$
\begin{aligned}
&A_{2s\uparrow}e^{-ik_{2z}h}\left[2\langle s|\hat{\alpha}_b|s\rangle + q_{2z} + q_{3z}\right] + A_{2p\uparrow}e^{-ik_{2z}h}\left[2\langle s|\hat{\alpha}_b|p\rangle\right] \\
&+A_{2s\downarrow}e^{ik_{2z}h}\left[2\langle s|\hat{\alpha}_b|s\rangle - q_{2z} + q_{3z}\right] + A_{2p\downarrow}e^{ik_{2z}h}\left[2\langle s|\hat{\alpha}_b|p\rangle\right] = 0
\end{aligned}
\tag{F.27}
$$

$$
\begin{aligned}
&A_{2s\uparrow}e^{-ik_{2z}h}\left[2\langle p|\hat{\alpha}_b|s\rangle\right]+A_{2p\uparrow}e^{-ik_{2z}h}\left[2\langle p|\hat{\alpha}_b|p\rangle+\frac{\varepsilon_2}{q_{2z}}+\frac{\varepsilon_3}{q_{3z}}\right]\\
&+A_{2s\downarrow}e^{ik_{2z}h}\left[2\langle p|\hat{\alpha}_b|s\rangle\right]+A_{2p\downarrow}e^{ik_{2z}h}\left[2\langle p|\hat{\alpha}_b|p\rangle-\frac{\varepsilon_2}{q_{2z}}+\frac{\varepsilon_3}{q_{3z}}\right]=0
\end{aligned}
\tag{F.28}
$$

where the explicit expression for the elements $\langle\beta|\hat{\alpha}_{t,b}|\gamma\rangle$ is given by:

$$
\langle s|\hat{\alpha}_i|s\rangle=\frac{1}{k_t^2}\left(-k_y,\,k_x\right)\begin{pmatrix}\alpha_{ixx}&\alpha_{ixy}\\\alpha_{ixy}&\alpha_{iyy}\end{pmatrix}\begin{pmatrix}-k_y\\k_x\end{pmatrix}=\frac{\alpha_{ixx}k_y^2-2\alpha_{ixy}k_xk_y+\alpha_{iyy}k_x^2}{k_t^2}
\tag{F.29}
$$

$$
\langle s|\hat{\alpha}_i|p\rangle=\frac{1}{k_t^2}\left(-k_y,\,k_x\right)\begin{pmatrix}\alpha_{ixx}&\alpha_{ixy}\\\alpha_{ixy}&\alpha_{iyy}\end{pmatrix}\begin{pmatrix}k_x\\k_y\end{pmatrix}=\frac{\alpha_{ixy}(k_x^2-k_y^2)+k_xk_y(\alpha_{iyy}-\alpha_{ixx})}{k_t^2}
\tag{F.30}
$$

$$
\langle p|\hat{\alpha}_i|s\rangle=\frac{1}{k_t^2}\left(k_x,\,k_y\right)\begin{pmatrix}\alpha_{ixx}&\alpha_{ixy}\\\alpha_{ixy}&\alpha_{iyy}\end{pmatrix}\begin{pmatrix}-k_y\\k_x\end{pmatrix}=\frac{\alpha_{ixy}(k_x^2-k_y^2)+k_xk_y(\alpha_{iyy}-\alpha_{ixx})}{k_t^2}
\tag{F.31}
$$

$$
\langle p|\hat{\alpha}_i|p\rangle=\frac{1}{k_t^2}\left(k_x,\,k_y\right)\begin{pmatrix}\alpha_{ixx}&\alpha_{ixy}\\\alpha_{ixy}&\alpha_{iyy}\end{pmatrix}\begin{pmatrix}k_x\\k_y\end{pmatrix}=\frac{\alpha_{ixx}k_x^2+2\alpha_{ixy}k_xk_y+\alpha_{iyy}k_y^2}{k_t^2}
\tag{F.32}
$$

Equations (F.25)–(F.28) can be thus expressed in matrix form:

$$
\begin{pmatrix}M_{11}&M_{12}&M_{13}&M_{14}\\M_{21}&M_{22}&M_{23}&M_{24}\\M_{31}&M_{32}&M_{33}&M_{34}\\M_{41}&M_{42}&M_{43}&M_{44}\end{pmatrix}\begin{pmatrix}A_{2s\uparrow}\\A_{2p\uparrow}\\A_{2s\downarrow}\\A_{2p\downarrow}\end{pmatrix}=\mathbf{0},
\tag{F.33}
$$

where the terms M_{ij} $(i,j=1,2,2,4)$ are given by:

$$
\begin{aligned}
M_{11}&=e^{ik_{2z}h}\left[2\left(\frac{\alpha_{txx}k_y^2-2\alpha_{txy}k_xk_y+\alpha_{tyy}k_x^2}{k_t^2}\right)+q_{1z}-q_{2z}\right]\\
M_{12}&=2e^{ik_{2z}h}\left[\frac{\alpha_{txy}(k_x^2-k_y^2)+k_xk_y(\alpha_{tyy}-\alpha_{txx})}{k_t^2}\right]\\
M_{13}&=e^{-ik_{2z}h}\left[2\left(\frac{\alpha_{txx}k_y^2-2\alpha_{txy}k_xk_y+\alpha_{tyy}k_x^2}{k_t^2}\right)+q_{1z}+q_{2z}\right]\\
M_{14}&=2e^{-ik_{2z}h}\left[\frac{\alpha_{txy}(k_x^2-k_y^2)+k_xk_y(\alpha_{tyy}-\alpha_{txx})}{k_t^2}\right]
\end{aligned}
$$

$$
\begin{aligned}
M_{21} &= 2e^{ik_{2z}h}\left[\frac{\alpha_{txy}(k_x^2-k_y^2)+k_xk_y(\alpha_{tyy}-\alpha_{txx})}{k_t^2}\right]\\
M_{22} &= e^{ik_{2z}h}\left[2\left(\frac{\alpha_{txx}k_x^2+2\alpha_{txy}k_xk_y+\alpha_{tyy}k_y^2}{k_t^2}\right)+\frac{\varepsilon_1}{q_{1z}}-\frac{\varepsilon_2}{q_{2z}}\right]\\
M_{23} &= 2e^{-ik_{2z}h}\left[\frac{\alpha_{txy}(k_x^2-k_y^2)+k_xk_y(\alpha_{tyy}-\alpha_{txx})}{k_t^2}\right]\\
M_{24} &= e^{-ik_{2z}h}\left[2\left(\frac{\alpha_{txx}k_x^2+2\alpha_{txy}k_xk_y+\alpha_{tyy}k_y^2}{k_t^2}\right)+\frac{\varepsilon_1}{q_{1z}}+\frac{\varepsilon_2}{q_{2z}}\right]\\
\\
M_{31} &= e^{-ik_{2z}h}\left[2\left(\frac{\alpha_{bxx}k_y^2-2\alpha_{bxy}k_xk_y+\alpha_{byy}k_x^2}{k_t^2}\right)+q_{2z}+q_{3z}\right]\\
M_{32} &= 2e^{-ik_{2z}h}\left[\frac{\alpha_{bxy}(k_x^2-k_y^2)+k_xk_y(\alpha_{byy}-\alpha_{bxx})}{k_t^2}\right]\\
M_{33} &= e^{ik_{2z}h}\left[2\left(\frac{\alpha_{bxx}k_y^2-2\alpha_{bxy}k_xk_y+\alpha_{byy}k_x^2}{k_t^2}\right)-q_{2z}+q_{3z}\right]\\
M_{34} &= 2e^{ik_{2z}h}\left[\frac{\alpha_{bxy}(k_x^2-k_y^2)+k_xk_y(\alpha_{byy}-\alpha_{bxx})}{k_t^2}\right]\\
\\
M_{41} &= 2e^{-ik_{2z}h}\left[\frac{\alpha_{bxy}(k_x^2-k_y^2)+k_xk_y(\alpha_{byy}-\alpha_{bxx})}{k_t^2}\right]\\
M_{42} &= e^{-ik_{2z}h}\left[2\left(\frac{\alpha_{bxx}k_x^2+2\alpha_{bxy}k_xk_y+\alpha_{byy}k_y^2}{k_t^2}\right)+\frac{\varepsilon_2}{q_{2z}}+\frac{\varepsilon_3}{q_{3z}}\right]\\
M_{43} &= 2e^{ik_{2z}h}\left[\frac{\alpha_{bxy}(k_x^2-k_y^2)+k_xk_y(\alpha_{byy}-\alpha_{bxx})}{k_t^2}\right]\\
M_{44} &= e^{ik_{2z}h}\left[2\left(\frac{\alpha_{bxx}k_x^2+2\alpha_{bxy}k_xk_y+\alpha_{byy}k_y^2}{k_t^2}\right)-\frac{\varepsilon_2}{q_{2z}}+\frac{\varepsilon_3}{q_{3z}}\right]
\end{aligned}
\tag{F.34}
$$

The system (F.33) has non-trivial solutions only when its determinant is equal to zero, yielding the dispersion relation:

$$\begin{vmatrix} M_{11} & M_{12} & M_{13} & M_{14} \\ M_{21} & M_{22} & M_{23} & M_{24} \\ M_{31} & M_{32} & M_{33} & M_{34} \\ M_{41} & M_{42} & M_{43} & M_{44} \end{vmatrix} = 0 \tag{F.35}$$

Appendix G
Phonon Polaritons in Twisted Biaxial Trilayers: Further Analytical Considerations

These calculations were carried out in collaboration with Kirill Voronin (Donostia International Physics Center, Spain) and published in "Multiple and spectrally robust photonic magic angles in reconfigurable α-MoO_3 trilayers" by Jiahua Duan, Gonzalo Álvarez-Pérez, Christian Lanza, Kirill Voronin et al. Nature Materials 22, 867–872 (2023) [28].

G.1 Dyadic Green's Function for Polaritons in Twisted Biaxial Trilayers

To calculate the field distribution created by a point source, we use the dyadic Green's function (DGF) formalism presenting a solution of the vectorial Maxwell's equations with a dipolar source. As was discussed in Appendix E, the physical meaning of the DGF, $\hat{G}(\mathbf{r}-\mathbf{r}')$, is the field created at the point $\mathbf{r}$ by a point source placed at the point $\mathbf{r}'$. The overall electric field created by the point dipole with the electric dipolar moment $\boldsymbol{\mu}$ along the multilayer structure can be calculated at any point above the top slab with the radius vector $\mathbf{r}$ according to the following expression [19]:

$$E(r) = 4\pi k_0^2 \hat{G}(\mathbf{r}-\mathbf{r}')\boldsymbol{\mu}.$$

Analogously to the derivations carried out in Appendix E, in the following discussion, we consider a vertically polarized electric point dipole $\boldsymbol{\mu} = (0, 0, \mu)^T$ and the z component of the electric field, as the latter presents the main interest in the s-SNOM experiments. Since the electric field of an s-polarized plane wave has a null projection on the z axis, we only consider the contribution from the p-polarized plane waves to the DGF, which we thus denote as $\hat{G}_{pp}$. Such dyadic provides the p-polarized component of the electromagnetic field excited by a p-polarized point

G. Álvarez Pérez, *Fundamentals of Nano-Optics in Hyperbolic van der Waals Materials*, Springer Theses, https://doi.org/10.1007/978-3-031-75767-9

source. The DGF for our heterostructure has the following explicit form [19]:

$$\hat{G}_{pp}(\mathbf{r}-\mathbf{r}') = \frac{1}{8\pi^2}\int d^2\mathbf{k}_{\|} e^{i\mathbf{k}_{\|}\Delta\mathbf{r}_{\|}} \left[\hat{M}_0 e^{ik_{4z}(z-z') - r_{pp}\hat{M}_r} e^{ik_{4z}(z+z')}\right],$$

where r_{pp} is the amplitude Fresnel's reflection coefficient for the heterostructure, $\mathbf{r}_{\|}$ is the in-plane position vector, $\mathbf{k}_{\|}$ is the in-plane component of the wavevector, and $k_{4z} = \sqrt{k_0\varepsilon_4^2 - k^2}$ stands for the z component of the wavevector in the isotropic superstrate, with dielectric permittivity ε_4. The explicit expression of the matrices $\hat{M}_0$ and $\hat{M}_r$ will not be derived here for space constraints but can be found in the Supplementary Materials of Ref. [28]. With the help of the DGF shown above, the z component of the electric field can be written as follows:

$$\hat{E}_z(\mathbf{r}) = iE_0 \int d^2\mathbf{k}_{\|} e^{i\mathbf{k}_{\|}\Delta\mathbf{r}_{\|}} \frac{k_{\|}^2}{k_0^3 k_{4z}} \left[e^{ik_{4z}(z-z')} + r_{pp} e^{ik_{4z}(z+z')}\right]. \tag{G.1}$$

where $E_0 = k^3\mu/2\pi\varepsilon_4$. Here we set the in-plane component of the dipole radius vector equal to 0, $\mathbf{r}' = 0$. To further simplify Eq. (G.1), we assume that the contribution from the electric field of polaritons dominates over the rest of the created fields. As the contribution from the polaritons comes from the poles of Fresnel's coefficient, r_{pp}, we neglect the non-singular first term in the integrand of Eq. (G.1). Additionally, as the polaritons have large wavevectors compared to the wavevector of propagating waves in free space (in the superstrate), we apply the large-k approximation to calculate r_{pp}. In this approximation, $k_{4z} = ik_{\|} = iqk_0$, where $q = k_{\|}/k_0$ is the normalized in-plane component of the wavevector. Converting the integral into polar coordinates $(q_x, q_y) \to (q, \varphi)$; $(x, y) \to (\rho, \theta)$, we obtain the final expression for the electric field:

$$E_z(r,\theta) = E_0 \int_0^{\infty} dq \int_0^{2\pi} d\varphi q^2 r_{pp}(q,\varphi) e^{-qk_0(e^z+z')} e^{iqk_0\rho\cos(\varphi-\theta)}, \tag{G.2}$$

where $E_0 = k^3\mu/2\pi\varepsilon_4$. The explicit expression for the reflection coefficient r_{pp} can also be found in the Supplementary Materials of Ref. [28].

G.2 Distribution of the Polariton Field in Twisted Biaxial Trilayers

We can calculate the electric field distribution of the polaritons excited by the point source and propagating along the trilayer biaxial slab using Eq. (G.2). Although Eq. (G.2) includes the integration up to infinity, in the numerical calculations the integration limit should be approximated by a finite value. In our calculations within the elliptic frequency range, we approximate the upper integration limit by the maximal value of q achieved by the fundamental mode. By doing this, we only take into account the contribution of the fundamental mode and *filter out* all higher-order modes. This approach is not applicable for the hyperbolic frequency range, therefore, for this range we use another approach based on the fact that the distance between the source and surface restricts the wavevector of exciting modes. In Eq. (G.2), this restriction is provided by the factor $e^{-qk_0(e^z+z')}$, vanishing the expression for high values of q. For example, in most of our calculations, unless otherwise specified, we use $z = z' = 0.1$ μm; therefore, the upper limit for polaritonic wavelength is 0.2 μm, which gives us $q_{\max} = 50$. To demonstrate the validity of Eq. (G.2), we compare the results of the performed full-wave simulations with the results based on Eq. (G.2). Figure G.1 shows the polar plots of the electric field amplitude extracted at the distance of two polariton wavelengths from the source for the bilayer (panels a–c), and for the trilayer (panels g–i). The solid curves and red points represent the result of analytical calculations and full-wave simulations, respectively. Also, using both methods we represent the dependence of the electric field as a function of the distance from the source along the canalization directions for the bilayer (Fig. G.1d–f) and for the trilayer (Fig. G.1j–l). According to Fig. G.1, the analytical results agree well with the numerical simulations.

G.3 Directivity of Anisotropic Polaritons and Canalization Criteria

In order to characterize the in-plane directivity of anisotropic polaritons we plot the directivity patterns for polaritons excited by a point source-that is, the dependence of the intensity $|E_z|^2$ as a function of the in-plane angle for a fixed distance from the source in polar coordinates (I, φ), as shown in Fig. G.2a. A similar representation is used in antenna theory.

To define the canalization regimes and evaluate their performance (i.e. the robustness against variations in the twist angles and the spectral broadbandness) we introduce a canalization criterion: (1) the propagation of canalized polaritons should be experimentally observable, i.e. the propagation length of the polariton modes should be larger than their wavelength, (2) the canalized polaritons should propagate with low angular spreading, i.e. the angular width of the beam (αw) should be smaller than a given critical angle α_{cr}, analogously to one of the directivity criteria in antenna theory. We chose the critical angle equal to 10° relatively arbitrarily, following the *beam* divergence to be at least five times smaller than its propagation length. We define the angular width of a polariton *beam*, α_w, as the angular distance between the points at which the value of intensity becomes half of the one along the canalization direction. Namely, α_w is the full width at the half maximum (FWHM) of the polariton directivity pattern. However, the angular width of a polariton beam depends on the distance from the source, and consequently, to unify the procedure we calculate the intensity at a distance ρ. We take the value for the distance to

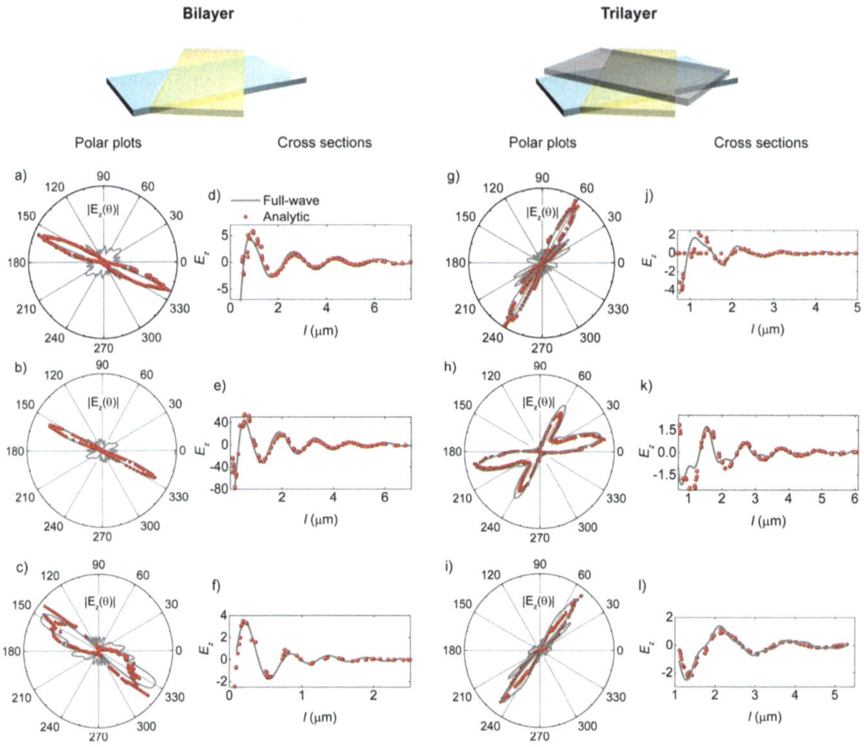

Fig. G.1 Comparison of the polar plots and field cross sections calculated with Eq. (G.2)**, and full-wave numerical simulations.** (**a–c**) polar plots of the field amplitude $|E_z(\theta)|$ at the distance of two polariton wavelengths from the source placed over the twisted bilayer, where θ is the in-plane angle. The thicknesses of the layers are $d_1 = d_2 = 200$ nm, the twisted angle is $\theta_{1-2} = 65°$, and the frequencies are (**a**) $\omega_0 = 901\ \text{cm}^{-1}$, (**b**) $\omega_0 = 909\ \text{cm}^{-1}$ (**c**) $\omega_0 = 930\ \text{cm}^{-1}$. (**d–f**) the field cross sections along the canalization direction for parameters chosen in panels **a–c** respectively. (**g–i**) polar plots of the field amplitude $|E_z(\varphi)|$ at the distance of two polariton wavelengths from the source placed over the twisted trilayer, where φ is the in-plane angle. For clarity, we omit the scale for the intensity, as the absolute values are normalized and the relevant magnitude is the angle. The thicknesses of the layers are $d_3 = 250$ nm, $d_2 = 240$ nm, $d_1 = 180$ nm; the frequencies and twisted angles are (**g**) $\omega_0 = 917\ \text{cm}^{-1}$, $\theta_{1-2} = 30°$, $\theta_{1-3} = 40°$, (**h**) $\omega_0 = 917\ \text{cm}^{-1}$, $\theta_{1-2} = 30°$, $\theta_{1-3} = 60°$ (**i**) $\omega_0 = 930\ \text{cm}^{-1}$, $\theta_{1-2} = 30°$, $\theta_{1-3} = -40°$. (**j–l**) the field cross sections along the canalization directions for the parameters chosen in panels (**g–h**) respectively

suppress the contribution from the other modes and polaritons propagating along different directions to preserve the canalized polaritons. Based on the fact that the propagation lengths of the polaritons propagating along the other directions are usually smaller than the propagation length of the canalized mode, we choose to match the propagation length, $L_p = 1/\text{Im}(k)$, along the canalization direction, φ_c. The real and imaginary parts of $\mathbf{k}_{\|}$ are extracted from the dispersion relation (see their isofrequency curves in Fig. G.2b). Note that the propagation direction is related to the group velocity, which is perpendicular to the IFC. Therefore, to calculate the

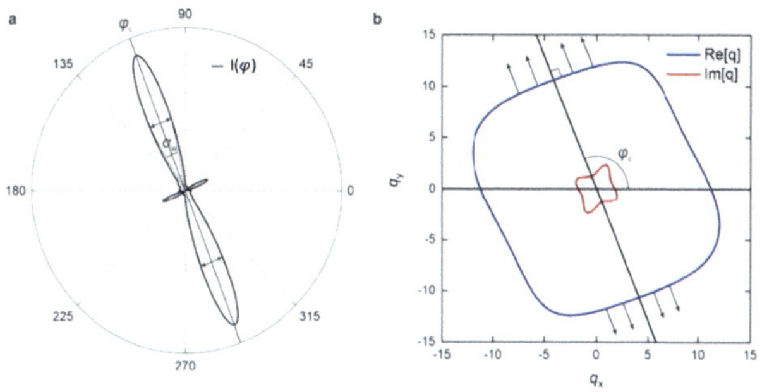

Fig. G.2 Canalization criteria. (**a**) Directivity plot of the field distribution with indicated canalization direction, φ_c, and angular width, α_w. For clarity, we omit the scale for the intensity, as the absolute values are normalized and the relevant magnitude is the angle. (**b**) IFC of the canalized mode shown in panel (**a**)

propagation length, we consider the direction at which the real IFC is orthogonal to the canalization direction. For such a purpose, we assume that the real and imaginary parts of the wavevector are collinear. Thus, the canalization condition is formally written as the following:

1. The propagation length must be larger than the polariton wavelength, i.e., $L_p > \lambda_p$.
2. The FWHM of the directivity pattern calculated at the distance $\rho = L_p$ from the source (where L_p is calculated along the canalization direction, φ_c) is smaller than 10°: $\alpha_w < 10°$

Note that the canalization criterion is developed for point sources and could not be applied to the distribution of field excited by an elongated antenna or other finite source. Consequently, in a general case, to verify the canalization regime one should perform calculations for the point source over the considered multilayer structure.

Appendix H
Active Tuning of Hyperbolic Polaritons in van der Waals Materials by Gated Graphene: Further Analytical and Numerical Considerations

These calculations were carried out in collaboration with Arturo González-Morán, Nathaniel Capote-Robayna, and Kirill V. Voronin (Donostia International Physics Center, Spain) and published in "Active Tuning of Highly Anisotropic Phonon Polaritons in Van der Waals Crystal Slabs by Gated Graphene" by Gonzalo Álvarez-Pérez, Arturo González-Morán et al. ACS Photonics 9 (2), 383–390 (2022) [29].

H.1 Hybridization Between Plasmon Polaritons in Graphene and HPhPs in α-MoO_3

In this section we study the effect of putting a graphene layer on top of an α-MoO_3 slab, as far as hybridization between PhPs and GPPs is concerned. To do so, we calculate the dispersion of HPhPs in a bare slab of α-MoO_3 surrounded by vacuum, as well as the dispersion of plasmon polaritons in a graphene sheet embedded between a semi-infinite air space and a semi-infinite dielectric environment of α-MoO_3 (with the *disabled* phonons in the dielectric permittivity).

In order to calculate the dispersion of plasmon polaritons in a graphene sheet embedded between a semi-infinite air space and a semi-infinite dielectric environment of α-MoO_3, we consider a graphene layer surrounded by media with permittivities ε_1 and ε_{eff}, where $\varepsilon_{\text{eff}} = \varepsilon_x^\infty$ when calculating the dispersion along the [100] crystal axis and $\varepsilon_{\text{eff}} = \varepsilon_y^\infty$ along the [001]. In this case, $\varepsilon_{x,y}^\infty$ is considered to take into account the dielectric environment of α-MoO_3 in which we *immerse* the graphene when calculating the dispersion of the GPPs, This dielectric environment will affect the confinement of GPPs, and therefore their coupling with HPhPs. The phonons of α-MoO_3 are not considered in this calculation, since we aim to study how GPPs in an *environment* of α-MoO_3 interact with PhPs in α-MoO_3.

The dispersion of plasmon polaritons in graphene placed between two media with arbitrary dielectric permittivities, ε_1 and ε_2 is given by [23]

G. Álvarez Pérez, *Fundamentals of Nano-Optics in Hyperbolic van der Waals Materials*, Springer Theses, https://doi.org/10.1007/978-3-031-75767-9

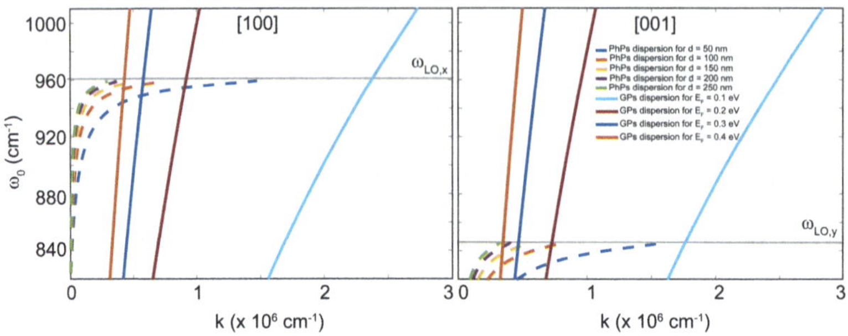

Fig. H.1 Analytical dispersion of HPhPs in bare slabs of α-MoO_3 of different thicknesses surrounded by vacuum (dashed lines), as well as the dispersion of plasmon polaritons in a graphene sheet surrounded by a semi-infinite air space and a semi-infinite dielectric environment of α-MoO_3 for different Fermi energies E_F. The dispersions are shown along both in-plane directions: [100] (left) and [001] (right panel)

$$k = \frac{\omega}{c} \cdot \frac{\varepsilon_1 + \varepsilon_2}{2} \cdot \frac{i}{\alpha},$$

with α being the normalized conductivity, given by Eq. (1.13).

The optimal distance for coupling depends on the inverse wavevector of the GPPs and PhPs, as they are evanescent waves along the out-of-plane direction. In our numerical simulations and analytical calculations, the distance between the graphene layer and the biaxial slab is zero. Hence, in this setting the overlap between the GPPs and HPhPs fields is maximal. Whether the GPPs and HPhPs couple or not is determined by whether there is a crossing point between the dispersion of GPPs (without PhPs) and the dispersion of HPhPs (without GPPs). This is, if there is a crossing point between the dispersions for a given configuration, there will be coupling.

Figure H.1 shows the dispersion for GPPs, derived using the equation above and taking $\varepsilon_1 = 1$ and $\varepsilon_2 = \varepsilon_i^{\infty}$ (with $i = x, y$)—with the suppressed HPhPs—for different Fermi energies E_F (continuous lines), as well as the dispersion of HPhPs in the slab for different thicknesses (dashed lines)—without the graphene layer. We can observe that the higher the Fermi energy, the wider the range of thicknesses of the biaxial slab for which there exists coupling between GPPs and HPhPs. If the slab is too thick, then a higher E_F is required to match the HPhPs wavevector and achieve the coupling regime. To give an example, for the range of thicknesses d between 50 and 200 nm and $E_F = 0.4$ eV we find a crossing of the dispersion curves and thus a coupling between HPhPs in α-MoO_3 and GPPs. On the contrary, for $d =$ 250 nm and $E_F = 0.4$ eV there is no crossing.

H.2 Derivation of the Dispersion Relation of Polaritons Propagating in a Heterostructure Made of a Graphene Sheet on Top of a Biaxial Slab

Here we derive the dispersion relation of polaritons propagating in a heterostructure made of a graphene sheet on top of a biaxial slab of thickness d, as depicted in Fig. H.1a in the main text. Let us consider a non-magnetic biaxial slab with dielectric permittivity tensor $\hat{\varepsilon}$. The coordinate system $\{x, y, z\}$ is chosen is such a way that $\hat{\varepsilon}$ is diagonal, so that:

$$\hat{\varepsilon} = \begin{pmatrix} \varepsilon_x & 0 & 0 \\ 0 & \varepsilon_y & 0 \\ 0 & 0 & \varepsilon_z \end{pmatrix}.$$

On the other hand, the graphene layer is modeled as a two-dimensional conductivity sheet of zero thickness. This avoids the calculation of the fields inside the graphene sheet [23], and has already been proven to be valid for graphene [23], transition layer polaritons [24], single layers of biaxial crystals [14, 25], polaritonic crystals of hexagonal boron nitride [26] (h-BN) and Moiré hyperbolic metasurfaces [30] with a layer thickness that is much smaller than the polariton wavelength ($d \gg \lambda_p$). Under this approximation, the conductivity of graphene, σ, is given by Eq. (1.12).

Dispersion Relation Under the Limit of Large-k Modes

As has been shown during this thesis, and described in detail in Chap. 3, the limit of large-k modes is of great importance for studying highly confined anisotropic polaritons in vdW biaxial crystal slabs. Under this limit, the dispersion relation of polaritons in a biaxial slab embedded between two semi-infinite isotropic media, derived in Chap. 3 and in Ref. [14] is given by:

$$q = \frac{\rho}{k_0 d}\left[\arctan\left(\frac{\varepsilon_1 \rho}{\varepsilon_z}\right) + \arctan\left(\frac{\varepsilon_3 \rho}{\varepsilon_z}\right) + \pi l\right], \qquad l = 0, 1, 2..., \tag{H.1}$$

where $q = k_t / k_0$ is the normalized in-plane wavevector ($k_t^2 = k_x^2 + k_y^2$), ε_1 and ε_3 are the permittivities of the superstrate (air) and substrate, respectively, $k_0 = \omega / c$ is the wavevector in free space, and

$$\rho = i\sqrt{\frac{\varepsilon_z q^2}{\varepsilon_x q_x^2 + \varepsilon_y q_y^2}} = i\sqrt{\frac{\varepsilon_z}{\varepsilon_x \cos^2\varphi + \varepsilon_y \sin^2\varphi}}, \tag{H.2}$$

with φ being the angle between the x-axis and the in-plane component of the wavevector.

To find out how the presence of graphene modifies the dispersion relation of the polaritons propagating along the biaxial slab, we calculate how the reflection and transmission coefficients for the biaxial slab surface covered by graphene are modified in comparison to those for the boundary between the biaxial and isotropic media without graphene. The transmission and reflection coefficients, t_{21} and r_{21}, respectively, can be calculated by applying the boundary conditions for **E** and **D** to the biaxial slab surface covered by graphene, which are given by:

$$\mathbf{E}_{1t} = \mathbf{E}_{2t} \qquad \text{and} \qquad \mathbf{D}_{1n} - \mathbf{D}_{2n} = 4\pi Q, \tag{H.3}$$

where the expression for these vectors of the electric fields are the ones derived for the "e" basis vectors in Chap. 3 (see Eq. (3.10)), reproduced here for completeness:

$$\mathbf{E}_1 = E_0|e\rangle_+ + E_r|e\rangle_- = \frac{E_0}{q}\begin{pmatrix} q_x \frac{\Delta_2 - q_y^2}{\Delta_x^e} \\ q_y \\ \frac{\Delta_2}{-q_{ez}} \end{pmatrix} e^{ik_x x + ik_y y} + \frac{E_r}{q}\begin{pmatrix} q_x \frac{\Delta_2 - q_y^2}{\Delta_x^e} \\ q_y \\ \frac{\Delta_2}{q_{ez}} \end{pmatrix} e^{ik_x x + ik_y y},$$

$$\mathbf{E}_2 = E_t|e\rangle_+ = \frac{E_t}{q}\begin{pmatrix} q_x \frac{\Delta_2 - q_y^2}{\Delta_x^e} \\ q_y \\ \frac{\Delta_2}{-q_{ez}} \end{pmatrix} e^{ik_x x + ik_y y}.$$

The factor $1/q$ [q is the normalized in-plane wavevector $q^2 = (k_x^2 + k_y^2)/k_0^2$] stands for the normalization, and:

$$\Delta_2 = \frac{\Delta_x^e \Delta_y^e - q_x^2 q_y^2}{\Delta_x^e - q_x^2}, \qquad \Delta_i^e = \varepsilon_i - q_x^2 - q_y^2 - q_z^2 + q_i^2.$$

Under the limit of large-k modes, the expressions for the z-components of the wave vectors inside the slab can be approximated as:

$$q_{oz}^2 = -q_x^2 - q_y^2, \quad q_{ez}^2 = -\frac{\varepsilon_x}{\varepsilon_z} q_x^2 - \frac{\varepsilon_y}{\varepsilon_z} q_y^2. \tag{H.4}$$

Using Eq. (H.4), we simplify the expressions for the basis "e"-vectors:

$$|e\rangle = \frac{1}{q}\begin{pmatrix} q_x \\ q_y \\ q_z \end{pmatrix} e^{ik_x x + ik_y y}.$$

Outside of biaxial media. $q_z = iq$ and

$$|p\rangle = \frac{1}{q}\begin{pmatrix} q_x \\ q_y \\ iq \end{pmatrix} e^{ik_x x + ik_y y}.$$

We define $t_{21} = \frac{E_t}{E_0}$ and $r_{21} = \frac{E_r}{E_0}$. Then Eqs. (H.3) lead to the following equations:

$$1 + r_{21} = t_{21} \qquad \text{and} \qquad i\varepsilon_1 q t_{21} - i q_{ez}\varepsilon_z (1 - r_{21}) = 4\pi \frac{Q_0}{E_0},$$

where Q is the surface charge, that can be calculated using the charge conservation law:

$$\frac{\partial Q}{\partial t} = -\operatorname{div} j \qquad \Rightarrow \qquad Q_0 = \frac{\mathbf{q}\mathbf{j}_0}{c} = \frac{\sigma_g}{c}\mathbf{q}\mathbf{E}_t.$$

Hence,

$$i\varepsilon_1 q t_{21} - i q_{ez}\varepsilon_z (1 - r_{21}) = 2\alpha_g q^2 t_{21} \qquad (\varepsilon_1 + 2i\alpha_g q) q t_{21} = q_{ez}\varepsilon_z (1 - r_{21}),$$

or, equivalently:

$$r_{21} = \frac{q_{ez}\varepsilon_z - q(\varepsilon_1 + 2i\alpha_g q)}{q_{ez}\varepsilon_z + q(\varepsilon_1 + 2i\alpha_g q)}, \qquad t_{21} = \frac{2q_{ez}\varepsilon_z}{q_{ez}\varepsilon_z + iq(\varepsilon_1 + 2i\alpha_g q)}. \tag{H.5}$$

The reflection and transmission coefficients for the boundary between the biaxial and isotropic media without graphene can also be obtained from Eqs. (H.5), by applying $\alpha_g = 0$:

$$r_{21} = \frac{q_{ez}\varepsilon_z - q\varepsilon_1}{q_{ez}\varepsilon_z + q\varepsilon_1}, \qquad t_{21} = \frac{2q_{ez}\varepsilon_z}{q_{ez}\varepsilon_z + iq\varepsilon_1}, \tag{H.6}$$

Therefore, to take into account graphene in the dispersion relation for high-q we should just change $\varepsilon_1 \to \varepsilon_1 + 2i\alpha q$ in the expression for the reflection coefficient. The dispersion relation given in Eq. (H.1) will then transform into:

$$q = \frac{\rho}{k_0 d}\left[\arctan\left(\frac{(\varepsilon_1 + 2i\alpha q)\rho}{\varepsilon_z}\right) + \arctan\left(\frac{\varepsilon_3 \rho}{\varepsilon_z}\right) + \pi l\right], \qquad l = 0, 1, 2..., \tag{H.7}$$

which can be solved numerically. In order to solve it, we define an implicit function in which the remaining variables act as arguments:

$$D(q) = \frac{\rho}{k_0 d}\left[\arctan\left(\frac{(\varepsilon_1 + 2i\alpha q)\rho}{\varepsilon_z}\right) + \arctan\left(\frac{\varepsilon_3 \rho}{\varepsilon_z}\right) + \pi l\right] - q.$$

For fixed values of the arguments, any complex number q that satisfies the condition $D(q) = 0$ is a solution of Eq. (H.7). The dispersion relation is then plotted giving a range of values for the frequency ω and the wavevector q following two main criteria.

- The values for the real part of q are taken in the range in which it is expected to find wave propagation, and always fulfilling $\mathrm{Re}(q) > 0$, since Eq. (H.1) is derived under this condition, as discussed in Chap. 3 and in Ref. [14].
- The sign of q (and thus of the phase velocity) is determined by $\mathrm{Im}(q)$, in such a way that if $\mathrm{Im}(q) < 0$, then we perform $\mathrm{Re}(q) \to -\mathrm{Re}(q)$.
- The complex number q must satisfy the condition $|\mathrm{Re}(q)| \geq |\mathrm{Im}(q)|$. This condition is imposed to consider propagating modes only.

Following these conditions, q and ω can be defined as an array of numbers q_i and ω_j. The implicit function D is then defined as a matrix D_{ij}. This matrix, taken as $\log_{10}(|D_{ij}|)$, is plotted as a colormap where the regions with the smallest values of $\log_{10}(|D_{ij}|)$ correspond to the values of q and ω that satisfy Eq. (H.7).

This same procedure can be followed in order to obtain the IFCs. In this case, it is given a range of values for q and φ (see Eq. (H.2)).

Dispersion Relation Under the Limit of an Ultrathin Biaxial Slab

On the other hand, the biaxial slab can also be modeled as a two-dimensional conductivity sheet of zero thickness. As already stated above, this avoids the calculation of the fields inside the slab, and has already been proven to be valid for several systems, including single layers of biaxial crystals, as discussed in Chap. 3. Under this approximation, the dispersion relation for a single biaxial conductivity sheet between two semi-infinite isotropic media (see Eq. (3.60)):

$$\left\{\alpha_x q_y^2 + \alpha_y q_x^2 + \frac{q_x^2 + q_y^2}{2}\left(iq_{1z} + iq_{3z}\right)\right\}\left\{\alpha_x q_x^2 + \alpha_y q_y^2 + \frac{q_x^2 + q_y^2}{2} \cdot \left(\frac{\varepsilon_1}{iq_{1z}} + \frac{\varepsilon_3}{iq_{3z}}\right)\right\} = q_x^2 q_y^2 \left(\alpha_x - \alpha_y\right)^2, \tag{H.8}$$

where $\alpha_{x,y} = k_o d \varepsilon_{x,y}/2i$ are the normalized 2D effective conductivity components.

In analogy to the case in Eq. (H.7), we should just modify the normalized conductivities $\alpha_i \to \alpha_i + \alpha_g$, where α_i is the normalized conductivity of the biaxial slab ($i = x, y$) and α_g is the graphene conductivity. Taking this into account, the dispersion relation given in Eq. (H.8) transforms into:

$$\left\{\alpha_x q_y^2 + \alpha_y q_x^2 + \frac{q_x^2 + q_y^2}{2}\left(iq_{1z} + iq_{3z} + 2\alpha_g\right)\right\}\left\{\alpha_x q_x^2 + \alpha_y q_y^2 + \frac{q_x^2 + q_y^2}{2}\right.$$
$$\left.\cdot\left(\frac{\varepsilon_1}{iq_{1z}} + \frac{\varepsilon_3}{iq_{3z}} + 2\alpha_g\right)\right\} = q_x^2 q_y^2 \left(\alpha_x - \alpha_y\right)^2. \quad \text{(H.9)}$$

Dispersion Relation Under the Limit of Large-k Modes and an Ultrathin Biaxial Slab

If we combine the approximations described in sections "Dispersion Relation Under the Limit of Large-k Modes" and "Dispersion Relation Under the Limit of an Ultrathin Biaxial Slab", i.e., we take the limit $q \gg 1$ in Eq. (H.9), we get:

$$\left\{\alpha_x q_y^2 + \alpha_y q_x^2 + i\frac{q^3}{2}\right\}\left\{\alpha_x q_x^2 + \alpha_y q_y^2 - i\frac{q}{2}\left(\varepsilon_1 + \varepsilon_2 + 2i\alpha_g q\right)\right\} = q_x^2 q_y^2 \left(\alpha_x - \alpha_y\right)^2, \quad \text{(H.10)}$$

From Eq. (H.10) or from Eq. (H.7) in the limit $qk_0 d/\rho \ll 1$, we obtain:

$$q = \frac{i(\varepsilon_1 + \varepsilon_2)}{2(\alpha_x \cos^2\varphi + \alpha_y \sin^2\varphi + \alpha_g)}. \quad \text{(H.11)}$$

H.3 Dispersion of hBN-Encapsulated Graphene

In this section we study to which extent our assumption of the graphene relaxation time $\tau = 0.5$ ps is reasonable for a single graphene layer. This value of τ is typical for encapsulated graphene used in optical experiments at ambient conditions, this is, a graphene layer sandwiched between two h-BN layers, as sketched in Fig. H.2.

To determine the validity of our assumption, we calculate via the transfer matrix method the dispersion of PhPs in heterostructures made of encapsulated graphene and α-MoO_3, (i.e., h-BN/graphene/h-BN/α-MoO_3) for different thicknesses. We then compare the as-obtained numerical solution with the corresponding analytical solution for non-encapsulated graphene on top of α-MoO_3.

Fig. H.2 Sketch encapsulated graphene on MoO3

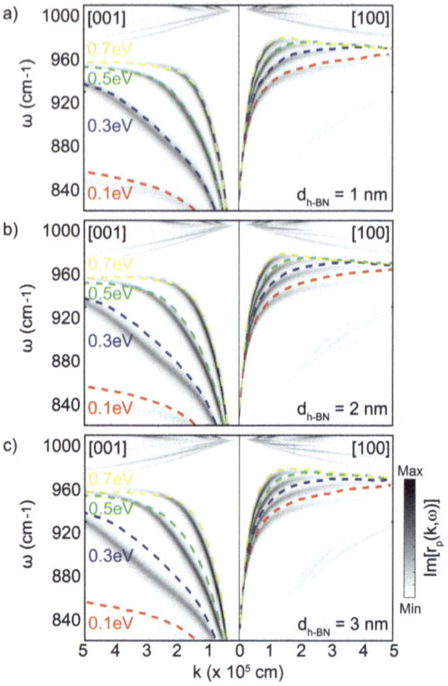

Fig. H.3 Dispersion of PhPs on an α-MoO_3 slab for different representative doping levels, i.e., different representative Fermi energies $E_F = 0.1$, 0.3, 0.5 and 0.7 eV. The false color plot shows transfer-matrix calculations displaying the imaginary part of Fresnel's reflection coefficient and the analytical dispersions are depicted as dashed lines; (**a**–**c**) correspond to different thicknesses of the h-BN layers

The results, shown in Fig. H.3, reveal that the h-BN layers do not introduce any significant changes to the dispersion nor any in-plane effects if they are thin enough. Hence, this justifies our assumption not to consider the encapsulating h-BN layers in our analytical study.

H.4 Interband and Intraband Contributions to the Conductivity and Landau Damping

As explained in Sect. 1.3, the conductivity of graphene can computed within the standard random-phase approximation. Under such approximation, it is found to depend on temperature (T), chemical potential (μ), and scattering time (τ), see Eq. (1.12). For relatively small frequencies $\hbar\omega/\mu \ll 1$ (where the intraband contribution to the conductivity dominates) the real part of the conductivity is sensitive to τ, while for higher frequencies $\hbar\omega/\mu \gtrsim 1$ (with a dominating interband contribution) the losses are predominantly temperature dependent [31]. By doping the graphene, the carrier concentration increases, and Landau damping can be eliminated, as the Pauli principle then prohibits plasmon-to-electron-hole-pair transitions in a large region of the $(\mathbf{k}, \omega)$ space.

The excitations in our devices lie well outside the Landau damping regime, as indicated in Fig. H.4 for different Fermi energies. The inset shows the intraband and

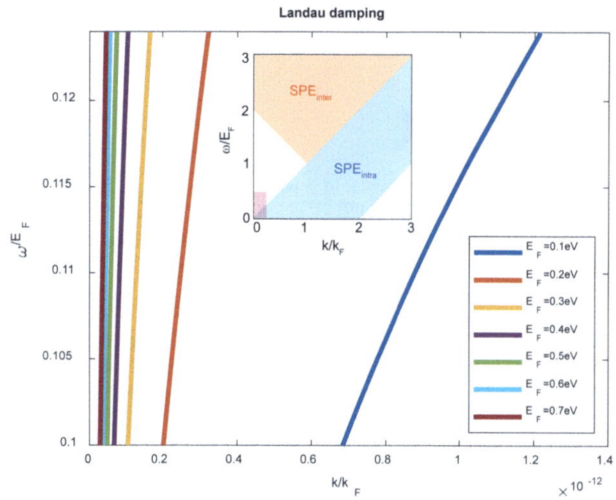

Fig. H.4 Dispersion of GPPs for different Fermi energies E_F (continuous lines). The inset shows the regions in which the interband and intraband Landau damping are significant. The area in which our calculations are located is shaded in pink

interband Landau damping regions in the (k, ω) space [32]. The pink-shaded region indicates the region in which we perform our calculations, namely in the spectral range 800–1000 cm^{-1} $\approx$0.1–0.125 eV, while the wavevectors span the range 0–8×10^5cm^{-1}. On the other hand, even for sufficiently small Fermi levels (providing large plasmon momenta), the confinement of the plasmon polaritons to graphene becomes substantial, so that the field virtually does not penetrate the α-MoO_3 layer. Therefore, the interaction between GPPs and HPhPs in α-MoO_3 slab is strongly suppressed and is not observable neither in the calculated hyperbolic filed patterns, nor in the potential near-field experiments.

H.5 Field Distributions and IFC of Polaritons Propagating in a Graphene/α-MoO_3 Heterostructure for Several Cases of Interest

Field Distributions and IFC of Polaritons Propagating in a Heterostructure Made of a Graphene Sheet on Top of an α-MoO_3 Slab of at a Doping Level $E_F = 0\,eV$

In this subsection we calculate the real-space electric near-field distribution and the corresponding numerical and analytical polaritonic IFCs of hybrid polaritons propagating in graphene/α-MoO_3 vdW heterostructures for a Fermi energy E_F = 0 eV. The incident frequency is fixed to $\omega_0 = 930\,\text{cm}^{-1}$ and the thickness of the

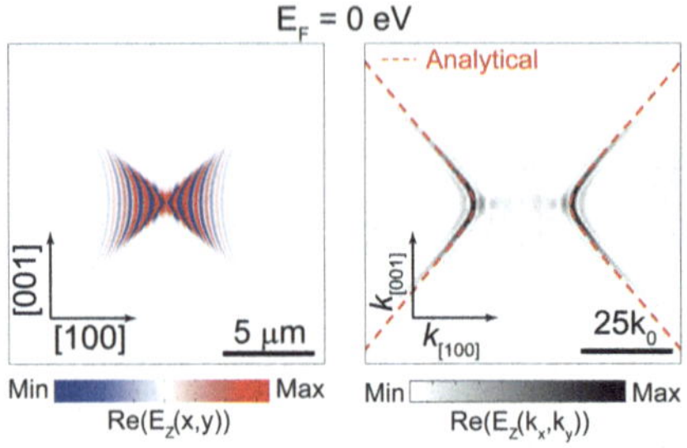

Fig. H.5 Simulated real-space electric near-field distribution (left panel) and corresponding numerical and analytical polaritonic IFCs (color plot and dashed red lines in the right panel, respectively) of hybrid polaritons propagating in graphene/α-MoO_3 vdW heterostructures for a Fermi energy $E_F = 0$ eV. The incident frequency is fixed to $\omega_0 = 930\,\mathrm{cm}^{-1}$ and the thickness of the slab is $d = 120$ nm

slab is $d = 120$ nm. The Fermi energy $E_F = 0$ eV is effectively modelled by setting the graphene conductivity α to zero. The results are shown in Fig. H.5, where we can observe propagation of PhPs within a hyperbolic sector centered along the [100] direction, in accordance with prior works [25, 33] and in excellent agreement with our analytical calculations.

Topological Transition in Graphene/α-MoO₃ Heterostructures

In this subsection we show the dependence of the open-to-closed topological transition of the IFCs in graphene/α-MoO_3 heterostructures as a function of the Fermi energy E_F. To do that, we perform full-wave numerical simulations, in which hybrid polaritons are launched by a point dipole on top of the heterostructures. Then, we FT the near electric field distribution to get the numerical IFCs. Since large-k waves do not contribute to the field distribution, we set a rather large value of the wavevector $\tilde{k} = 70k_0$. We will define the transition Fermi energy $\tilde{E}_F$ as the energy for which the curve is closed and entirely contained within a window of size $[-\tilde{k}, \tilde{k}] \times [-\tilde{k}, \tilde{k}]$. As such, the value at which the transition takes place is $\tilde{E}_F = 0.32$ eV (see Fig. H.6)

Canalization of Polaritons in Graphene/α-MoO₃ Heterostructures for Different Fermi Energies and Thicknesses of the Biaxial Slab

It is possible to manipulate the experimental design in order to achieve canalization for different values of E_F. Apart from the Fermi energy E_F, the dispersion relation

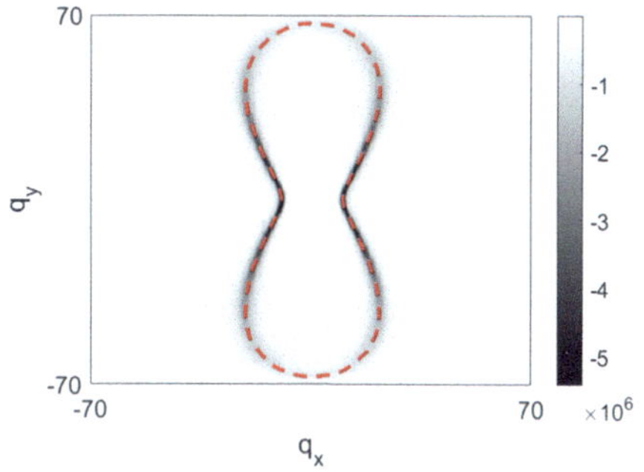

Fig. H.6 IFC of PhPs in graphene/α-MoO_3 at the transition Fermi energy $\tilde{E_F} = 0.32\,\text{eV}$ and $\omega_0 = 930\,\text{cm}^{-1}$

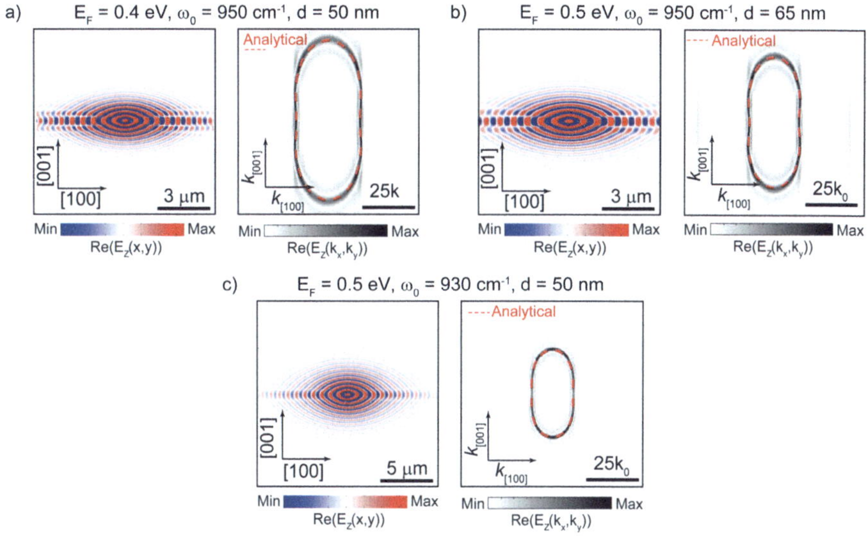

Fig. H.7 Canalization of polaritons in graphene/α-MoO_3 heterostructures for Fermi energies of (**a**) 0.4 and (**b**) 0.5 eV for biaxial slab thicknesses of $d = 50$ and 65 nm, respectively, at an incident frequency $\omega_0 = 950\,\text{cm}^{-1}$ and (**c**) for a Fermi energy of 0.5 eV, a slab thickness of $d = 50$ nm, and an incident frequency $\omega_0 = 930\,\text{cm}^{-1}$

for hybrid polaritons in heterostructures containing graphene/α-MoO_3 depends on the wavelength of the incident light and on the thickness of the biaxial slab (see Eq. (H.7)). Therefore, one can modify the polaritonic IFC by varying these parameters. For instance, it is possible to achieve canalization for experimentally easily accessible Fermi energies (0.4 and 0.5 eV) by considering biaxial slab thicknesses of $d = 50$ and 65 nm for fixed incident frequencies $\omega_0 = 930$ or $950\,\text{cm}^{-1}$ (see Fig. H.7).

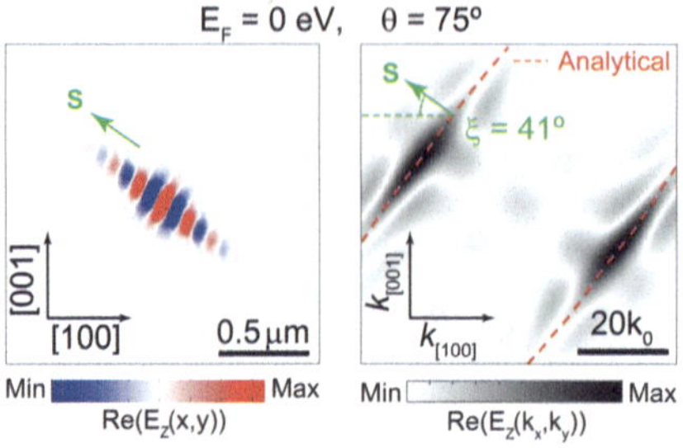

Fig. H.8 Simulated real-space electric near-field distribution (left panel) and corresponding numerical and analytical polaritonic IFCs (color plot and dashed red lines in the right panel, respectively) of polaritons propagating in a heterostructure made of two twisted slabs of α-MoO_3 with graphene on top for a Fermi energy $E_F = 0\,\mathrm{eV}$. The incident frequency is fixed to $\omega_0 = 930\,\mathrm{cm}^{-1}$ and the thickness of the slabs is $d = 50\,\mathrm{nm}$

Field Distributions and IFC of Polaritons Propagating in a Heterostructure Made of a Graphene Sheet on Top of a Two Twisted Slabs of α-MoO₃ at a Doping Level $E_F = 0\,eV$

In this subsection we calculate the real-space electric near-field distribution and the corresponding numerical and analytical polaritonic IFCs of polaritons propagating in a heterostructure made of two twisted slabs of α-MoO_3 with graphene on top for a Fermi energy $E_F = 0\,\mathrm{eV}$. The incident frequency is fixed to $\omega_0 = 930\,\mathrm{cm}^{-1}$ and the thickness of both slabs is set to $d = 50\,\mathrm{nm}$. The Fermi energy $E_F = 0\,\mathrm{eV}$ is effectively modelled by setting the graphene conductivity α to zero. The results are shown in Fig. H.8, where we can observe propagation of canalized PhPs for a twist angle $\theta = 75°$, in excellent agreement with our analytical calculations.

Field Distributions and IFC of Polaritons Propagating in a Heterostructure Made of a Graphene Sheet on Top of a Two Twisted Slabs of α-MoO₃ at a Doping Level $E_F = 0.5\,eV$

In this subsection we calculate the real-space electric near-field distribution and the corresponding numerical and analytical polaritonic IFCs of polaritons propagating in a heterostructure made of two twisted slabs of α-MoO_3 with graphene on top for a Fermi energy $E_F = 0.5\,\mathrm{eV}$. The incident frequency is fixed to $\omega_0 = 930\,\mathrm{cm}^{-1}$ and the thickness of both slabs is set to $d = 50\,\mathrm{nm}$. The results are shown in Fig. H.9, where we can observe propagation of canalized PhPs for a twist angle $\theta = 55°$, in excellent agreement with our analytical calculations.

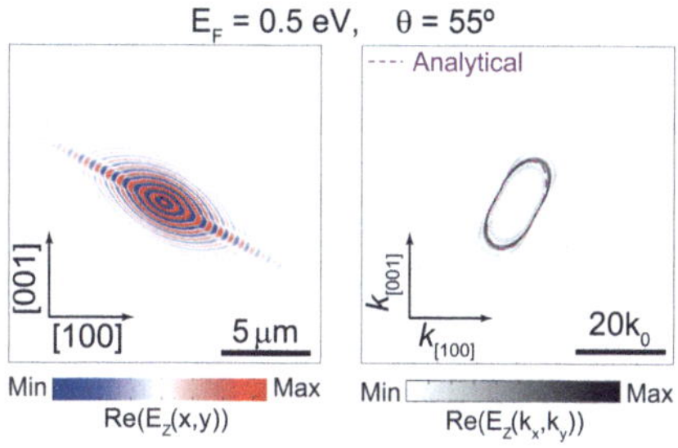

Fig. H.9 Simulated real-space electric near-field distribution (left panel) and corresponding numerical and analytical polaritonic IFCs (color plot and dashed red lines in the right panel, respectively) of polaritons propagating in a heterostructure made of two twisted slabs of α-MoO_3 with graphene on top for a Fermi energy $E_F = 0\,\text{eV}$. The incident frequency is fixed to $\omega_0 = 930\,\text{cm}^{-1}$ and the thickness of the slabs is $d = 50\,\text{nm}$

Influence of the Thickness of the Twisted α-MoO_3 Slabs on the Coupling with Graphene Plasmons

The strength of the interactions between the graphene plasmons and the HPhPs depends on the thicknesses of the two twisted layers of α-MoO_3. In addition, such interaction is also influenced by the incident frequency, the twist angle and the doping level, giving rise to a vast parametric space hosting many different particular cases which it is not practical to explore completely. Considering this, so far we have chosen a representative incident frequency ($930\,\text{cm}^{-1}$) and a representative thickness of the slabs (50 nm). To prove that our results are still valid beyond the chosen thickness, here we show full-wave numerical simulations and analytical calculations for other thicknesses.

We observe that for 100-nm-thick twisted α-MoO_3 slabs, hybrid polaritons can be canalized at a relatively low doping level of 0.3 eV (see Fig. H.10a,b,c), even if placing the graphene layer in between the two slabs (Fig. H.10b,c). This happens because HPhPs in the two twisted slabs couple and hence we can externally manipulate them by means of electrical doping. Therefore, our original inferences for 50-nm-thick α-MoO_3 slabs are still valid for 100-nm-thick α-MoO_3 slabs. However, this is not the case when we consider two slabs of very different thicknesses (a 250-nm-thick top α-MoO_3 slab and 50-nm-thick bottom α-MoO_3 slab) because HPhPs in the twisted heterostructure do not couple (see panels d,e,f in the figure below). We thus conclude that our choice of a representative thickness does not influence the generalization of the results reported and that the strength of the interactions between graphene and the twisted slabs is mainly determined by whether HPhPs in the twisted layers hybridize or not, while graphene can be already dielectrically affected by semi-infinite bulk α-MoO_3.

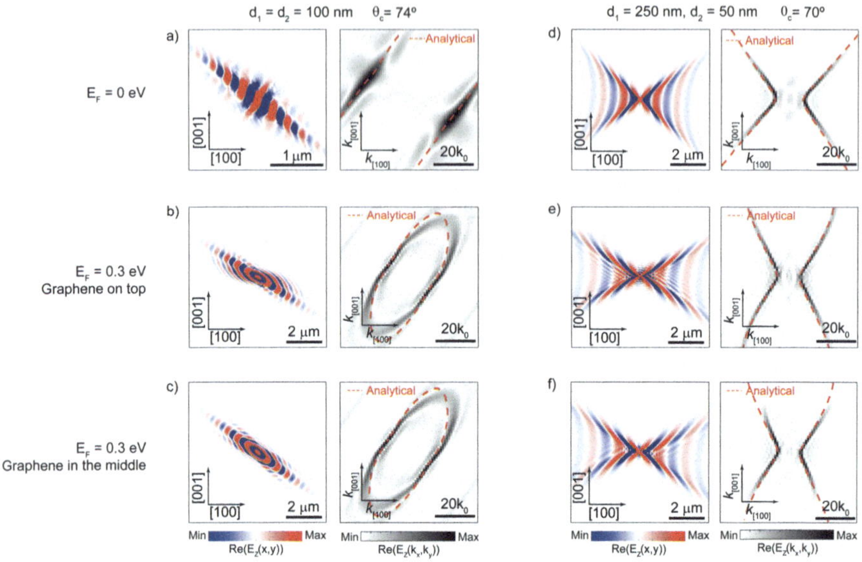

Fig. H.10 Simulated real-space electric near-field distribution (left panel) and corresponding numerical and analytical polaritonic IFCs (color plot and dashed red lines in the right panel, respectively) of polaritons propagating in a heterostructure made of two twisted slabs of α-MoO_3 with graphene on top for Fermi energies $E_F = 0$ and 0.3 eV. The incident frequency is fixed to $\omega_0 = 930\,\mathrm{cm}^{-1}$ and the thickness of the slabs is $d = 100\,\mathrm{nm}$ (left) and 250/50 nm (right). The graphene layer is placed on top (middle row) and in the middle (bottom row) of the twisted α-MoO_3 slabs

We also note that polaritons behave the same way no matter whether we consider the graphene layer on top of the twisted heterostructure or sandwiched between the two twisted slabs of α-MoO_3 (see the two bottom rows in Fig. H.10).

Dispersion Relation of Polaritons in a Heterostructure Made of Two Twisted Biaxial Thin Slabs and a Graphene Sheet In Between

In order to determine the influence of the position of the graphene layer, we perform full-wave numerical simulations considering the graphene monolayer sandwiched in between the two α-MoO_3 slabs. We then Fourier transform the electric near-field distribution to obtain the numerical IFCs in momentum space. Such numerical IFCs are compared to the analytical IFCs, derived considering the graphene monolayer on top of the two α-MoO_3 slabs. For the calculation of the analytical IFCs, we use Eq. (H.16) from Sect. H.6, which is sensitive to the position of graphene.

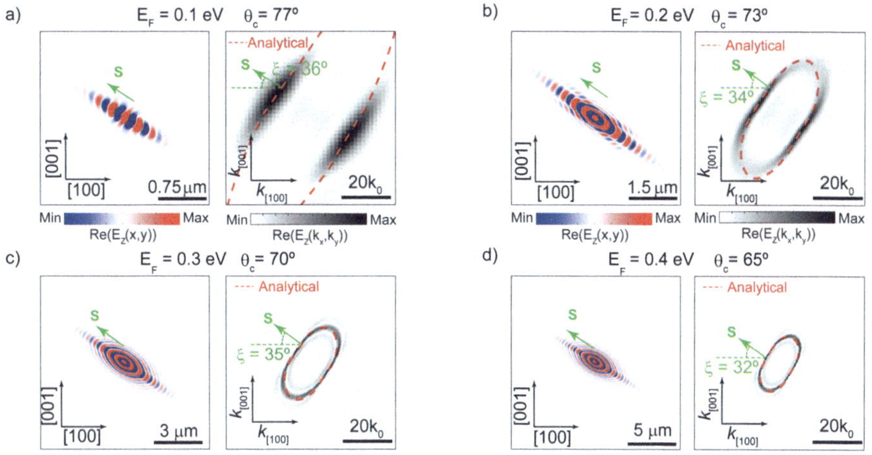

Fig. H.11 Simulated real-space electric near-field distribution (left panel) and corresponding numerical and analytical polaritonic IFCs (color plot and dashed red lines in the right panel, respectively) of polaritons propagating in a heterostructure made of two twisted slabs of α-MoO_3 with graphene in between (color plot) and on top (dashed red lines) of the twisted slabs. We illustrate the canalization regime for different for different Fermi energies E_F and twist angles θ_C. The incident frequency is fixed to $\omega_0 = 930\,\text{cm}^{-1}$ and the thickness of the slabs is $d = 50$ nm

The excellent agreement between the analytical expressions and the simulations allows us claiming that hybrid polaritons behave the same way no matter whether we consider the graphene layer on top of the twisted heterostructure or sandwiched between the two twisted slabs of α-MoO_3. In particular, we find canalization for the same twist angles θ_C, incident frequencies ω_0 and Fermi energies E_F for both configurations (see Fig. H.11 and compare with Fig. 3 in the main text).

H.6 Derivation of the Dispersion Relation of Polaritons in a Heterostructure Made of Two Twisted Biaxial Thin Slabs and a Graphene Sheet on Top

Here we derive the dispersion relation of polaritons propagating in a heterostructure made of graphene on top of two twisted biaxial slabs. Such relation can be directly obtained from the dispersion relation for the twisted bilayer with the change $\varepsilon_1 \to \varepsilon_1 + 2i\alpha q$. To find the dispersion relation for the polaritons propagating along the biaxial bilayer structure, we consider the field distribution, which is given by the following (see Fig. H.12):

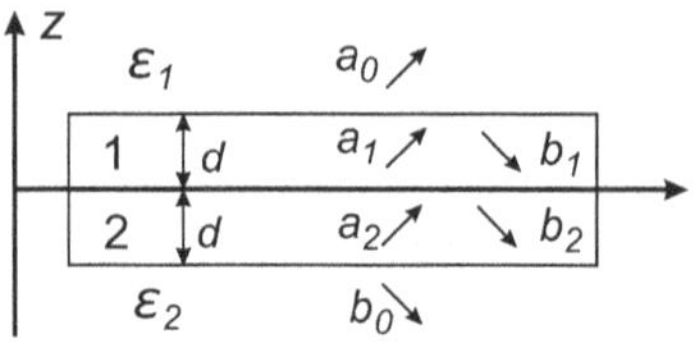

Fig. H.12 Geometry of the biaxial bilayer structure (1,2) and the field amplitudes and directions in each of the media

$$
\begin{aligned}
\mathbf{E}_0 &= a_0 \begin{pmatrix} q_x \\ q_y \\ iq \end{pmatrix} e^{ik_x x + ik_y y} e^{-qk_0 z} \qquad z > d \\
\mathbf{E}_1 &= a_1 \begin{pmatrix} q_x \\ q_y \\ q_{z1} \end{pmatrix} e^{ik_x x + ik_y y} e^{iq_{z1}k_0 z} + b_1 \begin{pmatrix} q_x \\ q_y \\ -q_{z1} \end{pmatrix} e^{ik_x x + ik_y y} e^{-iq_{z1}k_0 z} \qquad d > z > 0 \\
\mathbf{E}_2 &= a_2 \begin{pmatrix} q_x \\ q_y \\ q_{z2} \end{pmatrix} e^{ik_x x + ik_y y} e^{iq_{z2}k_0 z} + b_2 \begin{pmatrix} q_x \\ q_y \\ -q_{z2} \end{pmatrix} e^{ik_x x + ik_y y} e^{-iq_{z2}k_0 z} \qquad 0 > z > -d \\
\mathbf{E}_4 &= b_0 \begin{pmatrix} q_x \\ q_y \\ -iq \end{pmatrix} e^{ik_x x + ik_y y} e^{qk_0 z} \qquad -d > z
\end{aligned}
\tag{H.12}
$$

Using the boundary conditions for $\mathbf{E}_t$ and $\mathbf{D}_n$

$$
\mathbf{E}_{1t} = \mathbf{E}_{2t} \qquad \mathbf{D}_{1n} = \mathbf{D}_{2n}, \tag{H.13}
$$

we obtain the following system of equations:

$$
\begin{pmatrix}
-e^{-qk_0d} & e^{iq_{z1}k_0d} & e^{-iq_{z1}k_0d} & 0 & 0 & 0 \\
-i\varepsilon_1 q e^{-qk_0d} & \varepsilon_z q_{z1} e^{iq_{z1}k_0d} & -\varepsilon_z q_{z1} e^{-iq_{z1}k_0d} & 0 & 0 & 0 \\
0 & 1 & 1 & -1 & -1 & 0 \\
0 & q_{z1} & -q_{z1} & -q_{z2} & q_{z2} & 0 \\
0 & 0 & 0 & e^{-iq_{z2}k_0d} & e^{iq_{z2}k_0d} & -e^{-qk_0d} \\
0 & 0 & 0 & \varepsilon_z q_{z2} e^{-iq_{z2}k_0d} & -\varepsilon_z q_{z2} e^{iq_{z2}k_0d} & i\varepsilon_2 q e^{-qk_0d}
\end{pmatrix}
\times \begin{pmatrix} a_0 \\ a_1 \\ b_1 \\ a_2 \\ b_2 \\ b_0 \end{pmatrix} = 0. \tag{H.14}
$$

The homogeneous system (H.14) has non-trivial solutions only when its determinant is equal to zero. The zeros of the determinant yield the dispersion relation for the modes in the biaxial slab. As a result, we get the following expression

$$\tan(q_{z1}k_0d)\tan(q_{z2}k_0d)\varepsilon_z q\left(\varepsilon_1 q_{z2}^2+\varepsilon_2 q_{z1}^2\right)+\tan(q_{z1}k_0d)q_{z2}\left(\varepsilon_z^2 q_{z1}^2-\varepsilon_1\varepsilon_2 q^2\right)+ \\ +\tan(q_{z2}k_0d)q_{z1}\left(\varepsilon_z^2 q_{z2}^2-\varepsilon_1\varepsilon_2 q^2\right)=\varepsilon_1\varepsilon_z q q_{z1}q_{z2}. \tag{H.15}$$

In analogy to the procedure followed in Sect. H.2, by applying the transformation $\varepsilon_1 \to \varepsilon_1 + 2i\alpha q$, we obtain the dispersion relation for the polaritons propagating along the biaxial bilayer covered by graphene:

$$\begin{aligned}&\tan(q_{z1}k_0d)\tan(q_{z2}k_0d)\varepsilon_z q\left((\varepsilon_1+2i\alpha q)q_{z2}^2+\varepsilon_2 q_{z1}^2\right)\\&\quad+\tan(q_{z1}k_0d)q_{z2}\left(\varepsilon_z^2 q_{z1}^2-(\varepsilon_1+2i\alpha q)\varepsilon_2 q^2\right)\\&\quad+\tan(q_{z2}k_0d)q_{z1}\left(\varepsilon_z^2 q_{z2}^2-(\varepsilon_1+2i\alpha q)\varepsilon_2 q^2\right)\\&=(\varepsilon_1+2i\alpha q+\varepsilon_2)\varepsilon_z q q_{z1}q_{z2}.\end{aligned} \tag{H.16}$$

Equation (H.16) can be solved numerically or can be simplified in the limit of ultrathin slabs to find the fundamental transverse magnetic (TM_0) mode. To do this, we consider the Taylor expansion of the tangent function, $\tan(q_{zi}k_0d) \approx q_{zi}k_0d$, and neglect the terms with higher orders of d.

$$q=\frac{\varepsilon_1+\varepsilon_2}{k_0d\varepsilon_z\left(\beta_1^2+\beta_2^2\right)-2i\alpha}, \tag{H.17}$$

with

$$\beta_i=\sqrt{-\frac{\varepsilon_x}{\varepsilon_z}\cos^2\theta_i-\frac{\varepsilon_y}{\varepsilon_z}\sin^2\theta_i}. \tag{H.18}$$

where θ_i is the angle between $\mathbf{q}$ and the main axes of i-th slab. Equation (H.17) has the same form as Eq. (H.11), and if the layers are not twisted, then Eq. (H.17) reduces to Eq. (H.11). The explicit expression (H.17) allows us to easily estimate the wavevector, without the need of numerical calculations or simulations.

References

1. G. Álvarez-Pérez et al., Infrared permittivity of the biaxial van der Waals semiconductor α-MoO_3 from near- and far-field correlative studies. Adv. Mater. **32**, 1908176 (2020). https://onlinelibrary.wiley.com/doi/abs/10.1002/adma.201908176
2. R.O. Jones, O. Gunnarsson, The density functional formalism, its applications and prospects. Rev. Mod. Phys. **61**, 689–746 (1989). https://link.aps.org/doi/10.1103/RevModPhys.61.689
3. P. Hohenberg, W. Kohn, Inhomogeneous electron gas. Phys. Rev. **136**, B864–B871 (1964). https://link.aps.org/doi/10.1103/PhysRev.136.B864
4. W. Kohn, L.J. Sham, Self-consistent equations including exchange and correlation effects. Phys. Rev. **140**, A1133–A1138 (1965). https://link.aps.org/doi/10.1103/PhysRev.140.A1133

5. J.P. Perdew, K. Burke, M. Ernzerhof, Generalized gradient approximation made simple. Phys. Rev. Lett. **77**, 3865–3868 (1996). https://link.aps.org/doi/10.1103/PhysRevLett.77.3865
6. R. Cowley, The lattice dynamics of an anharmonic crystal. Adv. Phys. **12**, 421–480 (1963). https://doi.org/10.1080/00018736300101333
7. S. Baroni, S. de Gironcoli, A. Dal Corso, P. Giannozzi, Phonons and related crystal properties from density-functional perturbation theory. Rev. Mod. Phys. **73**, 515–562 (2001). https://link.aps.org/doi/10.1103/RevModPhys.73.515
8. P. Giannozzi et al., QUANTUM ESPRESSO: a modular and open-source software project for quantum simulations of materials. J. Phys. Condens. Matter **21**, 395502 (2009). https://dx.doi.org/10.1088/0953-8984/21/39/395502
9. P. Giannozzi et al., Advanced capabilities for materials modelling with Quantum ESPRESSO. J. Phys. Condens. Matter **29**, 465901 (2017). https://doi.org/10.1088/1361-648X/aa8f79
10. L. Paulatto, F. Mauri, M. Lazzeri, Anharmonic properties from a generalized third-order ab initio approach: Theory and applications to graphite and graphene. Phys. Rev. B **87**, 214303 (2013). https://link.aps.org/doi/10.1103/PhysRevB.87.214303
11. L. Paulatto, I. Errea, M. Calandra, F. Mauri, First-principles calculations of phonon frequencies, lifetimes, and spectral functions from weak to strong anharmonicity: The example of palladium hydrides. Phys. Rev. B **91**, 054304 (2015). https://link.aps.org/doi/10.1103/PhysRevB.91.054304
12. W. Li, J. Carrete, N.A. Katcho, N. Mingo, ShengBTE: A solver of the Boltzmann transport equation for phonons. Comput. Phys. Commun. **185**, 1747–1758 (2014). https://www.sciencedirect.com/science/article/pii/S0010465514000484
13. W. Li, L. Lindsay, D.A. Broido, D.A. Stewart, N. Mingo, Thermal conductivity of bulk and nanowire $Mg_2Si_xSn_{1-x}$ alloys from first principles. Phys. Rev. B **86**, 174307 (2012). https://link.aps.org/doi/10.1103/PhysRevB.86.174307
14. G. Álvarez-Pérez, K.V. Voronin, V.S. Volkov, P. Alonso-GonzáLlez, A.Y. Nikitin, Analytical approximations for the dispersion of electromagnetic modes in slabs of biaxial crystals. Phys. Rev. B **100**, 235408 (2019). https://link.aps.org/doi/10.1103/PhysRevB.100.235408
15. L.D. Landáu, Y.M. Lífshits, *Electrodynamics of Continuous Media* (Pergamon Press/Addison-Wesley, Oxford/Reading, MA, 1960), p. 324. https://www.elsevier.com/books/electrodynamics-of-continuous-media/landau/978-0-08-057060-0
16. J. Taboada-Gutiérrez et al., Broad spectral tuning of ultra-low-loss polaritons in a van der Waals crystal by intercalation. Nat. Mater. **19**, 964–968 (2020). https://doi.org/10.1038/s41563-020-0665-0
17. T.E. Tiwald et al., Carrier concentration and lattice absorption in bulk and epitaxial silicon carbide determined using infrared ellipsometry. Phys. Rev. B **60**, 11464–11474 (1999). https://link.aps.org/doi/10.1103/PhysRevB.60.11464
18. J. Duan et al., Enabling propagation of anisotropic polaritons along forbidden directions via a topological transition. Sci. Adv. **7**, eabf2690 (2021). https://www.science.org/doi/abs/10.1126/sciadv.abf2690
19. L. Novotny, B. Hecht, *Principles of Nano-optics* (Cambridge University Press, 2006)
20. F. Wijnands et al., Green's functions for Maxwell's equations: application to spontaneous emission. Opt. Quantum Electron. **29**, 199–216 (1997). https://doi.org/10.1023/A:1018506222632
21. J. Duan et al., Twisted nano-optics: manipulating light at the nanoscale with twisted phonon polaritonic slabs. Nano Lett. **20**, 5323–5329 (2020). https://doi.org/10.1021/acs.nanolett.0c01673
22. J.S. Gómez-Díaz, M. Tymchenko, A. Alù, Hyperbolic plasmons and topological transitions over uniaxial metasurfaces. Phys. Rev. Lett. **114**, 233901 (2015). https://link.aps.org/doi/10.1103/PhysRevLett.114.233901
23. A.Y. Nikitin, Graphene plasmonics, in *World Scientific Handbook of Metamaterials and Plasmonics*. World Scientific Series in Nanoscience and Nanotechnology (Cambridge University Press, 2017), pp. 307–338. https://www.worldscientific.com/worldscibooks/10.1142/10642-vol4#t=toc

24. V. Agranovich, D. Mills, *Surface Polaritons – Electromagnetic Waves at Surfaces and Interfaces* (Elsevier Science, 1982). https://www.elsevier.com/books/surface-polaritons/agranovich/978-0-444-86165-8
25. W. Ma et al., In-plane anisotropic and ultra-low-loss polaritons in a natural van der Waals crystal. Nature **562**, 557 (2018). https://www.nature.com/articles/s41586-018-0618-9
26. F.J. Alfaro-Mozaz et al., Deeply subwavelength phonon-polaritonic crystal made of a van der Waals material. Nat. Commun. **10**, 42 (2019). https://doi.org/10.1038/s41467-018-07795-6
27. H. Yu, G.-B. Liu, J. Tang, X. Xu, W. Yao, Moiré excitons: From programmable quantum emitter arrays to spin-orbit-coupled artificial lattices. Sci. Adv. **3**, e1701696 (2017). https://www.science.org/doi/abs/10.1126/sciadv.1701696
28. J. Duan et al., Multiple and spectrally robust photonic magic angles in reconfigurable α-MoO_3 trilayers. Nat. Mater. **22**, 867–872 (2023). https://doi.org/10.1038/s41563-023-01582-5
29. G. Álvarez-Pérez et al., Active tuning of highly anisotropic phonon polaritons in van der Waals crystal slabs by gated graphene. ACS Photon. **9**, 383–390 (2022). https://doi.org/10.1021/acsphotonics.1c01549
30. G. Hu, A. Krasnok, Y. Mazor, C.-W. Qiu, A. Alù, Moiré hyperbolic metasurfaces. Nano Lett. **20**, 3217–3224 (2020). https://doi.org/10.1021/acs.nanolett.9b05319
31. A.Y. Nikitin, F. Guinea, F.J. García-Vidal, L. Martín-Moreno, Edge and waveguide terahertz surface plasmon modes in graphene microribbons. Phys. Rev. B **84**, 161407 (2011). https://link.aps.org/doi/10.1103/PhysRevB.84.161407
32. E.H. Hwang, S. Das Sarma, Dielectric function, screening, and plasmons in two-dimensional graphene. Phys. Rev. B **75**, 205418 (2007). https://link.aps.org/doi/10.1103/PhysRevB.75.205418
33. Z. Zheng et al., A mid-infrared biaxial hyperbolic van der Waals crystal. Sci. Adv. **5**, eaav8690 (2019). https://advances.sciencemag.org/content/5/5/eaav8690

SPRINGER NATURE

GPSR Compliance

The European Union's (EU) General Product Safety Regulation (GPSR) is a set of rules that requires consumer products to be safe and our obligations to ensure this.

If you have any concerns about our products, you can contact us on ProductSafety@springernature.com

In case Publisher is established outside the EU, the EU authorized representative is:

Springer Nature Customer Service Center GmbH
Europaplatz 3
69115 Heidelberg, Germany

The manufacturer's authorised representative in the EU is Springer Nature Customer Service Centre GmbH, Europaplatz 3, 69115 Heidelberg, Germany. If you have any concerns regarding our products, please contact ProductSafety@springernature.com

Printed and bound by CPI Group (UK) Ltd, Croydon, CR0 4YY
30/09/2026
02225598-0008